VIBRATION
of PLATES

VIBRATION
of PLATES

S. Chakraverty

CRC Press
Taylor & Francis Group
Boca Raton London New York

CRC Press is an imprint of the
Taylor & Francis Group, an **informa** business

CRC Press
Taylor & Francis Group
6000 Broken Sound Parkway NW, Suite 300
Boca Raton, FL 33487-2742

First issued in paperback 2019

No claim to original U.S. Government works

ISBN-13: 978-0-367-45248-3 (pbk)
ISBN-13: 978-1-4200-5395-1 (hbk)

**Visit the Taylor & Francis Web site at
http://www.taylorandfrancis.com**

**and the CRC Press Web site at
http://www.crcpress.com**

Library of Congress Cataloging-in-Publication Data

Chakraverty, Snehashish.
    Vibration of plates / by Snehashish Chakraverty.
        p. cm.
    Includes bibliographical references and index.
    ISBN-13: 978-1-4200-5395-1 (alk. paper)
    ISBN-10: 1-4200-5395-7 (alk. paper)
    1. Plates (Engineering)--Vibration. I. Title.

TA660.P6C45 2009
624.1'7765--dc22   2008025635

Contents

Preface

This book is based on my experience as a researcher and a teacher for more than 15 years. It is intended primarily as a textbook as well as a reference for individual researchers in the study of vibration in general and dynamic behavior of structural members such as elastic plates in particular. The general theory of linear vibration is well established. In this regard, the book does not discuss the new theories of vibration of the structures such as plates. Moreover, there is a wealth of information on this subject given in the Reference section at the end of each chapter.

As of today, vibration analysis of complex-shaped structures is commonly encountered in engineering and architectural practice. In aeronautical, marine, mechanical, and civil structural designs, regular, irregular, and complex-shaped members are sometimes incorporated to reduce costly material, lighten the loads, provide ventilation, and alter the resonant frequencies of the structures. Accordingly, these shapes need to be accurately analyzed from an engineering perspective with easy and fast computational methods. In particular, plates and other structural members are integral parts of most engineering structures and their vibration analyses are needed for the safe design of structures. Analysis and design of such structures call for efficient computational tools. Finite element method (FEM), finite difference method (FDM), boundary element method (BEM), etc. are the standard industry methods to deal with such situations. But with irregular (complex) shapes of structural components, design is based on numerous approximations, which sometimes lead to inaccuracies and more computing time. In this book, very efficient shape functions are described, which result in far less computation time without compromising the accuracy of the results.

Vibration analysis of plates of various shapes and configurations has been studied extensively in the past. Dynamic behavior of these structures is strongly dependent on boundary conditions, geometrical shapes, material properties, different theories, and various complicating effects. In the initial stages, results were available for some simple cases, namely a limited set of boundary conditions and geometries, in which the analytical solution could be found. The lack of good computational facilities made it almost impossible to get accurate results even in these cases. With the advent of fast computers and various efficient numerical methods, there has been a tremendous increase in the amount of research done for getting better accuracy in the results. Although the discretization methods in terms of FEM, FDM, and BEM provide a general framework for general structures, they invariably result in problems with a large number of degrees of freedom. This deficiency may be overcome by using the Rayleigh–Ritz method. In recent times, a tremendous amount of work has been done throughout the globe by using the newly developed method of boundary characteristic orthogonal polynomials (BCOPs) with the Rayleigh–Ritz method. This method provides better accuracy of results, is more efficient, simple, and is easier for computer implementation.

The purpose of this book is to have a systematic understanding of the vibration and plate vibration theory, different plate geometries and their complicating effects along with their theory, analysis, and results. This may prove to be a benchmark for graduate and postgraduate students, teachers, and researchers in this field. The book provides comprehensive results, up-to-date and self-contained reviews along with application-oriented treatment of the use of newly developed method of BCOPs in vibration problems.

The linear vibration equations related to vibration analysis of plates and other vibration problems along with BCOPs are indeed very powerful tools for dynamical modeling, as will be illustrated throughout this book.

Chapter 1 introduces the theory, vibration basics, related equations of motion, and other fundamental details of single degree of freedom (SDF), multi-degree of freedom (MDF), and continuous systems. Chapter 2 provides the method of solution in simple cases of SDF, MDF, and continuous systems.

In Chapter 3, plate theory and equation of motion for plate vibration in different forms along with other vibration basics for plate members are discussed. General analysis in Cartesian, polar, and elliptic coordinates related to plate vibration is provided in order to give a perspective of the problem.

Various methods of solutions, namely exact, series type, and approximate methods for solution of transverse vibration of plates are included in Chapter 4. Exact solutions for circular and annular plates are well known and may also be found in other textbooks available in the market. So, in some cases only the final governing equations are presented. In this respect, for circular annular plates, there may have been some typographical mistakes, but here these have been corrected. Again solutions in terms of polar, rectangular, and elliptic coordinate systems are also given for a general idea. Moreover approximate solution methods are discussed in some detail, as these will be used in the solution of problems in later chapters.

The assumed deflection shapes used in the approximate methods such as in the Rayleigh–Ritz method were normally formulated by inspection and sometimes by trial and error, until recently, when a systematic method of constructing such a function in the form of characteristic orthogonal polynomials (COPs) was developed in 1985. Such developments of COPs in one and two dimensions along with preliminary details regarding orthogonal polynomials are discussed in Chapter 5. Here the first member of the COP is constructed so as to satisfy the essential boundary condition. Then the method of use of COPs in the solution of vibration of beams and plates with various shapes is presented.

In Chapter 6, the above COPs are modified to BCOPs to satisfy the essential boundary conditions in all the polynomials. Methods of generating the BCOPs in two and n dimensions are incorporated. Recurrence schemes to generate the BCOPs are discussed in detail for two dimensions and higher. Here, generalizations of the recurrence scheme along with their development in terms of grades are also presented.

Thus, Chapters 5 and 6 explain the new methodology of COPs and BCOPs in the analysis of vibration problems. One may generalize the procedure as discussed not only to other complex problems of vibration but the method may be extended to many diversified areas such as diffusion, potential theory, fluid mechanics, etc. where the physical quantity of interest can be approximated over a domain as a linear combination of these polynomials.

Chapters 7 through 10 treat the practical problems of plates of various geometries, namely circular, elliptic, rectangular, skew, triangular, and annular and their vibration analyses using the above-mentioned powerful method of BCOPs in the Rayleigh–Ritz method. Numerous results with different possible boundary conditions at the edges of the mentioned domains are presented. Getting the exact result in special cases where possible shows the efficacy and power of this methodology. In order to handle the BCOPs in a much easier form, Chapters 7 through 9 also deal with the mapping of general elliptic, triangular, and skew domain, respectively to circular, standard triangle, and unit square. This makes the analysis numerically efficient, and the unnecessary work of generating the BCOPs over each domain of a particular shape is reduced.

Chapter 10 demonstrates the analysis of vibration of circular annular and elliptic annular plates. Generation of BCOPs in annular domains and their use in the Rayleigh–Ritz method are discussed in detail. Numerical results for all possible boundary conditions in annular regions are cited.

Complicating effects in the plate members make the equation of motion complex and thus their analysis even more complicated. Accordingly, Chapters 11 through 14 introduce the concept of BCOPs in the study of vibration of plates with various complicating effects.

Nonhomogeneous material properties occur in the bodies especially due to imperfections of the materials. So, in Chapter 11, nonhomogeneous material properties in terms of Young's modulus and density are considered in the vibration analysis of plates. Plates with variable thickness along with different boundary conditions at the edges are analyzed in Chapter 12. Two types of variable thicknesses are considered to show the methodology about handling of this complicating effect in plate vibration.

In recent times, lightweight structures are widely used in a variety of engineering fields. Free vibration of orthotropic plates is an important area of such behavior. Analysis of vibration of plates with complicating effects such as orthotropic material properties are addressed in Chapter 13. Here, the method used is again the two-dimensional BCOPs, which shows the easy implementation of the method to extract the vibration characteristics of plates with respect to different boundary conditions at the edges. Example problems of elliptic orthotropic and annular elliptic orthotropic plates are considered in this chapter.

The last chapter (Chapter 14) deals with the effect of hybrid complicating effects in the study of vibration of plates. Simultaneous effects of two or more complicating effects on the vibration study of plates are termed here as hybrid effects that may invariably present in the materials of the plate.

Simultaneous behavior of nonhomogeneity, variable thickness, and ortho-tropy; generations of BCOPs in the plate domain with hybrid complicating materials; as well as a wide variety of results and analyses along with comparison in special cases are all presented in this chapter.

I sincerely hope that this book will help students, teachers, and researchers in developing an appreciation for the topic of vibration in general and vibration of plates in particular, especially with the use of the new method of BCOPs in the Rayleigh–Ritz method. Any errors, oversights, omissions, or other comments to improve the book can be communicated to S. Chakraverty, email: sne_chak@yahoo.com and will be greatly appreciated.

Acknowledgments

My sincere thanks are to Prof. Bani Singh, former professor, IIT Roorkee (Now JIIT, Noida) for initiating me in this field during my initial period of research; Prof. R.B. Bhat, Concordia University, Quebec, Canada and Prof. M. Petyt, ISVR, University of Southampton, Southampton, United Kingdom while working on further application as well as refinement of the BCOPs method; and to Prof. A. Sahu, Coppin State University, Baltimore, Maryland, for encouragement. Further, the credit of developing COPs by Prof. Bhat for the first time in 1985 is also greatly acknowledged.

I am thankful to the Director, Central Building Research Institute, Roorkee, for giving me the permission to publish this book and to Taylor & Francis Group/CRC Press for all the help during this project. I would like to thank Ragini Jindal for having computed numerical results of Chapter 14 and Sameer for drawing the figures in Word format. I am greatly indebted to all the contributors and to the authors of the books listed in the Reference section at the end of every chapter. In particular, I am indebted to Prof. A.W. Leissa, The Ohio State University, Columbus, Ohio, for his invaluable reference manual on "vibration of plates" (published by NASA, 1969; reprinted by Acoustical Society of America, 1993).

I would very much like to acknowledge the encouragement, patience, and support provided by all my family members, in particular, my parents Sh. Birendra K. Chakraborty and Smt. Parul Chakraborty, my wife Shewli, and my daughters Shreyati and Susprihaa.

1

Background of Vibration

1.1 Vibration Basics

Vibration is the mechanical oscillation of a particle, member, or a body from its position of equilibrium. It is the study that relates the motion of physical bodies to the forces acting on them. The basic concepts in the mechanics of vibration are space, time, and mass (or forces). When a body is disturbed from its position, then by the elastic property of the material of the body, it tries to come back to its initial position. In general, we may see and feel that nearly everything vibrates in nature; vibrations may be sometimes very weak for identification. On the other hand, there may be large devastating vibrations that occur because of manmade disasters or natural disasters such as earthquakes, winds, and tsunamis.

As already mentioned, natural and human activities always involve vibration in one form or the other. Recently, many investigations have been motivated by the engineering applications of vibration, such as the design of machines, foundations, structures, engines, turbines, and many control systems. Vibration is also used in pile-driving, vibratory testing of materials, and electronic units to filter out unwanted frequencies. It is also employed to simulate the complex earthquake phenomenon and to conduct studies in the design of nuclear reactors.

On the one hand, vibrations are of great help, while on the other, there are many cases of devastating effects of excessive vibration on engineering structures. Therefore, one of the important purposes of vibration study is to reduce vibration through proper and comparatively accurate design of machines and structures. In this connection, the mechanical, structural, and aerospace engineers need the information regarding the vibration characteristics of the systems before finalizing the design of the structures.

In a dynamics problem, the applied loadings (and hence the structural response such as deflection, internal forces, and stress) vary with time. Thus, unlike a statics problem, a dynamics problem requires a separate solution at every instant of time. The structure may be considered as subjected to two loadings, namely the applied load and the inertia forces. The inertia forces are the essential characteristics of a structural dynamics problem.

The magnitude of the inertia forces depends on (1) the rate of loading, (2) the stiffness of the structure, and (3) the mass of the structure.

If the loading is applied slowly, the inertia forces are small in relation to the applied loading and may be neglected, and thus the problem can be treated as static. If the loading is rapid, the inertia forces are significant and their effect on the resulting response must be determined by dynamic analysis.

Generally, structural systems are continuous and their physical properties or characteristics are distributed. However, in many instances, it is possible to simplify the analysis by replacing the distributed characteristics by discrete characteristics, using the technique of lumping. Thus, mathematical models of structural dynamics problems may be divided into two major types:

1. Discrete systems with finite degrees of freedom (DOFs)
2. Continuous systems with infinite DOFs

However, in the latter case, a good approximation to the exact solution can be obtained by using a finite number of appropriate shape functions. The main content of this book is the vibration of plates; however, to have a basic knowledge of vibration and the related terms to understand the plate vibration in detail, first the case of a discrete system with 1DOF will be described in this chapter, followed by a multi-degree-of-freedom (MDOF) system. Then, the continuous systems will be defined. However, methods of analysis and other details regarding these will be included in Chapter 2. Mainly free vibration will be taken into consideration, i.e., the vibration that takes place in the absence of external excitations, which is the foundation of other complex vibration studies.

Keeping this in mind, the following sections address the above in some detail.

1.1.1   Causes of Vibration

The main causes of vibration are as follows:

- Unequal distribution of forces in a moving or rotating machinery
- External forces like wind, tides, blasts, or earthquakes
- Friction between two bodies
- Change of magnetic or electric fields
- Movement of vehicles, etc.

1.1.2   Requirements for Vibration

The main requirements for the vibration are as follows:

- There should be a restoring force.
- The mean position of the body should be in equilibrium.
- There must be inertia (i.e., we must have mass).

As pointed out, in vibration study, we are having two types of systems, discrete and continuous. Accordingly, we next define the distinction between the two.

1.1.3 Discrete and Continuous Systems

Dynamic system models may be divided into two classes, discrete and continuous (or distributed). The systems do depend on system parameters such as mass, damping, and stiffness (these will be defined later). Discrete systems are described mathematically by the variables that depend only on time. On the other hand, continuous systems are described by variables that depend on time and space. As such, the equations of motion of discrete systems are described by ordinary differential equations (ODEs), whereas the equations of motion for continuous systems are governed by partial differential equations (PDEs). Because ODE contains only one independent variable, i.e. time, and PDE contains more than one independent variable, such as time and space coordinates. To describe a system, we need to know the variables or coordinates that describe the system and this follows a term known as degrees of freedom (DOF) and the DOF is defined as the minimum number of independent variables required to fully describe the motion of a system.

If the time dependence is eliminated from the equation of discrete system, then it will be governed by a set of simultaneous algebraic equations and the continuous system will be governed by boundary value problem. Most of the mechanical, structural, and aerospace systems can be described by using a finite number of DOFs. Continuous systems have infinite number of DOFs.

1.1.4 Glossary of Some Terms

The following are some terms related to vibration study that will be frequently used in the whole of this book:

Amplitude: The maximum distance to which the particle or system moves on either side of the mean position is called amplitude of the vibrating system.

Time period: The time taken by the particle (or mass) in making one complete oscillation is called time period. It is generally denoted by T.

Frequency: The number of oscillations performed by a vibrating system per second is termed as frequency of vibration and it is the reciprocal of the period, i.e., frequency $= 1/T$.

Angular frequency or circular frequency: The angular frequency of a periodic quantity, in radians per unit time, is 2π times the frequency.

Natural frequency: Natural frequency is the frequency of free vibration of a system.

Damping: The process of energy dissipation in the study of vibration is generally referred as damping.

Elasticity: A material property that causes it to return to its natural state after being deformed.

Spring: A spring is a flexible mechanical line between two particles in a mechanical system.

Stiffness: Force of elastic spring (string) per unit displacement. In other words, it is the ratio of change of force (or torque) to the corresponding change in translation (or rotation) deflection of an elastic element. This is usually denoted by k.

Potential energy: The potential energy of a body is the energy it possesses due to its position and is equal to the work done in raising it from some datum level. A body of mass m at a height h has the potential energy of mgh, where g is the acceleration due to gravity.

Kinetic energy: This energy is possessed due to the velocity. If a body of mass m attains a velocity v from rest subject to a force P and moves a distance say x, then the kinetic energy of the body is given by $(1/2)mv^2$.

Strain energy: The strain energy of a body is the energy stored when the body is deformed. If an elastic body of stiffness k is extended a distance x by a force P, then the strain energy is given by $(1/2)kx^2$.

Lagrangian: The Lagrangian or Lagrangian function is defined as the difference between the kinetic energy and the potential energy of a system.

Other terms related to vibration will be discussed in the places where it will be first used.

1.1.5 Basic Vibration Model

Basic vibration model of a simple oscillatory system consists of a mass m, a massless spring with stiffness k, and a damper with damping c.

The spring supporting the mass is assumed to be of negligible mass. Its force (F)–deflection (x) relationship is considered to be linear and so following the Hooke's law, we have

$$F = kx \tag{1.1}$$

The viscous damping, generally represented by a dashpot, is described by a force (f) proportional to the velocity. So, we have

$$f = c\dot{x} \tag{1.2}$$

1.2   One Degree of Freedom Systems

In this case, only one coordinate is required to define the configuration of the system and so we have 1 DOF system. Figure 1.1 shows a simple undamped spring mass system, which is assumed to move only along the vertical direction. Corresponding free-body diagram is shown in Figure 1.2. The system in Figure 1.1 has 1 DOF because a single coordinate x describes its motion. When this system is placed in motion, oscillation will take place at its natural frequency, f_n, which is a property of the system. The deformation of the spring in the static equilibrium position is δ and the spring force $k\delta$ (where k is stiffness of the spring) is equal to the gravitational force w acting on mass m. So, we can write

$$k\delta = w \tag{1.3}$$

By measuring the displacement x from the static equilibrium position, the forces acting on m are $k(\delta + x)$ and w. Here, x is chosen to be positive in the downward direction. So, all quantities, viz., force, velocity, and acceleration, will also be positive in the downward direction. We now apply Newton's second law of motion to the mass, and accordingly, we have

$$m\ddot{x} = \sum F = w - k(\delta + x) \tag{1.4}$$

where $\ddot{x} = d^2x/dt^2$. From Equations 1.3 and 1.4, we will have

$$m\ddot{x} = -kx \tag{1.5}$$

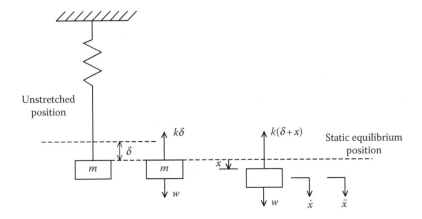

FIGURE 1.1
Undamped 1 DOF system (physical model).

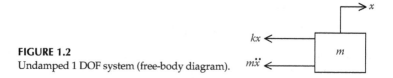

FIGURE 1.2

Undamped 1 DOF system (free-body diagram).

which shows that the resultant force on m is simply the spring force due to the displacement x. We will now write the standard form of the equation of motion for the above 1 DOF system from Equation 1.5 as

$$m\ddot{x} + kx = 0 \tag{1.6}$$

By defining the circular frequency ω_n by the equation

$$\omega_n^2 = k/m \tag{1.7}$$

Equation 1.6 may be written as

$$\ddot{x} + \omega_n^2 x = 0 \tag{1.8}$$

Vibratory motion represented by Equation 1.8 is called a harmonic motion. It is to be noted here that if there is a viscous damping (Equation 1.2) along with a force P, we can write the equation of motion of damped 1 DOF system as in Equation 1.6 by

$$m\ddot{x} + c\dot{x} + kx = P \tag{1.9}$$

The solution and other details regarding the 1 DOF systems will be discussed in Chapter 2.

Now we will discuss some simple problems of vibration such as simple pendulum, metal thin strip with a mass at one end, torsion of a rod having a pulley at one end, and an electric circuit, considering these as 1 DOF systems.

1.2.1 Simple Pendulum

Let us consider a simple pendulum as shown in Figure 1.3 where l is the length of the string and m is the mass of the bob of the pendulum. If O is the mean position of the pendulum, then restoring force will be given by $mg \sin \theta$, where acceleration due to gravity is denoted by g and θ is the angular displacement. The corresponding equation of circular path, i.e., S is written as

$$S = l\theta \tag{1.10}$$

If S is increasing along OP (Figure 1.3), then the equation of motion may be written as

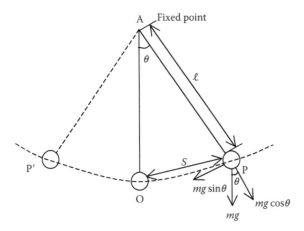

FIGURE 1.3
Simple pendulum.

$$m\frac{d^2s}{dt^2} = -mg\sin\theta \qquad (1.11)$$

Considering Equation 1.10 in Equation 1.11 and approximating $\sin\theta$ by θ for linear vibration, the equation of motion for a simple pendulum may be written as

$$\frac{d^2\theta}{dt^2} + \omega_n^2\theta = 0 \qquad (1.12)$$

where $\omega_n^2 = g/l$ is the natural frequency of simple pendulum.

1.2.2 Metal Thin Strip with a Mass at One End

Let us now consider the case of a mass m at the end of a massless thin metal strip of length l and flexural rigidity EI clamped at the other end as shown in Figure 1.4a. Here, E and I are, respectively, Young's modulus and polar moment of inertia of the cross section. Suppose the mass m is displaced to a distance x from its equilibrium position by a steady force. We have to first find the force acting on m due to the elasticity of the metal strip. The force acting on m must be equal and opposite to a steady force F, which would maintain a steady deflection x when applied transversely at the end of the strip as shown in Figure 1.4b. This restoring force F is proportional to displacement x and we will have the equation

$$F = \left(\frac{3EI}{l^3}\right)x \qquad (1.13)$$

Now, by Newton's second law, we may write

$$m\frac{d^2x}{dt^2} = -F \qquad (1.14)$$

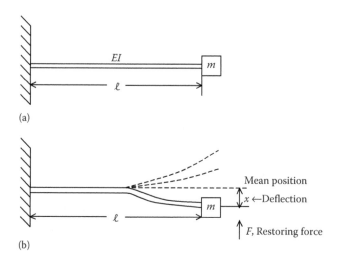

(a)

(b)

FIGURE 1.4
Metal thin strip with a mass m at one end.

Combining Equations 1.13 and 1.14, the equation of motion may now be written as

$$\frac{d^2x}{dt^2} + \omega_n^2 x = 0 \tag{1.15}$$

where $\omega_n^2 = 3EI/(ml^3)$ is the natural frequency of the 1 DOF system.

1.2.3 Torsion of a Rod Having a Pulley at One End

Here, we assume that a pulley of moment of inertia I is attached at the end of a rod of length l and torsional rigidity GJ, with the other end fixed, where G and J are, respectively, the modulus of rigidity and the polar moment of inertia of the cross section of the rod. The vibratory system is shown in Figure 1.5. The system exhibits a rotational oscillation when disturbed. So, a torque known as restoring torque will be created. We first consider the steady torque T that would maintain a steady angular deflection θ of the pulley. The restoring torque T in this case is proportional to the rate of angle of twist θ per unit length of the rod (i.e., $T \propto \theta/l$) and so we may have

$$T = GJ\frac{\theta}{l} \tag{1.16}$$

Again, for rotation we may write

$$I\frac{d^2\theta}{dt^2} = -T \tag{1.17}$$

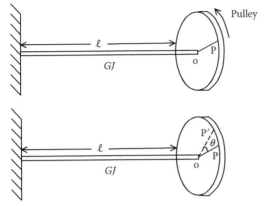

FIGURE 1.5
Torsion of a rod having a pulley at
one end.

Combining Equations 1.16 and 1.17, the equation of motion for torsional
vibrations of the system is given by

$$\frac{d^2\theta}{dt^2} + \omega_n^2\theta = 0 \tag{1.18}$$

where $\omega_n^2 = GJ/(lI)$ is the natural frequency of the above 1 DOF system.

1.2.4 Electric Circuit Having Current, Capacitance, Inductance, and Voltage

The circuit as shown in Figure 1.6 consists of a condenser of capacitance C,
discharging through an inductance L. Here, we neglect the electrical resistance of the circuit. At any time, let v be the voltage across the condenser,
i the current, and q the charge, then we have

$$i = -\frac{dq}{dt} \tag{1.19}$$

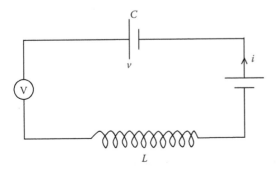

FIGURE 1.6
Electric circuit without resistance.

Faraday's law of induction gives

$$v = L\frac{di}{dt}$$

(1.20)

Moreover, charge q may be written as

$$q = Cv$$

(1.21)

Combining Equations 1.19 through 1.21 leads to the following equation of motion for the present system:

$$\frac{d^2q}{dt^2} + \omega_n^2 q = 0$$

(1.22)

where $\omega_n^2 = 1/(LC)$ is the natural frequency of the above 1 DOF system.

1.3 Two Degree of Freedom Systems

In the previous sections, only one coordinate was required to designate the system for a single mass attached to single spring as in Figure 1.1. In this section, we will consider systems where two masses are connected with two springs in series as depicted in Figure 1.7. Here, a mass m_1 is connected with a weightless spring of stiffness k_1 and a mass m_2 with a weightless

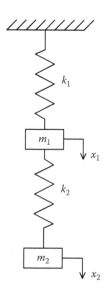

FIGURE 1.7
2 DOF system (displacement in vertical direction).

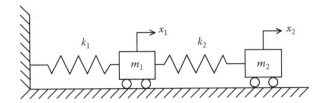

FIGURE 1.8
2 DOF system (displacement in horizontal direction).

spring of stiffness k_2. Two coordinates are needed to specify the configuration of the system and we come up with 2 DOFs.

1.3.1 Equation of Motion for Two Degree of Freedom System

For free vibration of systems with the 2 DOF, we may consider either Figure 1.7 or the system shown in Figure 1.8 where two masses m_1 and m_2 can slide without friction along a horizontal axis. The masses are connected by springs of stiffness k_1 and k_2 as shown. Either of the two systems is again depicted in Figure 1.9 where a free-body diagram with forces acting on masses m_1 and m_2 are given. From Figure 1.9, we can write the equation of motion for a 2 DOF system without damping as

$$m_1\ddot{x}_1 + k_1 x_1 = k_2(x_2 - x_1) \qquad (1.23)$$

$$m_2\ddot{x}_2 + k_2(x_2 - x_1) = 0 \qquad (1.24)$$

Equations 1.23 and 1.24 can now be written in matrix form as

$$\begin{bmatrix} m_1 & 0 \\ 0 & m_2 \end{bmatrix} \begin{Bmatrix} \ddot{x}_1 \\ \ddot{x}_2 \end{Bmatrix} + \begin{bmatrix} k_1 + k_2 & -k_2 \\ -k_2 & k_2 \end{bmatrix} \begin{Bmatrix} x_1 \\ x_2 \end{Bmatrix} = \begin{Bmatrix} 0 \\ 0 \end{Bmatrix} \qquad (1.25)$$

1.3.2 Example of Two Degree of Freedom System (with Damping and Force)

Consider a two-storey shear building idealized as a discrete system. If the floor girders are assumed to be rigid and the axial deformation of the columns is neglected, then this system will have 2 DOF, one degree of translation per floor, as shown in Figure 1.10.

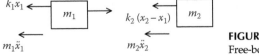

FIGURE 1.9
Free-body diagram for 2 DOF system.

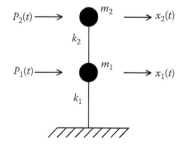

FIGURE 1.10
Two-storey shear building.

Let the lumped masses at the first and second storey be m_1 and m_2. If the lateral stiffnesses of the first and second storey are k_1 and k_2 and the corresponding viscous damping coefficients are c_1 and c_2, then this physical system can be represented by the mechanical system shown in Figure 1.11. The corresponding two equations of motion, one for each mass, can be written. These two equations may be expressed in matrix form as follows:

$$[M]\{\ddot{x}\} + [C]\{\dot{x}\} + [K]\{x\} = \{P\} \qquad (1.26)$$

where

$\{P\} = \begin{Bmatrix} P_1 \\ P_2 \end{Bmatrix}$, the force vector

$\{x\} = \begin{Bmatrix} x_1 \\ x_2 \end{Bmatrix}$, the displacement vector

$[M] = \begin{bmatrix} m_1 & 0 \\ 0 & m_2 \end{bmatrix}$, the mass matrix

$[C] = \begin{bmatrix} c_1 + c_2 & -c_2 \\ -c_2 & c_2 \end{bmatrix}$, the damping matrix

$[K] = \begin{bmatrix} k_1 + k_2 & -k_2 \\ -k_2 & k_2 \end{bmatrix}$, the stiffness matrix

In the governing equation, the mass matrix is diagonal because of the lumped mass idealization adopted. The stiffness and damping matrices are symmetrical with positive diagonal terms and with the largest term being in

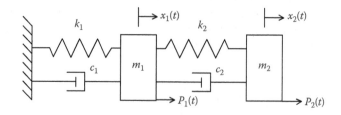

FIGURE 1.11
Mechanical system corresponding to the two-storey shear building.

the diagonal. Methods of solution and other details will be included in the subsequent chapters.

1.3.3 Coordinate Coupling

The differential equation of motion for discrete systems is, in general, coupled, in that all coordinates appear in each equation. For a 2 DOF system, in the most general case, the two equations for the undamped system have the form

$$m_{11}\ddot{x}_1 + m_{12}\ddot{x}_2 + k_{11}x_1 + k_{12}x_2 = 0 \qquad (1.27)$$

$$m_{21}\ddot{x}_1 + m_{22}\ddot{x}_2 + k_{21}x_1 + k_{22}x_2 = 0 \qquad (1.28)$$

These equations can be expressed in matrix form as

$$\begin{bmatrix} m_{11} & m_{12} \\ m_{21} & m_{22} \end{bmatrix} \begin{Bmatrix} \ddot{x}_1 \\ \ddot{x}_2 \end{Bmatrix} + \begin{bmatrix} k_{11} & k_{12} \\ k_{21} & k_{22} \end{bmatrix} \begin{Bmatrix} x_1 \\ x_2 \end{Bmatrix} = \begin{Bmatrix} 0 \\ 0 \end{Bmatrix} \qquad (1.29)$$

which immediately reveals the type of coupling present. Mass or dynamical coupling exists if the mass matrix is non-diagonal, whereas stiffness or static coupling exists if the stiffness matrix is non-diagonal.

It is possible to find a coordinate system that has neither form of coupling. The two equations are then decoupled and each equation can be solved independently of the other (as described in Chapter 2 for 1 DOF system). Such coordinates are called principal coordinates (also called normal coordinates). Although it is always possible to decouple the equations of motion for the undamped system, this is not always the case for a damped system.

The following matrix equation has zero dynamic and static coupling, but the damping matrix couples the coordinates.

$$\begin{bmatrix} m_{11} & 0 \\ 0 & m_{22} \end{bmatrix} \begin{Bmatrix} \ddot{x}_1 \\ \ddot{x}_2 \end{Bmatrix} + \begin{bmatrix} c_{11} & c_{12} \\ c_{21} & c_{22} \end{bmatrix} \begin{Bmatrix} \dot{x}_1 \\ \dot{x}_2 \end{Bmatrix} + \begin{bmatrix} k_{11} & 0 \\ 0 & k_{22} \end{bmatrix} \begin{Bmatrix} x_1 \\ x_2 \end{Bmatrix} = \begin{Bmatrix} 0 \\ 0 \end{Bmatrix} \qquad (1.30)$$

If in the foregoing equation $c_{12} = c_{21} = 0$, then the damping is said to be proportional (to the stiffness or mass matrix), and the system equations become uncoupled.

1.4 Multi-Degree of Freedom Systems

In the previous section, we have discussed the systems with 2 DOF. Systems with more than 2 DOFs may be studied in a similar fashion. Although the difficulty increases with the number of equations that govern the system,

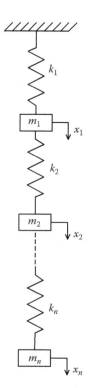

FIGURE 1.12
n DOF system (displacement in vertical direction).

in this case we may understand that the system is having several or MDOF. If the MDOF system is having *n* DOF, then it will have *n* independent coordinates designating the structural system as shown in Figure 1.12. In Figure 1.12, a series of masses m_1, m_2,..., m_n are connected in series by massless springs with stiffnesses k_1, k_2,..., k_n.

1.4.1 Equation of Motion for Multi-Degree of Freedom System

As in 2 DOF system, *n* DOF system is depicted in Figure 1.13 where *n* masses can slide without friction along a horizontal axis. The corresponding free-body diagram is given in Figure 1.14. Thus, we can write the equations

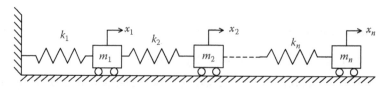

FIGURE 1.13
n DOF system (displacement in horizontal direction).

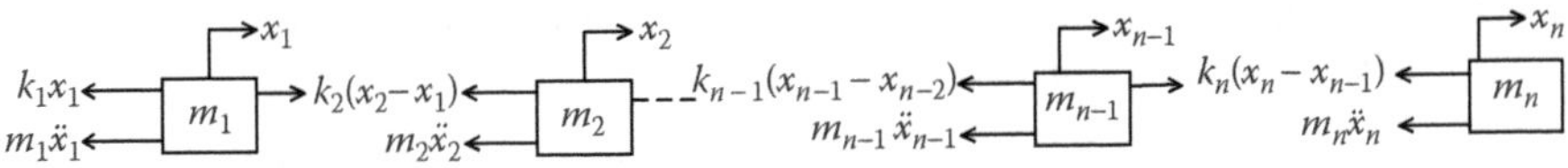

FIGURE 1.14
Free-body diagram for n DOF system (undamped).

of motion for n DOF system with no damping and no force as the following n equations:

$$k_1 x_1 - k_2(x_2 - x_1) + m_1 \ddot{x}_1 = 0$$

$$k_2(x_2 - x_1) - k_3(x_3 - x_2) + m_2 \ddot{x}_2 = 0$$

$$k_3(x_3 - x_2) - k_4(x_4 - x_3) + m_3 \ddot{x}_3 = 0$$

$$\cdots\cdots\cdots\cdots\cdots\cdots$$
$$\cdots\cdots\cdots\cdots\cdots\cdots$$

$$k_{n-1}(x_{n-1} - x_{n-2}) - k_n(x_n - x_{n-1}) + m_{n-1}\ddot{x}_{n-1} = 0$$

$$k_n(x_n - x_{n-1}) + m_n \ddot{x}_n = 0$$

The above n equations designating the n DOF system may be written in matrix form as

$$[M]\{\ddot{x}\} + [K]\{x\} = \{0\} \tag{1.31}$$

where

$$[M] = \begin{bmatrix} m_1 & 0 & \cdots & \cdots & 0 \\ 0 & m_2 & 0 & \cdots & 0 \\ \cdots & \cdots & \cdots & \cdots & \cdots \\ \cdots & \cdots & 0 & m_{n-1} & 0 \\ 0 & \cdots & \cdots & 0 & m_n \end{bmatrix}$$

$$[K] = \begin{bmatrix} k_1 + k_2 & -k_2 & 0 & \cdots & 0 \\ -k_2 & k_2 + k_3 & -k_3 & \cdots & 0 \\ \cdots & \cdots & & & \cdots \\ 0 & \cdots & -k_{n-1} & k_{n-1} + k_n & -k_n \\ 0 & & \cdots & -k_n & k_n \end{bmatrix}$$

$$\{\ddot{x}\} = \begin{Bmatrix} \ddot{x}_1 \\ \ddot{x}_2 \\ \vdots \\ \ddot{x}_{n-1} \\ \ddot{x}_n \end{Bmatrix}, \quad \{x\} = \begin{Bmatrix} x_1 \\ x_2 \\ \vdots \\ x_{n-1} \\ x_n \end{Bmatrix} \quad \text{and} \quad \{0\} = \begin{Bmatrix} 0 \\ 0 \\ \vdots \\ 0 \\ 0 \end{Bmatrix}$$

1.4.2 Equation of Motion for Multi-Degree of Freedom System with Damping and Force

We may now refer to the free-body diagram in Figure 1.15 for the mass m_i of a MDOF system with damping and force. The corresponding equation of motion in matrix form is written as

$$[M]\{\ddot{x}\} + [C]\{\dot{x}\} + [K]\{x\} = \{P\} \tag{1.32}$$

where the mass matrix $[M]$ and stiffness matrix $[K]$ are defined earlier whereas the damping matrix $[C]$ is written as

$$[C] = \begin{bmatrix} c_1 + c_2 & -c_2 & 0 & \cdots & 0 \\ -c_2 & c_2 + c_3 & -c_3 & \cdots & 0 \\ \cdots & \cdots & & & \cdots \\ 0 & \cdots & -c_{n-1} & c_{n-1} + c_n & -c_n \\ 0 & & \cdots & -c_n & c_n \end{bmatrix}$$

$$\{P\} = \begin{Bmatrix} P_1 \\ P_2 \\ \vdots \\ P_{n-1} \\ P_n \end{Bmatrix}$$

and

$$\{\dot{x}\} = \begin{Bmatrix} \dot{x}_1 \\ \dot{x}_2 \\ \vdots \\ \dot{x}_{n-1} \\ \dot{x}_n \end{Bmatrix}$$

For a system with n DOF, the governing equation of motion, viz., Equation 1.32, consists of matrices of size $n \times n$ representing n coupled simultaneous second-order differential equations. The solution is obtained through a step-by-step integration procedure with certain assumptions about the variation of acceleration, velocity, and displacement over the selected time step. However, in case of the linear system, a more convenient and elegant method to

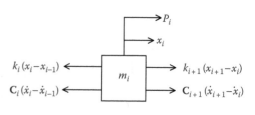

FIGURE 1.15
Free-body diagram for n DOF system (damped and forced).

solve this set of equations is through modal analysis, where the solution is obtained as the superposition of the contribution from different modes of vibrations. This method will be elaborated in later chapters.

1.5 Continuous Systems

Till now, we described the discrete systems, which are defined by finite number of DOFs and the corresponding differential equation is ordinary. But in the structural or mechanical systems such as strings, beams, rods, membranes, plates, and shells, elasticity and mass are considered to be distributed and hence these are called distributed or continuous systems. The continuous systems are designated by infinite number of DOFs. Displacement of continuous systems is described by a continuous function of position and time and consequently will be governed by partial differential equations (PDEs). In the following sections, we will address the vibration equation of simple continuous systems, viz., string, rod, beam, and membrane. Other complex continuous systems such as plates, which are the main topic of this book, will be described in detail in the subsequent chapters.

1.5.1 Transverse Vibration of a String

Consider a uniform elastic string stretched tightly between two fixed points O and A (Figure 1.16a) under tension T. Taking O as origin, OA as the axis of X and a line OY perpendicular to OX as the axis of Y, let $y(x, t)$ denote the transverse displacement of any point of string at distance x from O at time t.

To study the motion, the following assumptions are made:

1. Entire motion takes place in the XY-plane, i.e., each particle of the string moves in a direction perpendicular to X-axis.
2. String is perfectly flexible and offers no resistance to bending.
3. Tension in the string is large enough so that the weight of the string can be neglected.
4. Transverse displacement y and the slope $\partial y / \partial x$ are small so that their squares and higher powers can be considered negligible.

Now, let m be the mass per unit length of the string and consider a differential element PQ $(= ds)$ at a distance x from O. Then $ds = \sqrt{1 + (\partial y / \partial x)^2} \approx dx$.

The equation of motion for this element would be (from Figure 1.16b)

$$(m\,dx)\frac{\partial^2 y}{\partial t^2} = T_2 \sin \beta - T_1 \sin \alpha \text{ (along the vertical direction)} \qquad (1.33)$$

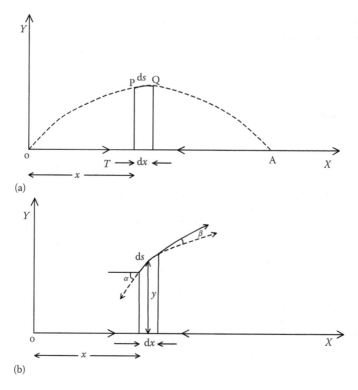

FIGURE 1.16
Transverse vibration of string.

$$0 = T_2 \cos\beta - T_1 \cos\alpha \text{ (along the horizontal direction)} \quad (1.34)$$

Equation 1.34 may be written as

$$T_2 \cos\beta = T_1 \cos\alpha = T \quad (1.35)$$

From Equations 1.33 and 1.35, we will arrive at

$$\frac{mdx}{T}\frac{\partial^2 y}{\partial t^2} = \tan\beta - \tan\alpha$$

$$= \left(\frac{\partial y}{\partial x}\right)_{x+dx} - \left(\frac{\partial y}{\partial x}\right)_x$$

Expanding the first term in the above equation by Taylor's series, the equation of motion for vibration of string may be written as

$$\frac{\partial^2 y}{\partial t^2} = c^2 \frac{\partial^2 y}{\partial x^2} \quad (1.36)$$

where $c^2 = \frac{T}{m}$. The above equation of motion is also called the one-dimensional wave equation.

1.5.2 Longitudinal Vibration of a Rod

Here, we study the longitudinal vibrations of a rod as shown in Figure 1.17 where the following assumptions are made for deriving the equation of motion:

1. The rod is thin and has a uniform cross section (R).
2. During vibration, a plane section of the rod normal to its axis remains plane and normal to the axis.
3. Each particle on a section undergoes axial displacement only.
4. A section of the rod can be specified by its x-coordinate only, where x denotes the distance of the section from the origin O (Figure 1.17).

Figure 1.17 also shows the free-body diagram of a differential element of this rod of length dx. The equilibrium position of the element is denoted by x and deformed position is u. So, if u is displacement at x, the displacement at $x + dx$ will be $u + \frac{\partial u}{\partial x} dx$. Let ρ, E, and σ be, respectively, the mass density of the material of the rod, the Young's modulus, and stress. Then, applying Newton's second law to the differential element gives

$$-\sigma R + \left(\sigma + \frac{\partial \sigma}{\partial x} dx\right) R = \rho R \, dx \frac{\partial^2 u}{\partial t^2} \tag{1.37}$$

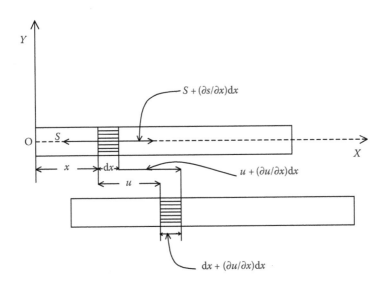

FIGURE 1.17
Displacement of element of the rod.

For elastic deformations, Hooke's law gives the modulus of elasticity E as the ratio of the unit stress to unit strain and so we get

$$\sigma = E\frac{\partial u}{\partial x} \tag{1.38}$$

Combining Equations 1.37 and 1.38 leads to

$$E\frac{\partial^2 u}{\partial x^2} = \rho\frac{\partial^2 u}{\partial t^2} \tag{1.39}$$

Finally, the equation of motion for the rod may be written from Equation 1.39 as

$$\frac{\partial^2 u}{\partial t^2} = c^2\frac{\partial^2 u}{\partial x^2} \tag{1.40}$$

where $c^2 = E/\rho$, which is called the velocity of propagation of the displacement.

1.5.3 Transverse Vibration of an Elastic Beam

Let us consider an elastic beam of modulus of rigidity EI, density ρ, and cross-section area A as shown in Figure 1.18. Here, E is the modulus of elasticity of the beam material and I is the moment of inertia of the beam about the axis of bending. Let q be the external normal force per unit length and Q and M be the shear force and bending moment at a distance x from the

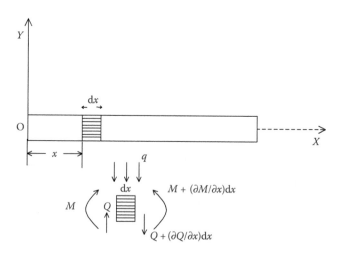

FIGURE 1.18
Transverse vibration of beam.

origin O as shown. Considering the forces acting downwards, we get the equation of motion as

$$\rho A\, dx \frac{\partial^2 y}{\partial t^2} = Q + \frac{\partial Q}{\partial x} dx - Q + q\, dx \tag{1.41}$$

We also have the relations for the shear force and bending moment, respectively, as

$$Q = \frac{\partial M}{\partial x} \tag{1.42}$$

and

$$M = -EI \frac{\partial^2 y}{\partial x^2} \tag{1.43}$$

Using Equations 1.41 and 1.42, we write

$$\rho A \frac{\partial^2 y}{\partial t^2} dx = \frac{\partial^2 M}{\partial x^2} dx + q\, dx \tag{1.44}$$

Now, substituting M from Equation 1.43 in Equation 1.44, the equation of motion may be obtained in this case as

$$EI \frac{\partial^4 y}{\partial x^4} + \rho A \frac{\partial^2 y}{\partial t^2} = q \tag{1.45}$$

If there is no external load, then putting $q = 0$ in Equation 1.45, the equation of motion for free vibration of elastic beam is written as

$$\frac{\partial^4 y}{\partial x^4} + \frac{1}{a^2} \frac{\partial^2 y}{\partial t^2} = 0 \tag{1.46}$$

where $a^2 = \dfrac{EI}{\rho A}$.

1.5.4 Vibration of Membrane

A membrane is a perfectly flexible infinitely thin lemna of uniform tension. It is further assumed that fluctuations in the tension of the membranes due to small deflections during vibrations can be neglected. Suppose the plane of the membrane coincides with the XY-plane as shown in Figure 1.19, we will consider forces acting on the elementary membrane $\Delta x \Delta y$. Before proceeding further, the following assumptions are made:

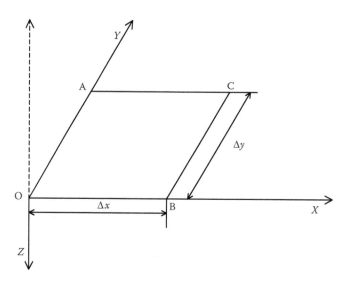

FIGURE 1.19
Elementary membrane in the XY-plane.

1. Displacement z of any point of the membrane is perpendicular to the plane of the lemna during vibration.
2. Tension on the membrane is uniform, and therefore, we let the tension per unit length along the boundary OB, BC, CA, and AO of the membrane be P.
3. Let m be the mass of the membrane per unit area so that the mass of the elementary portion of the membrane is $m\Delta x\Delta y$.
4. As we are assuming that the displacements are perpendicular to the XY-plane during vibrations, we also assume that there is no sideway motion.

By referring to Figure 1.20, the horizontal component of tensions gives

$$T_0 \cos \psi_0 = T_1 \cos \psi_1 = P\Delta y \tag{1.47}$$

where T_0 and T_1 are the tensions acting on the edges OA and BC, respectively.

As the displacements are assumed to be small, by taking $\cos \psi_0 = \cos \psi_1 = 1$ in Equation 1.47, we obtain

$$T_0 = T_1 = P\Delta y \tag{1.48}$$

Next, considering vertical components (downward) of tension for the faces OA and BC due to motion, one may arrive at the expression

$$V_t \approx T_1 \sin \psi_1 - T_0 \sin \psi_0 \tag{1.49}$$

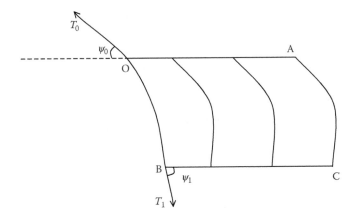

FIGURE 1.20
Tension T_0 and T_1 acting on the edges OA and BC of the membrane.

We have already assumed that $\cos \psi_0 = \cos \psi_1 = 1$ for small displacement, so we may write $\sin \psi_1 = \tan \psi_1$ and $\sin \psi_0 = \tan \psi_0$. Substituting these in Equation 1.49 along with the values of T_0 and T_1 from Equation 1.48, the expression for V_t turns out to be

$$V_t \approx P\Delta y \left\{ \left(\frac{\partial z}{\partial x} \right)_{\psi_1} - \left(\frac{\partial z}{\partial x} \right)_{\psi_0} \right\} \tag{1.50}$$

Rewriting Equation 1.50 by expanding $\left(\frac{\partial z}{\partial x} \right)_{\psi_1}$ and considering the first two terms of the expansion, we will arrive at the following expression for vertical component V_t for the faces OA and BC

$$V_t \approx P\Delta x \Delta y \frac{\partial^2 z}{\partial x^2} \tag{1.51}$$

Similarly, considering the vertical component of tension for the faces OB and AC, the expression may be written as

$$\overline{V}_t \approx P\Delta x \Delta y \frac{\partial^2 z}{\partial y^2} \tag{1.52}$$

From Equations 1.51 and 1.52, we may now write the equation of motion for vibration of membrane as

$$m\Delta x \Delta y \frac{\partial^2 z}{\partial t^2} = P\Delta x \Delta y \left(\frac{\partial^2 z}{\partial x^2} + \frac{\partial^2 z}{\partial y^2} \right) \tag{1.53}$$

Finally, the equation of motion in standard form becomes

$$\frac{\partial^2 z}{\partial t^2} = c^2\left(\frac{\partial^2 z}{\partial x^2} + \frac{\partial^2 z}{\partial y^2}\right) \tag{1.54}$$

where $c^2 = P/m$.

Equation 1.54 can also be expressed as

$$c^2\nabla^2 z = \frac{\partial^2 z}{\partial t^2} \tag{1.55}$$

where

$$\nabla^2 = \frac{\partial^2}{\partial x^2} + \frac{\partial^2}{\partial y^2} \tag{1.56}$$

is known as the two-dimensional Laplacian operator. The derived equation as above is in Cartesian coordinate in *XY*-plane. Similarly, the equation of motion for vibration of a circular membrane in polar coordinates may be derived. The following is the final equation that may be obtained for circular membrane with radius *r*.

$$\frac{\partial^2 z}{\partial r^2} + \frac{1}{r}\frac{\partial z}{\partial r} + \frac{1}{r^2}\frac{\partial^2 z}{\partial \theta^2} = \frac{m}{P}\frac{\partial^2 z}{\partial t^2} \tag{1.57}$$

1.6 Initial and Boundary Conditions

We should now understand that for most part, discrete and continuous systems represent different mathematical models of the same physical system. It has already been mentioned that motion of discrete system is governed by only time variable, whereas continuous systems are governed by variables that depend not only on time but also on the space coordinates. Accordingly, PDE is needed to describe the continuous system model and, on the other hand, ODES describe the discrete systems.

As such, owing to the solution of the ODE in the discrete case, only initial conditions, viz., the conditions of the dependent variable and possibly its derivatives at a single point (time), are prescribed and the problem solution of the ODE with the initial conditions is known as the initial value problem. Initial value problems are generally time-dependent problems in which the initial values (i.e., values at time $t = 0$) of the dependent variable and its time derivatives are specified. On the other hand, for continuous system governed by PDE, we need the value of the dependent variable and possibly their derivatives at more than one point or on the boundary. These problems are referred to as boundary value problems. For example, a bar in vibration may have three types of boundary conditions, viz., clamped, simply supported, and free. We will define here what the conditions for these boundary conditions are:

Clamped boundary condition: In this case, both the displacement and slope are considered to be zero.

Simply supported boundary condition: Here the displacement and bending moment must be zero.

Free boundary condition: In this boundary condition, the bending moment and shear force must be zero.

The details of these conditions in the solution of the vibration problems will be discussed in the subsequent chapters.

It is worth mentioning that 1 DOF behavior of the system may sometimes correspond very closely to the real situation, but often it is merely an assumption based on the consideration that only a single vibration pattern is developed. But in certain circumstances, a 1 DOF system may be good enough for practical purposes. For continuous systems, the success of this procedure does depend on assumed modes method, which depends on certain assumptions and on the appropriate choice of a shape or trial function, which in particular depends on the physical characteristics of the system. The selected shape functions should also satisfy specified boundary conditions of the problem. Thus, we encountered two classes of boundary condition: essential or geometric boundary conditions and natural or force (or dynamic) boundary condition. Now, we will define these two classes of boundary conditions.

Essential or geometric boundary condition: These boundary conditions are demanded by the geometry of the body. This is a specified condition placed on displacements or slopes on the boundary of a physical body. Essential boundary conditions are also known as Dirichlet boundary conditions.

Natural or force (or dynamic) boundary condition: These boundary conditions are demanded by the condition of shearing force and bending moment balance. Accordingly, this is a condition on bending moment and shear. Natural boundary conditions are also known as Neumann boundary conditions.

The above classification of boundary conditions has great implications in the analysis and solution of boundary value problems, particularly in continuous systems such as plates, by approximate methods. The boundary conditions for a given end-point can be of any type and of any combination. It may be seen that for a clamped end, there are only geometric boundary conditions and for a free end, there are only natural boundary conditions. But for a simply supported end, there is one geometric boundary condition and one natural boundary condition. However, the following three possible combinations of boundary conditions exist for any problem:

1. All are of essential type.
2. All are of natural type.
3. Some of them are of natural type and the remaining are of essential type.

Problems in which all the boundary conditions are of essential type are called Dirichlet problems (or boundary value problem of the first kind) and those in which all the boundary conditions are of natural type are called Neumann problems (or boundary value problem of the second kind). Another type of problem known as mixed type (or boundary value problem of the third kind) consists of those in which both essential and natural boundary conditions are satisfied. Examples of the three kinds of boundary value problems may again be given, respectively, by a beam clamped at both ends, a beam that is free at both ends, and a cantilever beam, i.e., having one end clamped and one end free.

For higher-order differential equations governing the equation of motion of vibration, such as plates (having fourth-order differential equation), the essential and natural boundary conditions may be specified accordingly. As mentioned, the details regarding plates will be discussed in later chapters. In general, let us consider that $Lu = f$ is a differential equation where L is the differential operator and u is the dependent variable. Suppose the differential equation is of order $2n$, then the essential boundary conditions are associated with the conditions of the given function and its derivatives of orders at most $n - 1$:

$$u = h_0, \ u' = h_1, \ u'' = h_2, \ldots, \ u^{n-1} = h_{n-1}$$

where superscripts denote various order derivatives.

The natural boundary conditions consist of derivatives of orders higher than $n - 1$, viz.,

$$u^n = g_n, \ u^{n+1} = g_{n+1}, \ldots, \ u^{2n-1} = g_{2n-1}$$

1.7 Equation of Motion through Application of Energy Method

Equation of motion of vibrating systems can sometimes be advantageously obtained by using the law of conservation of energy provided that damping is negligible. This section includes some of the vibration systems whose equation of motion are derived by the application of energy method.

1.7.1 Massless Spring Carrying a Mass *m*

Let us consider a mass m, which is attached to a massless spring with stiffness k as shown in Figure 1.21. Let the position of equilibrium be O and x be the extension for the motion. Then, the potential energy V of the system may be written as

$$V = \int_0^x kx\,dx = \frac{1}{2}kx^2 \tag{1.58}$$

and the kinetic energy T of the system becomes

$$T = \frac{1}{2}m\dot{x}^2 \tag{1.59}$$

By the principle of conservation of energy, sum of the kinetic and potential energies remains constant, and therefore, from Equations 1.58 and 1.59, one may obtain

$$\frac{1}{2}kx^2 + \frac{1}{2}m\dot{x}^2 = \text{constant} \tag{1.60}$$

Differentiating Equation 1.60 with respect to t, it becomes

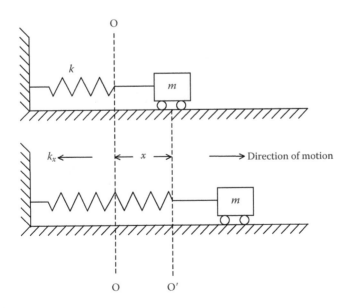

FIGURE 1.21
Motion of massless spring carrying a mass *m*.

$$m\ddot{x} + kx = 0 \qquad (1.61)$$

which is the same equation as given in Equation 1.6. We define again the circular frequency ω_n by the following relation:

$$\omega_n^2 = k/m \qquad (1.62)$$

and then Equation 1.61 may be written as

$$\ddot{x} + \omega_n^2 x = 0 \qquad (1.63)$$

which is the standard form of the equation of motion as obtained in Equation 1.8. Thus, the same vibration model (equation) may be obtained by energy consideration.

1.7.2 Simple Pendulum

We may refer to Figure 1.22, in which m is the mass of the bob of the pendulum, l is the length of the string, and θ is the angular displacement. Then, potential energy V is written in this case as

$$V = mg(\text{AB})$$
$$= mg(\text{OA} - \text{OB})$$

Then, writing the above in terms of l and θ, we have the potential energy as

$$V = mgl(1 - \cos\theta) \qquad (1.64)$$

The kinetic energy T of the system may be written as

$$T = \frac{1}{2}m(\dot{x}^2 + \dot{y}^2) \qquad (1.65)$$

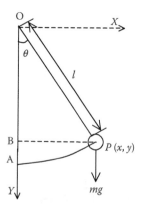

FIGURE 1.22
Motion of simple pendulum.

Writing Equation 1.65 in terms of l and θ and from Figure 1.22, we have the kinetic energy as

$$T = \frac{1}{2}ml^2\dot{\theta}^2 \tag{1.66}$$

where $x = l\sin\theta$, $\dot{x} = l\cos\theta\dot{\theta}$, $y = l\cos\theta$, $\dot{y} = -l\sin\theta\dot{\theta}$.

The equation of motion may be obtained by the principle of conservation of energy, i.e., $T + V = $ constant. Accordingly, we get

$$\frac{1}{2}ml^2\dot{\theta}^2 + mgl(1 - \cos\theta) = \text{constant} \tag{1.67}$$

Differentiating Equation 1.67 with respect to t, the final equation of motion for the simple pendulum is written as

$$\frac{d^2\theta}{dt^2} + \omega_n^2\theta = 0 \tag{1.67a}$$

where $\omega_n^2 = g/l$ is the natural frequency of simple pendulum. Thus, we obtain the same equation of motion as derived in Equation 1.12.

1.7.3 Spring of Mass m_s Carrying a Mass m

Let us consider a spring of length l, stiffness k, and mass m_s, which carries a mass m as shown in Figure 1.23. Potential energy V in this case is written as

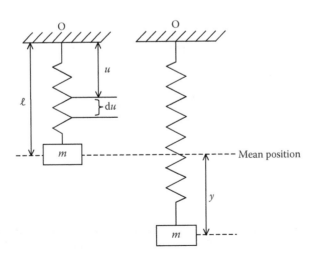

FIGURE 1.23
Motion of spring of mass m_s carrying a mass m.

$$V = \int_0^y ky \, dy = \frac{1}{2}ky^2 \tag{1.68}$$

Here, the total kinetic energy T is the addition of kinetic energy of mass m (say, T_m) and kinetic energy of spring of mass m_s (say, T_s). The kinetic energy due to mass m is given by

$$T_m = \frac{1}{2}m\dot{y}^2 \tag{1.69}$$

We can also write the mass of an element du of the spring as $\frac{m_s}{l}du$. Accordingly, kinetic energy of the element du of spring becomes

$$T_s^e = \frac{1}{2}\left(\frac{m_s}{l}du\right)\left(\frac{\dot{y}u}{l}\right)^2 \tag{1.70}$$

Also, the kinetic energy for the spring turns out to be

$$T_s = \int_0^l T_s^e = \frac{1}{6}m_s\dot{y}^2 \tag{1.71}$$

Combining Equations 1.69 and 1.71, we may have the total kinetic energy T as

$$T = \frac{1}{2}\left(m + \frac{1}{3}m_s\right)\dot{y}^2 \tag{1.72}$$

Using the principle of conservation of energy, i.e., using $T+V=$ constant, and then differentiating with respect to t, the equation of motion is finally written as

$$\frac{d^2y}{dt^2} + \omega_n^2 y = 0 \tag{1.73}$$

where

$$\omega_n^2 = \frac{k}{\left(m + \frac{1}{3}m_s\right)} \tag{1.74}$$

which is the natural frequency of the system. It is worth mentioning that if the mass of the spring is negligible, i.e., if $m_s = 0$, then the natural frequency from Equation 1.74 turns out to be $\omega_n^2 = k/m$ as was obtained in Equation 1.62 for a massless spring.

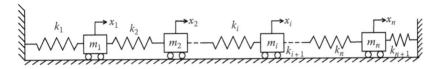

FIGURE 1.24
n DOF system.

In the following sections, only the expressions for potential and kinetic energies for vibration of MDOF system, string, and membranes will be provided for the sake of completeness for the benefit of serious students/researchers. They may derive the equation of motion in these cases from the expressions of the energies.

1.7.4 Multi-Degree of Freedom System

Figure 1.24 depicts *n* DOF system with *n* masses $m_1, m_2, \ldots, m_n$ without friction, which can slide along a horizontal axis. Here, $k_1, k_2, \ldots, k_n$ are the stiffnesses of the springs connected as shown. Then, the potential energy V and the kinetic energy T can, respectively, be written as

$$V = \frac{1}{2}\left[k_1 x_1^2 + k_2(x_2 - x_1)^2 + \cdots + k_n(x_n - x_{n-1})^2 + k_{n+1}x_n^2\right] \tag{1.75}$$

$$T = \frac{1}{2}\left[m_1 \dot{x}_1^2 + m_2 \dot{x}_2^2 + \cdots + m_n \dot{x}_n^2\right] \tag{1.76}$$

1.7.5 Vibration of String

If S and m denote the tensile force and mass per unit length of the string, then the potential energy V and the kinetic energy T for vibration of string are, respectively, given as

$$V = \frac{S}{2}\int_0^l \left(\frac{dy}{dx}\right)^2 dx \tag{1.77}$$

$$T = \frac{m}{2}\int_0^l (\dot{y})^2 dx \tag{1.78}$$

1.7.6 Vibration of Membrane

We may write the potential energy V and the kinetic energy T in this case, respectively, as

$$V = \frac{S}{2} \int \int \left[\left(\frac{\partial u}{\partial x} \right)^2 + \left(\frac{\partial u}{\partial y} \right)^2 \right] dx\ dy \qquad (1.79)$$

$$T = \frac{m}{2} \int \int (\dot{u})^2 dx\ dy \qquad (1.80)$$

where
    S is the tensile force
    m is the mass per unit area

Further Reading

Anderson, R.A., *Fundamental of Vibration*, Macmillan, New York, 1967.

Bhat, R.B. and Dukkipati, R.V., *Advanced Dynamics*, Narosa Publishing House, New Delhi, India, 2001.

Bishop, R.E.D., *Vibration*, Cambridge University Press, Cambridge, England, 1979.

Inman, D.J., *Engineering Vibrations*, Prentice Hall, Englewood Cliffs, NJ, 1994.

Jacobsen, L.S. and Ayre, R.S., *Engineering Vibrations*, McGraw-Hill, New York, 1958.

Meirovitch, L., *Elements of Vibration Analysis*, 2nd ed., McGraw-Hill, New York, 1986.

Ramamurti, V., *Mechanical Vibration Practice with Basic Theory*, CRC Press, Boca Raton, FL, 2000.

Rao, J.S., *Advanced Theory of Vibration*, Wiley Eastern Limited, New Delhi, 1991.

Shabana, A.A., *Theory of Vibration: Discrete and Continuous Systems*, Springer, New York, 1991.

Timoshenko, S., Young, D.H. and Weaver, W., *Vibration Problems in Engineering*, 5th ed., Wiley, New York, 1990.

2

Methods of Analysis for Vibration Problems

In Chapter 1, we derived the equation of motion for a variety of vibration problems. It has been seen that mathematical modeling of vibrating systems leads to ordinary and partial differential equations (PDEs). Moreover, it is also clear that the one-degree-of-freedom (1 DOF) systems are governed by ordinary differential equations (ODEs), the multi-degree-of-freedom (MDOF) system leads to a system of ODEs, and the continuous systems are prescribed by PDEs. Overall complex behavior of a vibrating system may be described by the corresponding simplified mathematical model of the system. The governing equations of motion of the vibrating systems may be exactly solved for simple cases only. With the advent of fast computers and numerical algorithms, there is a tremendous amount of rise in using approximate and numerical methods. Methods like finite element methods, boundary integral equation methods, finite difference methods, and the method of weighted residuals have made it possible to handle a variety of vibration problems. In the following sections, few methods of solution for 1 DOF, MDOF, and continuous vibrating systems are addressed for creating a base to encounter the problem of vibration of plates.

2.1 Single Degree of Freedom System

Let us write the physical problem represented by the model shown in Figure 2.1. In this model, the spring with stiffness k and the dashpot with the viscous damping c (the energy-loss mechanisms of a structure) are connected with a mass m. The displacement and the exciting force are denoted by $x(t)$ and $P(t)$, respectively. The equation of motion of the system may be written by the procedure discussed in Chapter 1 as

$$m\ddot{x} + c\dot{x} + kx = P(t) \tag{2.1}$$

With this model of 1 DOF system, we will now proceed with the solution of various cases of the system.

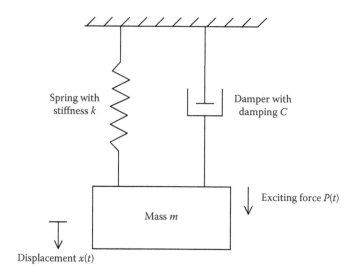

FIGURE 2.1
Single degree of freedom vibrating system.

2.1.1 Free Vibration without Damping

Let us consider that m and k are the mass and stiffness of the 1 DOF system. The corresponding equation of motion without damping may be written as

$$m\ddot{x} + kx = 0 \tag{2.2}$$

Putting $x = A\cos\omega t + B\sin\omega t$, we get

$$-m\omega^2 x + kx = 0$$

$$\Rightarrow \omega^2 = \frac{k}{m}$$

Then, the solution may be written as

$$x = A\cos\omega t + B\sin\omega t$$

which gives

$$\dot{x} = -A\omega\sin\omega t + B\omega\cos\omega t \tag{2.3}$$

$$\ddot{x} = -A\omega^2\cos\omega t - B\omega^2\sin\omega t$$

$$= \omega^2 x \tag{2.4}$$

This ω is called the undamped natural frequency and will be written as

$$\omega_n^2 = \frac{k}{m} \tag{2.5}$$

For the following initial conditions of the above problem

$$x(0) = x_0, \quad \dot{x}(0) = \dot{x}_0$$

the constants may be found as

$$x(0) = A \Rightarrow A = x_0$$
$$x = x_0 \cos \omega_n t + B \sin \omega_n t$$
$$\dot{x} = -x_0 \omega_n \sin \omega_n t + B\omega_n \cos \omega_n t$$

Plugging into the second initial condition $\dot{x}(0) = \dot{x}_0$ leads to

$$\dot{x}(0) = B\omega_n \Rightarrow B = \frac{\dot{x}_0}{\omega_n}$$

Thus, the final solution with the above initial conditions turns out to be

$$x = x_0 \cos \omega_n t + \frac{\dot{x}_0}{\omega_n} \sin \omega_n t \tag{2.6}$$

2.1.2 Free Vibration with Damping without Force

Equation of motion in this case may be written as

$$m\ddot{x} + c\dot{x} + kx = 0 \tag{2.7}$$

where c is the damping of the 1 DOF system.

Now, putting $\qquad\qquad x = Ae^{qt}$ $\qquad\qquad\qquad$ (2.8)

we can get from Equation 2.7

$$mq^2 + cq + k = 0 \tag{2.9}$$

Solving Equation 2.9, value of q is written as

$$q = \frac{-c \pm \sqrt{(c^2 - 4mk)}}{2m} \tag{2.10}$$

$$= -\frac{c}{2m} \pm \left[\left(\frac{c}{2m}\right)^2 - \left(\frac{k}{m}\right) \right]^{1/2} \tag{2.11}$$

Now, the following cases will arise in this case:

Case (i)

If

$$\left(\frac{c}{2m}\right)^2 > \frac{k}{m}$$

then the solution will be of the form

$$x = A_1 \exp(q_1 t) + A_2 \exp(q_2 t) \tag{2.12}$$

where q_1 and q_2 are given by Equation 2.11 (for $+$ and $-$)
Substituting the following initial conditions in Equation 2.12

$$x(0) = x_0, \quad \dot{x}(0) = \dot{x}_0$$

constants A_1 and A_2 may be obtained as

$$A_1 = \frac{(\dot{x}_0 - x_0 q_2)}{(q_1 - q_2)} \tag{2.13}$$

$$A_2 = \frac{(x_0 q_1 - \dot{x}_0)}{(q_1 - q_2)} \tag{2.14}$$

Putting values of A_1 and A_2 from Equations 2.13 and 2.14 in Equation 2.12, we get the required response for this case.

Case (ii)

If

$$\left(\frac{c}{2m}\right)^2 < \frac{k}{m}$$

The roots of Equation 2.11 are complex and can be written as

$$q = -\frac{c}{2m} \pm i \left[\frac{k}{m} - \left(\frac{c}{2m}\right)^2 \right]^{1/2} \tag{2.15}$$

and the corresponding solution may be written as

$$x = e^{-\left(\frac{c}{2m}t\right)} [A_1 \sin \omega_1 t + A_2 \cos \omega_1 t] \tag{2.16}$$

where

$$\omega_1 = \frac{k}{m} - \left(\frac{c}{2m}\right)^2$$

Again, using initial conditions

$$x(0) = x_0, \quad \dot{x}(0) = \dot{x}_0$$

the constant A_2 is obtained as

$$X(0) = x_0 = A_2 \tag{2.17}$$

Consequently, putting Equation 1.17 in Equation 2.16, we get the value of A_1

$$A_1 = \frac{\dot{x}_0}{\omega_1} + \frac{c}{2m\omega_1} x_0 \tag{2.18}$$

Plugging the values of A_1 and A_2 from Equations 2.18 and 2.17 in Equation 2.16, we have the response solution in this case. It is convenient to express the damping C in terms of the critical damping $C_c = 2m\omega_n$, where ω_n is undamped natural frequency.

This is actually the case when the roots are equal in Equation 2.11. So from Equation 2.11, we may write

$$\left(\frac{C}{2M}\right)^2 = \frac{k}{m} \Rightarrow \frac{c}{2m} = \omega_n \Rightarrow c \approx C_c = 2m\omega_n$$

The damping ratio is then defined by

$$\zeta = \frac{c}{C_c} = \frac{c}{2m\omega_n}$$

Therefore, Equation 2.15 reduces to

$$q = -\omega_n \zeta \pm i\omega_n \sqrt{(1 - \zeta^2)} \tag{2.19}$$

Now, three cases corresponding to $\zeta > 1$, $\zeta < 1$, and $\zeta = 1$ are obtained.

Case (a): When $\zeta > 1$

Both terms in Equation 2.19 will be real and this implies a steadily decaying response with no oscillation (Figure 2.2). This is termed as overdamped system (heavy damping) and the response is given as in Equation 2.12 of Case (i).

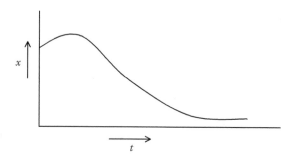

FIGURE 2.2
Response in 1 DOF system for $\zeta > 1$.

Case (b): When $\zeta < 1$

The square root term will be real, positive, and q in this case will be complex with a negative real part. Thus, the response will oscillate (Figure 2.3) at a damped natural frequency,

$$\omega_d = \omega_n \sqrt{(1 - \zeta^2)} \tag{2.20}$$

and will decay in amplitude with increasing time; this is called light damping or underdamped system. The system oscillates, but it is not periodic because the motion never repeats itself.

Using the initial conditions

$$x(0) = x_0, \quad \dot{x}(0) = \dot{x}_0$$

we have response as

$$x = e^{-\omega_n \zeta t} \left[x_0 \cos \omega_d t + \frac{\dot{x}_0 + \zeta \omega_n x_0}{\omega_d} \sin \omega_d t \right] \tag{2.21}$$

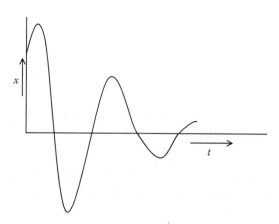

FIGURE 2.3
Response in 1 DOF system for $\zeta < 1$.

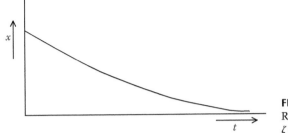

FIGURE 2.4
Response in 1 DOF system for $\zeta = 1$.

This is called transient solution (without force). The response is also written by Equation 2.16 of Case (ii).

Case (c): When $\zeta = 1$ (critical damping)

Here, the system does not oscillate and it comes to rest in the shortest period of time (Figure 2.4).

$$x = (A + Bt)e^{-\omega_n \zeta t} \tag{2.22}$$

The corresponding response is as follows:
   Putting initial conditions

$$x(0) = x_0, \quad \dot{x}(0) = \dot{x}_0$$

in Equation 2.22, we get

$$A = x_0, \quad B = (\dot{x}_0 + x_0 \omega_n \zeta)$$

Finally, substituting the values of A and B in Equation 2.22, the response in this case becomes

$$x = [x_0 + (\dot{x}_0 + x_0 \omega_n \zeta)t]e^{-\omega_n \zeta t} \tag{2.23}$$

2.1.3 Forced Vibration in Single Degree of Freedom System

In this section, forcing functions, viz., harmonic distributing force with and without damping, will be considered to have an idea of handling the forced vibration problems in 1 DOF systems.

2.1.3.1 Harmonic Distributing Force (with Damping)

Let the external forcing function, i.e., the harmonic distributing force be of the form

$$P(t) = P_0 \cos \omega t$$

where

P_0 is a constant

$\omega/2\pi$ is the frequency of the distributing force

The equation of motion in this case is written as

$$m\ddot{x} + c\dot{x} + kx = P_0 \cos \omega t \tag{2.24}$$

From the theory of differential equation, the solution will be written as the summation of transient solution (called complimentary function, C.F.) and the steady-state solution (called particular integral, P.I.)

C.F., i.e., the transient solution is obtained earlier with right-hand side of Equation 2.24 = 0. So P.I. remains to be found, which becomes

$$\text{P.I.} = \frac{P_0 \cos(\omega t - \phi)}{\left[(k - m\omega^2)^2 + c^2\omega^2\right]^{1/2}}$$

where

$$\tan \phi = \frac{c\omega}{(k - m\omega^2)}$$

and so

$$x = e^{-\frac{ct}{2m}}(A_1 \sin \omega_d t + A_2 \cos \omega_d t + \frac{P_0 \cos(\omega t - \phi)}{\left[(k - m\omega^2)^2 + c^2\omega^2\right]^{1/2}} \tag{2.25}$$

$$\omega_d = \omega_n \sqrt{(1 - \zeta^2)}, \quad \omega_n = \sqrt{\frac{k}{m}}, \quad \zeta = \frac{c}{2m\omega_n}$$

In the response expression of Equation 2.25, A_1 and A_2 may be determined from the initial conditions.

NOTE: Transient solution dies out because of damping, but the steady state will soon be there.

Therefore, transient solution is significant in many practical cases.

Now, using initial conditions in Equation 2.25 as

$$x(0) = x_0, \quad \dot{x}(0) = \dot{x}_0$$

we get

$$A_2 = x_0 - \frac{P_0 \cos \phi}{\left[(k - m\omega^2)^2 + c^2\omega^2\right]^{1/2}} \quad \text{and} \tag{2.26}$$

$$A_1 = \frac{\dot{x}_0}{\omega_d} + \frac{c}{2m\omega_d} \left[x_0 - \frac{P_0 \cos \phi}{\sqrt{(k^2 - m\omega^2)^2 + c^2\omega^2}} \right]$$
$$- \frac{P_0\omega \sin \phi}{\omega_d \sqrt{(k^2 - m\omega^2) + c^2\omega^2}} \tag{2.27}$$

Putting A_1 and A_2 from Equations 2.27 and 2.26 in Equation 2.25, we may get the response of the system.

2.1.3.2 Undamped System with Sinusoidal Force

Taking $c = 0$ in Equation 2.25, the undamped solution is written as

$$x = A_1 \sin \omega_n t + A_2 \cos \omega_n t + \frac{P_0 \cos \omega t}{(k - m\omega^2)} \tag{2.28}$$

Again, using the initial conditions

$$x(0) = x_0, \quad \dot{x}(0) = \dot{x}_0$$

in Equation 2.28, we have the constants as follows:

$$A_1 = \frac{\dot{x}_0}{\omega_n}, \quad A_2 = x_0 - \frac{P_0}{(k - m\omega^2)}$$

As before, substituting A_1 and A_2 in Equation 2.28, we obtain the response as

$$x = \left[\frac{\dot{x}_0}{\omega_n} \sin \omega_n t + \left(x_0 - \frac{P_0}{(k - m\omega^2)} \right) \cos \omega_n t \right] + \frac{P_0 \cos \omega t}{(k - m\omega^2)} \tag{2.29}$$

On the other hand, if we have the equation of motion as

$$m\ddot{x} + kx = P_0 \sin \omega t$$

where the forcing function is $P_0 \sin \omega t$, then the solution becomes

$$x = A_1 \sin \omega_n t + A_2 \cos \omega_n t + \frac{P_0 \sin \omega t}{(k - m\omega^2)} \tag{2.30}$$

Using again the initial conditions in Equation 2.30 as

$$x(0) = x_0, \quad \dot{x}(0) = \dot{x}_0$$

the constants turn out to be

$$A_1 = \frac{1}{\omega_n}\left(x_0 - \frac{P_0\omega}{(k - m\omega^2)}\right), \quad A_2 = x_0$$

Plugging in the constants A_1 and A_2 in Equation 2.30, we get the response as

$$x = \left[\frac{1}{\omega_n}\left(\dot{x}_0 - \frac{P_0\omega}{(k - m\omega^2)}\right)\sin\omega_n t\right] + x_0\cos\omega_n t + \frac{P_0\sin\omega t}{(k - m\omega^2)} \qquad (2.31)$$

2.2 Two Degree of Freedom System

Solution of 2 DOF system will be discussed in this section. For this, let us consider a two-storey shear frame structure as shown in Figure 2.5, where the system needs two coordinates, viz., x_1 and x_2, to designate the motion as shown in the figure. We will find the natural frequency and mode shapes for this 2 DOF system subject to ambient vibration (i.e., without force). We designate the height of each storey as h and the elastic modulus as E. I_b and I_c are the second moment of the cross-sectional area about the axis of bending of beam and column, respectively. The flexural rigidities for beam and column are shown in Figure 2.5. Beams and floors are considered to be rigid. Only the lateral displacement x_1 and x_2 are considered with no axial deformation, and masses of the first and second storey are supposed to be $2m$ and m. Then, stiffness of the two lower columns will be given by

$$k_1 = k_2 = \frac{12(2EI_c)}{h^3}$$

The total stiffness for the columns of the first storey is written as (because in parallel)

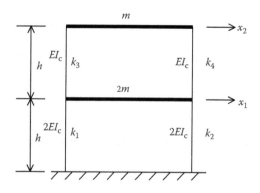

FIGURE 2.5
Two-storey shear frame (2 DOF system).

$$K_1 = k_1 + k_2 = \frac{2(12 \times 2EI_c)}{h^3} \tag{2.32}$$

Similarly, for the second floor the total stiffness may be obtained as

$$K_2 = \frac{2(12EI_c)}{h^3} \tag{2.33}$$

Given (from Figure 2.5) $m_1 = 2m$ and $m_2 = m$, the mass (M) and stiffnes (K) matrices for this 2 DOF system may be written in the form

$$M = \begin{bmatrix} m_1 & 0 \\ 0 & m_2 \end{bmatrix} = m \begin{bmatrix} 2 & 0 \\ 0 & 1 \end{bmatrix} \tag{2.34}$$

$$K = \begin{bmatrix} k_1 + k_2 & -k_2 \\ -k_2 & k_2 \end{bmatrix} = \frac{24EI_c}{h^3} \begin{bmatrix} 3 & -1 \\ -1 & 1 \end{bmatrix} \tag{2.35}$$

The governing equation of motion from Equation 1.25 will become

$$m \begin{bmatrix} 2 & 0 \\ 0 & 1 \end{bmatrix} \begin{Bmatrix} \ddot{x}_1 \\ \ddot{x}_2 \end{Bmatrix} + \frac{24EI_c}{h^3} \begin{bmatrix} 3 & -1 \\ -1 & 1 \end{bmatrix} \begin{Bmatrix} x_1 \\ x_2 \end{Bmatrix} = \begin{Bmatrix} 0 \\ 0 \end{Bmatrix} \tag{2.36}$$

The above equations are uncoupled and we will rewrite it as

$$M\ddot{x} + Kx = 0 \tag{2.37}$$

We will solve the above free vibration equation for vibration characteristics, viz., for frequency and mode shapes of the said structural system. Accordingly, putting $x = \{\phi\}e^{i\omega t}$ in free vibration equation 2.37, we get

$$(K - M\omega^2)\{\phi\} = \{0\} \tag{2.38}$$

where ω and $\{\phi\}$ denote the natural frequency and mode shapes of the system, respectively. Substituting elements of M and K from Equations 2.34 and 2.35 in Equation 2.38, it will become

$$\left(\begin{bmatrix} 3k & -k \\ -k & k \end{bmatrix} - \omega_n^2 \begin{bmatrix} 2m & 0 \\ 0 & m \end{bmatrix} \right) \begin{Bmatrix} \phi_1 \\ \phi_2 \end{Bmatrix} = \begin{Bmatrix} 0 \\ 0 \end{Bmatrix} \tag{2.39}$$

where

$$k = 24EI_c/h^3 \tag{2.40}$$

For nontrivial solution, the following determinant must be zero,

$$\begin{vmatrix} (3k - 2m\omega^2) & -k \\ -k & (k - m\omega^2) \end{vmatrix} = 0 \tag{2.41}$$

Writing $\omega^2 = \lambda$, we have the following equation called the charateristic equation as

$$(3k - 2m\lambda)(k - m\lambda) - k^2 = 0 \tag{2.42}$$

Solution of the above quadratic equation becomes

$$\lambda = \omega^2 = \frac{k}{2m}, \frac{2k}{m} \tag{2.43}$$

Thus, the natural frequencies of the above system may be obtained in the form

$$\omega_1 = 3.464\sqrt{\frac{EI_c}{mh^3}}, \quad \omega_2 = 6.928\sqrt{\frac{EI_c}{mh^3}} \tag{2.44}$$

From Equation 2.39 after plugging in $\omega_1^2 = k/(2m)$, we have

$$\begin{bmatrix} (3k - k) & -k \\ -k & (k - \frac{1}{2}k) \end{bmatrix} \begin{Bmatrix} \phi_1 \\ \phi_2 \end{Bmatrix}_{(1)} = \begin{Bmatrix} 0 \\ 0 \end{Bmatrix} \tag{2.45}$$

After putting $\phi_2 = 1$ in the above, one can obtain ϕ_1 as $1/2$ and then we may find the first mode shape, which is called first mode as

$$\begin{Bmatrix} \phi_1 \\ \phi_2 \end{Bmatrix}_{(1)} = \begin{Bmatrix} 1/2 \\ 1 \end{Bmatrix} \tag{2.46}$$

Similarly, substituting $\omega_2^2 = 2k/m$ and $\phi_2 = 1$ in Equation 2.39, the second mode shape or the second mode may be obtained as

$$\begin{Bmatrix} \phi_1 \\ \phi_2 \end{Bmatrix}_{(2)} = \begin{Bmatrix} -1 \\ 1 \end{Bmatrix}$$

Figure 2.6 depicts the mode shapes, viz., the first and second modes, geometrically for clear understanding of the deflected shapes for these modes. The response computation will be included after the understanding of the solution methodology for MDOF system. Also, the discussion regarding the damping and force will be undertaken in the section of MDOF system.

2.3 Multi-Degree of Freedom System

2.3.1 Reduction to an Eigenvalue Problem for General System (Conservative)

The equation of motion of a conservative system with n DOF may be written, in matrix form, as follows:

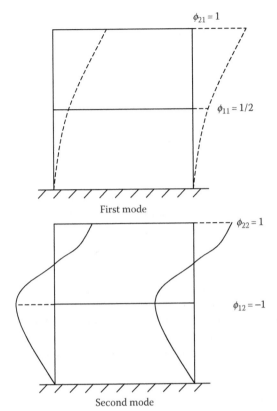

$\phi_{21} = 1$

$\phi_{11} = 1/2$

First mode

$\phi_{22} = 1$

$\phi_{12} = -1$

Second mode

FIGURE 2.6
First and second modes of the 2 DOF system.

$$[M]\{\ddot{x}\} + [K]\{x\} = \{0\} \tag{2.47}$$

where
 　[M] is the mass or inertia matrix
 　[K] the stiffness matrix
 　{x} is the displacement vector

In general, [M] and [K] are symmetric matrices. Now, we define the inverse dynamical matrix [W] by $[W] = [M]^{-1}[K]$ and rewrite Equation 2.47 as follows (i.e., pre-multiply Equation 2.47 by $[M]^{-1}$):

$$\{\ddot{x}\} + [W]\{x\} = \{0\} \tag{2.48}$$

We search for oscillatory solutions of this equation of the form

$$\{x\} = \{A\}e^{iwt} \tag{2.49}$$

where
> $\{A\}$ is a column matrix of n unknown amplitudes
> ω is a circular frequency to be determined

Putting Equation 2.49 in Equation 2.48, we get

$$-\omega^2\{A\} + [w]\{A\} = \{0\} \tag{2.50}$$

Let $\omega^2 = \lambda$, then the above equation may be written as

$$([W] - \lambda[I])\{A\} = \{0\} \tag{2.51}$$

where $[I]$ is the unit matrix of nth order. This equation represents a set of linear homogeneous equations. For nontrivial solution, the determinant of the coefficients of $\{A\}$ must vanish. We, therefore, have

$$\det([W] - \lambda[I]) = 0 \tag{2.52}$$

which is the characteristic equation of the inverse dynamical matrix $[W]$.

Let us assume that its n roots are distinct and that they have values $\lambda_1, \lambda_2, \ldots, \lambda_n$. These roots are the eigenvalues of $[W]$. To each eigenvalue corresponds a natural frequency of oscillation ω_i given by $\omega_i = (\lambda_i)^{1/2}$; ω_i normally will be real. The corresponding eigenvectors and modal columns will be discussed next.

Eigenvectors and Modal Columns

From Equation 2.51, it may be seen that for each λ_i, there corresponds a column vector $\{A\}_i$ or eigenvector of the matrix $[W]$. This must satisfy the equation

$$[W]\{A\}_i = \lambda_i\{A\}_i, \quad i = 1, 2, \ldots, n \tag{2.53}$$

The eigenvectors $\{A\}_i$ are sometimes called modal columns, as each column matrix $\{A\}_i$ corresponds to a particular mode, viz., the ith-mode of the system.

2.3.2   Orthogonality of the Eigenvectors

From Equation 2.53, we may write

$$[W]\{A\}_i = \lambda_i\{A\}_i$$
$$\Rightarrow [K]\{A\}_i = \lambda_i[M]\{A\}_i \tag{2.54}$$

Similarly, for the jth mode, Equation 2.53 becomes

$$[K]\{A\}_j = \lambda_j[M]\{A\}_j \tag{2.55}$$

Pre-multiplying Equations 2.54 by $\{A\}_j^T$ and 2.55 by $\{A\}_i^T$, we get

$$\{A\}_j^T[K]\{A\}_i = \lambda_i\{A\}_j^T[M]\{A\}_i \tag{2.56}$$

$$\{A\}_i^T[K]\{A\}_j = \lambda_j\{A\}_i^T[M]\{A\}_j \tag{2.57}$$

Now taking transpose on both sides of Equation 2.57, we obtain

$$\{A\}_j^T[K]\{A\}_i = \lambda_j\{A\}_j^T[M]\{A\}_i \tag{2.58}$$

{because $(AB)^T = B^T A^T$ and K and M are symmetric $=> [K]^T = [K]$}
   Subtracting Equation 2.58 from Equation 2.56, we obtain

$$0 = (\lambda_i - \lambda_j)\{A\}_j^T[M]\{A\}_i$$

Since we have assumed that the λ's are distinct, i.e., $\lambda_i \neq \lambda_j$, we must have

$$\{A\}_j^T[M]\{A\}_i = 0 \tag{2.59}$$

Therefore, putting Equation 2.59 on the right-hand side of Equation 2.56, we arrive at

$$\{A\}_j^T[K]\{A\}_i = 0 \quad \text{for } i \neq j \tag{2.60}$$

Equations 2.59 and 2.60 are "generalized" or weighted orthogonality relationships satisfied by the eigenvectors of the inverse dynamical matrix $[W]$.
   The simple type of orthogonality, i.e., $\{A\}_j^T\{A\}_i = 0$ may also occur, but for this $[W]$ itself must be symmetric. In this connection, it is important to recall that the product of two symmetric matrices need not be symmetric, that is why while $[M]$ and $[K]$ are symmetric, $[W]$ generally is not symmetric.

2.3.3 Modal Matrix

The modal matrix $[A]$ is a partitioned matrix made up of the modal columns, or eigenvectors, placed side by side such that

$$[A] = [\{A\}_1\{A\}_2 \cdots \{A\}_n] \tag{2.61}$$

Hence, the set of Equations 2.53 may be written in the form

$$[W][\{A\}_1\{A\}_2 \cdots \{A\}_n] = [\lambda_1\{A\}_1\lambda_2\{A\}_2 \cdots \lambda_n\{A\}_n] \Rightarrow [W][A] = [A][\lambda] \tag{2.62}$$

where $[\lambda]$ is a diagonal matrix made up of the eigenvalues λ_i.
   (Note that $[\lambda][A]$ does not produce the Equation 2.53, but $[A][\lambda]$ does.)

If the eigenvalues λ_i are distinct, then $[A]$ may be shown to be nonsingular; therefore, pre-multiplying Equation 2.62 by $[A]^{-1}$, we obtain

$$[\lambda] = [A]^{-1}[W][A] \qquad (2.63)$$

Therefore, the modal matrix $[A]$, by the above operation, diagonalizes the matrix $[W]$.

We next discuss some further properties of the modal matrix, which arise from the orthogonality of the eigenvectors and let us define a new matrix $[P]$ by

$$[P] = [A]^{T}[M][A] = \begin{bmatrix} (A)_1^T[M](A)_1 & (A)_1^T[M](A)_2 & \cdots \\ (A)_2^T[M](A)_1 & (A)_2^T[M](A)_2 & \cdots \\ \vdots & \vdots & \vdots \end{bmatrix}$$

$$= \begin{bmatrix} (A)_1^T[M](A)_1 & 0 & \cdots & 0 \\ 0 & (A)_2^T[M](A)_2 & \cdots & 0 \\ 0 & 0 & \cdots & (A)_n^T[M](A)_n \end{bmatrix} \qquad (2.64)$$

which in view of the orthogonality of the eigenvectors becomes

$$\{A\}_j^T[K]\{A\}_i = 0 \quad \text{for } i \neq j$$

Accordingly, $[P]$ is a diagonal matrix and is called the generalized mass matrix.

In a similar fashion, we can show that

$$[S] = [A]^{T}[K][A] \qquad (2.65)$$

is also diagonal and $[S]$ is called the spectral matrix.

2.3.4 Relationship between [P], [S], and [λ]

Equation 2.62 gives the relationship between $[W]$, $[A]$, and $[\lambda]$, where $[\lambda]$ matrix is diagonal,

$$[W][A] = [A][\lambda] \qquad (2.66]$$

Pre-multiplying the above by $[M]$, we obtain

$$[M][W][A] = [M][A][\lambda]$$
$$\Rightarrow [M][M]^{-1}[K][A] = [M][A][\lambda] \text{ (since, } [W] = M^{-1}K)$$
$$\Rightarrow [K][A] = [M][A][\lambda]$$

Again pre-multiplying by $[A]^T$ yields

$$[A]^T[K][A] = [A]^T[M][A][\lambda]$$
$$\Rightarrow [S] = [P][\lambda] \qquad (2.67)$$

Since $[S]$, $[P]$, $[\lambda]$ are all diagonal, Equation 2.67 represents a set of n equations of the type

$$S_{ii} = \lambda_i P_{ii} \qquad (2.68)$$

2.3.5 Solution of the Dynamical Problem (Free Vibration)

The dynamical problem is expressed as (discussed previously)

$$[M]\{\ddot{x}\} + [K]\{x\} = \{0\} \qquad (2.69)$$

Once we have found the eigenvalues and eigenvectors of the inverse dynamical matrix $[W]$, we may certainly write $[A]$ and $[\lambda]$ as above. We next introduce a new set of coordinates $\{y\}$ related to the coordinates $\{x\}$ by the transformation

$$\{x\} = [A]\{y\} \qquad (2.70)$$

where $y_1, y_2, \ldots, y_n$ are called normal or principal coordinates. Then, Equation 2.69 becomes

$$[M][A]\{\ddot{y}\} + [K][A]\{y\} = \{0\} \qquad (2.71)$$

Pre-multiplying by $[A]^T$, we obtain

$$[A]^T[M][A]\{\ddot{y}\} + [A]^T[K][A]\{y\} = \{0\}$$

Utilizing the definitions of matrices $[P]$ and $[S]$ in the above, one may get

$$[P]\{\ddot{y}\} + [S]\{y\} = \{0\} \qquad (2.72)$$

and then by the relation (Equation 2.67), one obtains

$$[P](\{\ddot{y}\} + [\lambda]\{y\}) = \{0\} \qquad (2.73)$$

Since $[P]$ is not generally zero (the problem would then be a static one), we have

$$\{\ddot{y}\} + [\lambda]\{y\} = \{0\} \qquad (2.74)$$

which is equivalent to the set of equations

$$\ddot{y}_i + \lambda_i y_i = 0, \quad i = 1, 2, \ldots, n \qquad (2.75)$$

The whole object of the analysis, so far, was to separate the coordinates (de-couple the equations), so that each of the resulting equations contains only one coordinate. The procedure was aimed at finding the necessary transformation operator, i.e., the modal matrix. Each normal coordinate performs a simple harmonic oscillation. So, we have

$$y_i = a_i \cos(\omega_i t) + b_i \sin(\omega_i t), \quad i = 1, 2, \ldots, n \qquad (2.76)$$

This may be expressed in the form

$$\{y\} = [C]\{a\} + [D]\{b\} \qquad (2.77)$$

where

$$[C] = \begin{bmatrix} \cos(\omega_1 t) & 0 & 0 & \cdots \\ 0 & \cos(\omega_2 t) & 0 & \cdots \\ \vdots & \vdots & \vdots & \vdots \\ & \vdots & \vdots & \vdots \end{bmatrix}$$

and

$$[D] = \begin{bmatrix} \sin(\omega_1 t) & 0 & 0 & \cdots \\ 0 & \sin(\omega_2 t) & 0 & \cdots \\ \vdots & \vdots & \vdots & \vdots \\ & \vdots & \vdots & \vdots \end{bmatrix}$$

The final solution may be expressed in terms of the original coordinates $\{x\}$ by pre-multiplying Equation 2.77 by $[A]$, i.e.,

$$[A]\{y\} = [A][C]\{a\} + [A][D]\{b\}$$
$$\Rightarrow \{x\} = [A][C]\{a\} + [A][D]\{b\} \qquad (2.78)$$

where $\{a\}$ and $\{b\}$ are column vectors, which are to be determined by initial conditions.

Suppose the initial displacements and velocities of the system are given by $\{x_0\}$ and $\{v_0\}$ at $t = 0$.

Therefore, Equation 2.78 at $t = 0$ gives

$$\{x_0\} = [A]\{a\} \qquad (2.79)$$

since $[C(0)] = [I]$ and $[D(0)] = [0]$ (because $\cos 0 = 1$ and $\sin 0 = 0$).

Similarly, differentiating Equation 2.78, we obtain at $t = 0$

$$(v_0) = [A][\omega]\{b\} \tag{2.80}$$

where $[\omega]$ is a diagonal matrix, the diagonal terms being $\omega_1, \omega_2, \ldots, \omega_n$. The vectors $\{a\}$ and $\{b\}$ are obtained from Equations 2.79 and 2.80 as

$$\{a\} = [A]^{-1}\{x_0\}$$

and

$$\{b\} = [\omega]^{-1}[A]^{-1}\{v_0\}$$

Substituting the above vectors $\{a\}$ and $\{b\}$ in Equation 2.78, one can get

$$\{x\} = [A][C][A]^{-1}\{x_0\} + [A][D][\omega]^{-1}[A]^{-1}\{v_0\} \tag{2.81}$$

which is the general form of the solution.

2.3.6 Classical Solution for Forced Vibration without Damping

Let us consider a system with n DOF acted upon by forces not derivable from a potential. The matrix equation of motion is

$$[M]\{\ddot{x}\} + [K]\{x\} = \{Q(t)\} \tag{2.82}$$

We suppose that the eigenvalues and eigenvectors of the homogeneous system have already been found. Now, proceeding by transforming into normal coordinates (i.e., $\{x\} = [A]\{y\}$),

$$[M][A]\{\ddot{y}\} + [K][A]\{y\} = \{Q\} \tag{2.83}$$

pre-multiplication of Equation 2.83 by $[A]^T$ yields

$$[A]^T[M][A]\{\ddot{y}\} + [A]^T[k][A]\{y\} = [A]^T\{Q\} \tag{2.84}$$

Writing the above equation in terms of generalized mass and spectral matrix gives

$$[P]\{\ddot{y}\} + [S]\{y\} = [A]^T\{Q\} \tag{2.85}$$

where
    $[P]$ is the generalized mass matrix
    $[S]$ is the spectral matrix

The column matrix $[A]^T\{Q\}$ is called the generalized force matrix and denoted by $\{F\}$, so that

$$[P]\{\ddot{y}\} + [S]\{y\} = \{F\} \tag{2.86}$$

Again pre-multiplying Equation 2.86 by $[P]^{-1}$, we obtain

$$\{\ddot{y}\} + [P]^{-1}[S]\{y\} = [P]^{-1}\{F\} \tag{2.87}$$

Now, since $[S] = [P][\lambda]$ and $[P]$ and $[S]$ are diagonal matrices, we have

$$\{\ddot{y}\} + [\lambda]\{y\} = [P]^{-1}\{F\} \tag{2.88}$$

which represents the following set of linear equations:

$$\ddot{y}_k + \omega_k^2 y_k = (1/p_{kk})F_k, \quad k = 1, 2, \ldots, n \tag{2.89}$$

The above equations can be solved for y and again the original coordinate $\{x\}$ can be found from the transformed equation

$$\{x\} = [A]\{y\}$$

2.3.7 Modal Damping in Forced Vibration

The equation of motion of n DOF system with viscous damping and arbitrary excitation $F(t)$ can be presented in matrix form as

$$[M]\{\ddot{x}\} + [C]\{\dot{x}\} + [K]\{x\} = \{F\} \tag{2.90}$$

It is generally a set of n coupled equations. The classical method of solving these equations in the absence of damping is to find the normal modes of oscillations of the homogeneous equation

$$[M]\{\ddot{x}\} + [K]\{x\} = \{0\}$$

and to determine the normal coordinates.

In this case again, introducing normal coordinates ($\{x\} = [A]\{y\}$) we have, by pre-multiplying Equation 2.90 with $[A]^T$,

$$[A]^T[M][A]\{\ddot{y}\} + [A]^T[C][A]\{\dot{y}\} + [A]^T[K][A]\{y\} = [A]^T\{F\} \tag{2.91}$$

Rewriting Equation 2.91 in terms of generalized mass and spectral matrix, we obtain

$$[P]\{\ddot{y}\} + [A]^T[C][A]\{\dot{y}\} + [S]\{y\} = [A]^T\{F\} \tag{2.92}$$

In general, $[A]^T[C][A]$ is not diagonal, whereas $[P]$ and $[S]$ are diagonal and so the preceding equation is coupled by the damping matrix. If $[C]$ is proportional to $[M]$ or $[K]$, then $[A]^T[C][A]$ becomes diagonal, in which case we can say that the system has proportional damping and Equation 2.92 is then completely uncoupled. Thus, instead of n coupled equations, we would have n uncoupled equations similar to that in a 1 DOF system.

To solve Equation 2.92 for y_i, we will write it, for example, for a 3 DOF system as follows:

$$
\begin{bmatrix} P_1 & 0 & 0 \\ 0 & P_2 & 0 \\ 0 & 0 & P_3 \end{bmatrix} \begin{Bmatrix} \ddot{y}_1 \\ \ddot{y}_2 \\ \ddot{y}_3 \end{Bmatrix} + \begin{bmatrix} \overline{C}_1 & 0 & 0 \\ 0 & \overline{C}_2 & 0 \\ 0 & 0 & \overline{C}_3 \end{bmatrix} \begin{Bmatrix} \dot{y}_1 \\ \dot{y}_2 \\ \dot{y}_3 \end{Bmatrix}
$$
$$
+ \begin{bmatrix} S_1 & 0 & 0 \\ 0 & S_2 & 0 \\ 0 & 0 & S_3 \end{bmatrix} \begin{Bmatrix} y_1 \\ y_2 \\ y_3 \end{Bmatrix} = \begin{Bmatrix} \overline{F}_1 \\ \overline{F}_2 \\ \overline{F}_3 \end{Bmatrix} \tag{2.93}
$$

where $[A]^T\{F\} = \{\overline{F}\}$.

Thus, we get three uncoupled equations as

$$
P_i\ddot{y}_i + \overline{C}_i\dot{y}_i + S_iy_i = \overline{F}_i, \quad i = 1, 2, 3 \tag{2.94}
$$

For $i = 1$, 2, 3, the above differential equations can be solved as in 1 DOF system and the final solution may be given by $\{x\} = [A]\{y\}$.

2.3.8 Normal Mode Summation

The forced vibration equation for the n DOF system can be written as discussed

$$
[M]\{\ddot{x}\} + [C]\{\dot{x}\} + [K]\{x\} = \{F\} \tag{2.95}
$$

and this can be solved by a digital computer.

However, for systems of large number of DOFs, the computation happens to be costly. But it is possible to cut down the size of the computation by a procedure known as the mode summation method. Essentially, the displacement of the structure under forced excitation is approximated by the sum of a limited number of normal modes of the system multiplied by normal coordinates. The procedure will now be discussed by an example.

Consider a 30-storey building with 30 DOF. The solution of its undamped homogeneous equation will lead to 30 eigenvalues and 30 eigenvectors that describe the normal modes of the system. If we know that the excitation of the building centers around the lower frequencies, the higher modes will not be excited and we would be justified in assuming the forced response to be the superposition of only a few of the lower modes. Let us assume that $\{A\}_1 (x_i), \{A\}_2(x_i)$, and $\{A\}_3(x_i)$ are sufficient number of lower modes.

Then, the deflection under forced excitation can be written as

$$
\{x\} = [A]\{y\}
$$

which may be written in the form

$$x_i = \{A\}_1(x_i)\{y_1\}(t) + \{A\}_2(x_i)\{y_2\}(t) + \{A\}_3(x_i)\{y_3\}(t) \qquad (2.96)$$

The position of all n floors in matrix notation may be expressed in terms of the modal matrix $[A]$, which is composed of only three modes

$$\begin{Bmatrix} x_1 \\ x_2 \\ \vdots \\ x_n \end{Bmatrix}_{n \times 1} = \begin{bmatrix} (A)_1(x_1) & (A)_2(x_1) & (A)_3(x_1) \\ (A)_1(x_2) & (A)_2(x_2) & (A)_3(x_2) \\ \vdots & \vdots & \vdots \\ (A)_1(x_n) & (A)_2(x_n) & (A)_3(x_n) \end{bmatrix}_{n \times 3} \begin{Bmatrix} y_1 \\ y_2 \\ y_3 \end{Bmatrix}_{3 \times 1} \qquad (2.97)$$

The use of the limited modal matrix then reduces the system to that equal to the number of modes used. For the present example of the 30-storey building with three normal modes, we may have the three equations represented by the following instead of the 30 coupled equations:

$$[P]_{3 \times 3}\{\ddot{y}\} + ([A]^T[C][A])_{3 \times 3}\{\dot{y}\} + [S]_{3 \times 3}\{y\} = [A]^T\{F\} \qquad (2.98)$$

Again, as discussed earlier, if $[C]$ is proportional to $[M]$ and $[K]$, then $[A]^T[C][A]$ becomes diagonal and Equation 2.98 would totally be uncoupled. Thus, the solution may be achieved easily. After getting $\{y\}$ from above, the displacement x_i of any floor may be obtained by using again the equation

$$\{x\} = [A]\{y\}$$

2.3.9 Response Computation

We have now studied the solution methodology for MDOF system. Accordingly, the response computation will be included now of the example discussed in Section 2.2. Let us assume that F_1 and F_2 are the forces that are acting on the 2 DOF system as shown in Figure 2.7. The equation of motion will be (without damping)

$$[M]\{\ddot{x}\} + [K]\{x\} = \{F(t)\} \qquad (2.99)$$

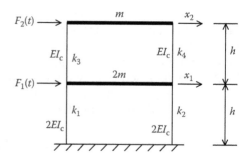

FIGURE 2.7
Two-storey shear frame subject to forces.

where

$$\{F(t)\} = \begin{Bmatrix} F_1(t) \\ F_2(t) \end{Bmatrix} \qquad (2.100)$$

By transforming into normal coordinates (i.e., $\{x\} = [A]\{y\}$), we may write as in Equation 2.88

$$\{\ddot{y}\} + [\lambda]\{y\} = [P]^{-1}[A]^T\{F(t)\} = \{\bar{F}(t)\} \qquad (2.101)$$

where $[P]$ is the generalized mass matrix.

We had already computed the two eigenvectors, for example, described in Section 2.2 as

$$\phi_1 = \begin{Bmatrix} 1/2 \\ 1 \end{Bmatrix}, \quad \phi_2 = \begin{Bmatrix} -1 \\ 1 \end{Bmatrix}$$

From these two eigenvectors, the modal matrix $[A]$ for the 2 DOF system are written as

$$[A] = \begin{bmatrix} 1/2 & -1 \\ 1 & 1 \end{bmatrix} \qquad (2.102)$$

The generalized mass matrix P may next be obtained as

$$[P] = [A]^T[M][A] = \begin{bmatrix} 1/2 & 1 \\ -1 & 1 \end{bmatrix} \begin{bmatrix} 2m & 0 \\ 0 & m \end{bmatrix} \begin{bmatrix} 1/2 & -1 \\ 1 & 1 \end{bmatrix}$$

$$= \begin{bmatrix} 3m/2 & 0 \\ 0 & 3m \end{bmatrix} \qquad (2.103)$$

The inverse of $[P]$ will have the form

$$[P]^{-1} = \begin{bmatrix} \frac{2}{3m} & 0 \\ 0 & \frac{1}{3m} \end{bmatrix} \qquad (2.104)$$

Equations 2.101, 2.102, and 2.104 then yield

$$\{\bar{F}(t)\} = \begin{Bmatrix} \bar{F}_1 \\ \bar{F}_2 \end{Bmatrix} = [P]^{-1}[A]^T \begin{Bmatrix} F_1 \\ F_2 \end{Bmatrix} = \begin{Bmatrix} \frac{1}{3m}(F_1 + 2F_2) \\ \frac{1}{3m}(F_2 - F_1) \end{Bmatrix} \qquad (2.105)$$

From Equations 2.101 and 2.105, we may now write

$$\ddot{y}_i + \omega_i^2 y_i = \bar{F}_i, \quad i = 1, 2 \qquad (2.106)$$

Then, the solution for each DOF becomes

$$y_i(t) = \frac{\dot{y}_i(0)}{\omega_i} \sin \omega_i t + y_i(0) \cos \omega_i t + \frac{1}{\omega_i} \int_0^t \overline{F}_i(\tau) \sin \omega_i(t - \tau) d\tau, \quad i = 1, 2 \quad (2.107)$$

After finding y_i from above, we obtain x_i (the original coordinate) from the relation

$$\{x\} = [A]\{y\} \tag{2.108}$$

The initial conditions in Equation 2.107 are found from Equation 2.108 at $t = 0$ by the following procedure.

Denoting the initial conditions as $\{x\}_{t=0} = \{x_0\}$ and $\{\dot{x}\}_{t=0} = \{\dot{x}_0\}$, we may then write from Equation 2.108

$$\{y\}_{t=0} = [A]^{-1}\{x_0\} \tag{2.109}$$

and

$$\{\dot{y}\}_{t=0} = [A]^{-1}\{\dot{x}_0\} \tag{2.110}$$

One can obtain the final response of each DOF, i.e., of each floor, by substituting Equation 2.107 along with 2.109 and 2.110 in Equation 2.108.

2.4 Continuous Systems

2.4.1 Vibration of a Taut String

Let us consider a string of mass per unit length m, which is stretched tightly between two fixed points O and A distance l apart under tension T as shown in Figure 2.8. It has been derived in Chapter 1 that the differential equation that governs the vibration is given by

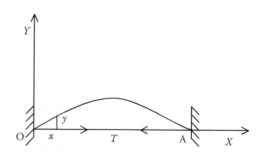

FIGURE 2.8
Vibration of a taut string.

$$\frac{\partial^2 y}{\partial t^2} = c^2 \frac{\partial^2 y}{\partial x^2} \tag{2.111}$$

where $c^2 = T/m$ and y denotes the transverse displacement at time t of a particle distant x from O, which is taken as origin, OA is the axis of X and the line OY through O perpendicular to OA as the axis of Y. We will solve Equation 2.111 by the separation of variables. For this, let us assume the solution as

$$y(x, t) = X(x)f(t) \tag{2.112}$$

Substituting Equation 2.112 into Equation 2.111, we get

$$X\frac{d^2 f}{dt^2} = c^2 f \frac{d^2 X}{dx^2}$$

Dividing the above equation on both sides by $X(x)f(t)$ gives

$$\frac{1}{f}\frac{d^2 f}{dt^2} = c^2 \frac{1}{X}\frac{d^2 X}{dx^2}$$

Since left-hand side of the above equation is a function of t alone while right-hand side is a function of x alone, their common value must be a constant, say a. Then, we have

$$\frac{1}{f}\frac{d^2 f}{dt^2} = c^2 \frac{1}{X}\frac{d^2 X}{dx^2} = a, \quad a > 0 \tag{2.113}$$

From Equation 2.113, we can write

$$\frac{d^2 f}{dt^2} - af = 0$$

Its solution will contain exponential function, which increases continuously as t increases, which is not possible for a vibratory system. So, we suppose $a = -\omega^2$. Hence, Equation 2.113 becomes

$$\frac{d^2 f}{dt^2} + \omega^2 f = 0 \tag{2.114}$$

and

$$\frac{d^2 X}{dx^2} + \frac{\omega^2}{c^2}X = 0 \tag{2.115}$$

Solution of the differential equations (Equations 2.114 and 2.115) are given by

$$f(t) = A \cos \omega t + B \sin \omega t \tag{2.116}$$

and

$$X(x) = C \cos \frac{\omega}{c} x + D \sin \frac{\omega}{c} x \tag{2.117}$$

Since the ends of the string are fixed, we have the following conditions:

$$y(0, t) = 0, \quad y(l, t) = 0$$

The above conditions imply $X(0) = 0$, $X(l) = 0$, which yield

$$C = 0$$

and $D \sin \frac{\omega}{c} l = 0$ reduces to $\frac{\omega l}{c} = n\pi$ $(n = 1, 2, 3, \dots)$.
Thus, we get $\omega = \frac{n\pi c}{l}$ $(n = 1, 2, 3, \dots)$.
This is the frequency equation, and the natural frequencies are given by

$$\omega_n = \frac{n\pi c}{l}, \quad n = 1, 2, 3, \dots \tag{2.118}$$

Corresponding to the natural frequency ω_n, there will correspond a characteristic mode written as

$$X_n = D_n \sin \frac{\omega_n}{c} x = D_n \sin \frac{n\pi x}{l} \tag{2.119}$$

When the string vibrates with frequency ω_n, the transverse displacement $y_n(x, t)$ will be given by

$$
\begin{aligned}
y_n(x,t) &= X_n(x) f_n(t) \\
&= D_n \sin \frac{n\pi x}{l} (A_n \cos \omega_n t + B_n \sin \omega_n t) \\
&= \sin \frac{n\pi x}{l} \left(A'_n \cos \frac{n\pi c}{l} t + B'_n \sin \frac{n\pi c}{l} t \right) \tag{2.120}
\end{aligned}
$$

Hence, the general solution for a string with its ends fixed will be given by

$$y(x, t) = \sum_{n=1}^{\infty} y_n(x, t) = \sum_{n=1}^{\infty} \sin \frac{n\pi x}{l} \left(A'_n \cos \frac{n\pi c}{l} t + B'_n \sin \frac{n\pi c}{l} t \right) \tag{2.121}$$

Next, we will describe few modes of vibration of the system.

Fundamental Mode: Here, we will put $n = 1$ in Equation 2.121, and accordingly we will get

$$y_1(x, t) = \sin \frac{\pi x}{l} \left(A'_1 \cos \frac{\pi c}{l} t + B'_1 \sin \frac{\pi c}{l} t \right) \tag{2.122}$$

Circular frequency is given by

$$\omega_1 = \frac{\pi c}{l} = \frac{\pi}{l}\sqrt{\frac{T}{m}}$$

Period of vibration is written as

$$T_1 = \frac{2\pi}{\omega_1} = \frac{2l}{c}$$

Mode shape will become

$$X_1 = \sin\frac{\pi x}{l}$$

The form of the deflection shape is shown in Figure 2.9a.

Second Mode: In this case, we will put $n=2$ in Equation 2.121 and accordingly we obtain

$$y_2(x,t) = \sin\frac{2\pi x}{l}\left(A_2'\cos\frac{2\pi c}{l}t + B_2'\sin\frac{2\pi c}{l}t\right) \tag{2.123}$$

Circular frequency is given by $\omega_2 = \frac{2\pi c}{l} = \frac{2\pi}{l}\sqrt{\frac{T}{m}}$, the period of vibration will be $T_1 = \frac{2\pi}{\omega_2} = \frac{l}{c}$, and the mode shape may be written as $X_2 = \sin\frac{2\pi x}{l}$.
The form of the deflection shape is shown in Figure 2.9b.

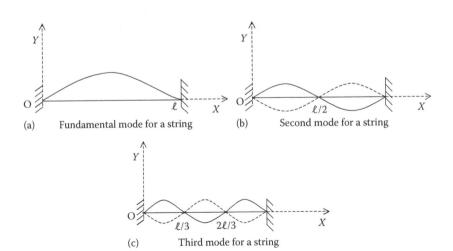

(a)   Fundamental mode for a string

(b)   Second mode for a string

(c)   Third mode for a string

FIGURE 2.9
(a) Fundamental mode for a string; (b) second mode for a string; and (c) third mode for a string.

Third Mode: Here, we will put $n = 3$ in Equation 2.121, which gives

$$y_3(x, t) = \sin\frac{3\pi x}{l}\left(A_3' \cos\frac{3\pi c}{l}t + B_3' \sin\frac{3\pi c}{l}t\right) \tag{2.124}$$

Circular frequency is given by $\omega_3 = \frac{3\pi c}{l} = \frac{3\pi}{l}\sqrt{\frac{T}{m'}}$, the period of vibration is written as $T_3 = \frac{2\pi}{\omega_3} = \frac{2l}{3c'}$, and the mode shape will become $X_1 = \sin\frac{3\pi x}{l}$.

The form of the deflection shape is shown in Figure 2.9c.

For a particular solution, we must be given the initial conditions. Let these conditions be

$$y(x, 0) = u(x); \quad \frac{\partial y}{\partial t}(x, 0) = v(x), \quad 0 \leq x \leq l \tag{2.125}$$

Applying the initial conditions (Equation 2.125) to the general solution (Equation 2.121) yields

$$u(x) = \sum_{n=1}^{\infty} A_n' \sin\frac{n\pi x}{l}$$

and

$$v(x) = \sum_{n=1}^{\infty} \sin\frac{n\pi x}{l} B_n'\left(\frac{n\pi c}{l}\right)$$

The above equations of $u(x)$ and $v(x)$ represent Fourier series representation of the functions $u(x)$ and $v(x)$ in the interval $0 \leq x \leq l$, and we have

$$\int_0^l u(x) \sin\frac{n\pi x}{l}dx = A_n'\frac{l}{2}$$

This gives

$$A_n' = \frac{2}{l}\int_0^l u(x) \sin\frac{n\pi x}{x}dx \tag{2.126}$$

Similarly, we can write

$$\int_0^l v(x) \sin\frac{n\pi x}{l}dx = B_n'\frac{n\pi c}{l}\frac{l}{2}$$

and so the constant B'_n is found to be

$$B'_n = \frac{2}{n \pi c} \int_0^l v(x) \sin \frac{n \pi x}{l} \, dx \qquad (2.127)$$

Thus, Equation 2.121 is the required solution where A'_n and B'_n are given by Equations 2.126 and 2.127, respectively.

2.4.2 Transverse Vibration of an Elastic Beam

We have the equation of motion for transverse vibration of a beam as (Chapter 1)

$$\frac{\partial^4 y}{\partial x^4} + \frac{1}{a^2} \frac{\partial^2 y}{\partial t^2} = 0 \qquad (2.128)$$

where
ρ is the density
A is the cross-sectional area
E is the modulus of elasticity of the beam material and
I is the moment of inertia of the beam about the axis of bending

Accordingly,

$$a^2 = \frac{EI}{\rho A} \qquad (2.129)$$

The differential equation (Equation 2.128) is solved by the separation of variables, and for this let us take

$$y = X(x)T(t) \qquad (2.130)$$

Substituting the expression for y from Equation 2.130 in Equation 2.128, we have the form

$$T \frac{d^4 X}{dx^4} + \frac{X}{a^2} \frac{d^2 T}{dt^2} = 0 \qquad (2.131)$$

which may be written as

$$-\frac{a^2}{X} \frac{d^4 X}{dx^4} = \frac{1}{T} \frac{d^2 T}{dt^2} = -\omega^2 \qquad (2.132)$$

Equation 2.132 yields two ODEs as

$$\frac{d^2 T}{dt^2} + \omega^2 T = 0 \qquad (2.133)$$

and

$$\frac{d^4 X}{dx^4} - k^4 X = 0 \tag{2.134}$$

where

$$k^4 = \frac{\omega^2}{a^2}, \quad k_n^4 = \frac{\omega_n^2}{a^2} \tag{2.135}$$

Solution of the ODE 2.133 is

$$T = \overline{E} \sin \omega t + \overline{F} \cos \omega t \tag{2.136}$$

and solution of the differential equation (Equation 2.134) is written as

$$X = \overline{A} \cos kx + \overline{B} \sin kx + \overline{C} \cosh kx + \overline{D} \sinh kx \tag{2.137}$$

Then, the complete solution turns out to be

$$y = \sum_{n=1}^{\infty} (\overline{A}_n \cos k_n x + \overline{B}_n \sin k_n x + \overline{C}_n \cosh k_n x + \overline{D}_n \sinh k_n x)$$
$$\times (\overline{E}_n \sin \omega_n t + \overline{F}_n \cos \omega_n t) \tag{2.138}$$

We write the above as

$$y = \sum_{n=1}^{\infty} (A \cos kx + B \sin kx + C \cosh kx + D \sinh kx)$$
$$\times (E \sin \omega t + F \cos \omega t) \tag{2.139}$$

For the sake of convenience while applying the boundary conditions on X, the form of X is taken as

$$X = A_1(\sin kx - \sinh kx) + A_2(\sin kx + \sinh kx) + A_3(\cos kx - \cosh kx)$$
$$+ A_4(\cos kx + \cosh kx) \tag{2.140}$$

where we have from Equations 2.139 and 2.140

$$A_1 + A_2 = B, \quad A_2 - A_1 = D, \quad A_3 + A_4 = A, \quad A_4 - A_3 = C$$

Various boundary conditions for this elastic beam are defined next:
Fixed or clamped end: In this case, deflection and slope are zero i.e.,

$$y = 0, \quad \frac{dy}{dx} = 0 \tag{2.141}$$

Simply supported end: Here, deflection and bending moment are zero and so we have

$$y = 0, \quad \frac{d^2y}{dx^2} = 0 \tag{2.142}$$

Free end: Free end is defined when we have bending moment and shear force as zero and accordingly we write

$$\frac{d^2y}{dx^2} = 0, \quad \frac{d^3y}{dx^2} = 0 \tag{2.143}$$

Next, the solution of transverse vibration of beam is described when both the ends are simply supported, and for this Equation 2.142 is used.

Considering the end $x = 0$, the boundary condition to be satisfied are $x(0) = 0$ and $\frac{d^2y}{dx^2}(0) = 0$, which give

$$A_4 = 0, \quad A_3 = 0 \tag{2.144}$$

To satisfy the simply supported boundary condition at the end $x = l$, we have $x(l) = 0$ and $\frac{d^2y}{dx^2}(l) = 0$, which yields

$$A_1(\sin kl - \sinh kl) + A_2(\sin kl + \sinh kl) = 0 \tag{2.145}$$

Rearranging the above equation, we get

$$-A_1(\sin kl + \sinh kl) + A_2(-\sin kl + \sinh kl) = 0 \tag{2.146}$$

Equations 2.145 and 2.146 provide

$$A_1 = A_2 \tag{2.147}$$

Thus, from Equations 2.144 and 2.147, solution for X may be obtained as $X = A \sin kx$, and in general this is written in the form

$$X_n = A_n \sin k_n x \tag{2.148}$$

Then, the characteristic equation of simply supported beam in general form is given by

$$y = XT = \sum_{n=1}^{\infty} A_n \sin k_n x (E_n \sin \omega_n t + F_n \cos \omega_n t) \tag{2.149}$$

The remaining constants may be found by imposing the given initial conditions. Similarly, the other boundary conditions may be handled for the general solution.

2.4.3 Vibration of Membrane

2.4.3.1 Rectangular Membrane

We have derived the equation of motion for vibration of a membrane in Chapter 1, which is given by

$$\frac{\partial^2 z}{\partial t^2} = c^2 \left(\frac{\partial^2 z}{\partial x^2} + \frac{\partial^2 z}{\partial y^2} \right) \tag{2.150}$$

where

$$c^2 = \frac{P}{m} \tag{2.151}$$

This is first written in the form

$$\frac{\partial^2 z}{\partial x^2} + \frac{\partial^2 z}{\partial y^2} - \frac{1}{c^2} \frac{\partial^2 z}{\partial t^2} = 0 \tag{2.152}$$

Let us take the solution of Equation 2.152 as

$$z(x, y, t) = X(x)Y(y)\sin(\omega t + \varepsilon) \tag{2.153}$$

Substituting expression of z from Equation 2.153 in Equation 2.152, we get

$$\frac{1}{X} \frac{d^2 X}{dx^2} + \frac{1}{Y} \frac{d^2 Y}{dy^2} + \frac{\omega^2}{c^2} = 0 \tag{2.154}$$

Now, the above is written in the form of two differential equations:

$$\frac{1}{X} \frac{d^2 X}{dx^2} = -k_1^2 \tag{2.155}$$

and

$$\frac{1}{Y} \frac{d^2 Y}{dy^2} = -k_2^2 \tag{2.156}$$

where

$$k_1^2 + k_2^2 = \frac{\omega^2}{c^2} \tag{2.157}$$

These differential equations are written next in the following form:

$$\frac{d^2X}{dx^2} + k_1^2 X = 0 \tag{2.158}$$

$$\frac{d^2Y}{dy^2} + k_2^2 Y = 0 \tag{2.159}$$

The solutions of these two differential equations are given by

$$X = A\cos k_1 x + B\sin k_1 x \tag{2.160}$$

$$Y = C\cos k_2 y + D\sin k_2 y \tag{2.161}$$

Therefore, the solution for the membrane may be written as

$$z = (A\cos k_1 x + B\sin k_1 x)(C\cos k_2 y + D\sin k_2 y)\sin(\omega t + \varepsilon) \tag{2.162}$$

As the membrane is in XY-plane as shown in Figure 2.10, we have

at

$$z = 0, \quad x = 0, \quad x = a \tag{2.163}$$

Also at

$$z = 0, \quad y = 0, \quad y = b \tag{2.164}$$

Applying the conditions at $z=0$ with $x=0$ and $y=0$ in Equation 2.162, the value of the constants A and C are found to be zero. Now using the condition $z=0$ at $x=a$ gives

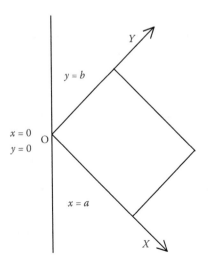

FIGURE 2.10
Membrane in Cartesian coordinate system.

$\sin k_1 a = 0 = \sin m\pi$, which gives

$$k_1 = \frac{m\pi}{a}, \quad m = 1, 2, 3, \ldots \tag{2.165}$$

Similarly, introducing the condition $z = 0$ at $y = b$ leads to $\sin k_2 b = 0 = \sin n\pi$, and we obtain

$$k_2 = \frac{n\pi}{b}, \quad n = 1, 2, 3, \ldots \tag{2.166}$$

Thus, the solution may be written as

$$z = BD \sin\frac{m\pi}{a}x \sin\frac{n\pi}{b}y \sin(\omega t + \varepsilon)$$

$$= A_{mn} \sin\frac{m\pi}{a}x \sin\frac{n\pi}{b}y \sin(\omega t + \varepsilon)$$

We will write the above by some rearrangement of the constants with $\cos \varepsilon$ and $\sin \varepsilon$ as

$$z = \sin\frac{m\pi}{a}x \sin\frac{n\pi}{b}y[B_{mn} \sin \omega t + C_{mn} \cos \omega t]$$

Therefore, the general solution finally would be written as

$$z = \sum_{m=1}^{\infty}\sum_{n=1}^{\infty} \sin\frac{m\pi}{a}x \sin\frac{n\pi}{b}y[B_{mn} \sin \omega_{mn}t + C_{mn} \cos \omega_{mn}t] \tag{2.167}$$

where the frequency parameter is given by

$$\omega_{mn}^2 = c^2\left[\left(\frac{m\pi}{a}\right)^2 + \left(\frac{n\pi}{b}\right)^2\right], \quad m = 1, 2, \ldots \quad \text{and} \quad n = 1, 2, \ldots \tag{2.168}$$

Constants B_{mn} and C_{mn} in general solution (Equation 2.167) may again be found using the prescribed initial conditions of the problem.

2.4.3.2 Circular Membrane

The equation of motion in Cartesian coordinate system is given by Equation 2.152. Taking the solution in the form

$$z = u(x, y) \sin(\omega t + \varepsilon) \tag{2.169}$$

the equation of motion may be obtained as

$$\frac{\partial^2 u}{\partial x^2} + \frac{\partial^2 u}{\partial y^2} + \frac{\omega^2}{c^2}u = 0 \tag{2.170}$$

Transforming the above equation in polar form by substituting $x = r \cos \theta$ and $y = r \sin \theta$, we get

$$\frac{\partial^2 u}{\partial r^2} + \frac{1}{r}\frac{\partial u}{\partial r} + \frac{1}{r^2}\frac{\partial^2 u}{\partial \theta^2} + \frac{\omega^2}{c^2}u = 0 \qquad (2.171)$$

Let us now take

$$u = R(r)V(\theta) \qquad (2.172)$$

And substituting the above form of u in Equation 2.171, we have

$$\frac{1}{R}\left(r^2\frac{d^2R}{dr^2} + r\frac{dR}{dr} + \frac{\omega^2}{c^2}r^2R\right) + \frac{1}{V}\frac{d^2V}{d\theta^2} = 0 \qquad (2.173)$$

In view of the above, it may be written in the form of two differential equations as

$$\frac{1}{R}\left(r^2\frac{d^2R}{dr^2} + r\frac{dR}{dr} + \frac{\omega^2}{c^2}r^2R\right) = n^2 \qquad (2.174)$$

and

$$\frac{1}{V}\frac{d^2V}{d\theta^2} = -n^2 \qquad (2.175)$$

Equation 2.174 is a standard differential equation known as Bessel's equation, which may be written in a simpler form as

$$r^2\frac{d^2R}{dr^2} + r\frac{dR}{dr} + \left(\frac{\omega^2}{c^2}r^2 - n^2\right)R = 0 \qquad (2.176)$$

and whose solution may be obtained as

$$R = DJ_n\left(\frac{\omega r}{c}\right) + EJ_{-n}\left(\frac{\omega r}{c}\right) \qquad (2.177)$$

Since z is infinite at $r = 0$, therefore $E = 0$ (as J_{-n} is not defined). Thus we have

$$R = DJ_n\left(\frac{\omega r}{c}\right) \qquad (2.178)$$

The differential Equation 2.175 is then written in the form

$$\frac{d^2V}{d\theta^2} + n^2V = 0 \qquad (2.179)$$

whose solution may be given as

$$V = A \cos n\theta + B \sin n\theta \qquad (2.180)$$

Therefore, the solution of membrane vibration in polar coordinates is found to be

$$z = RV \sin(\omega t + \varepsilon) = DJ_n\left(\frac{\omega r}{c}\right)\{A \cos n\theta + B \sin n\theta\} \sin(\omega t + \varepsilon) \qquad (2.181)$$

2.5 Approximate Methods for Vibration Problems

In the previous sections, we have given the exact solutions for vibration of simple continuous systems. These solutions are in the form of infinite series of principal modes. In a variety of vibration problems, exact solutions may not be obtained, and in those cases, one has to employ approximate methods. In this regard, many methods exist. But here, we give only a few of them. This section is dedicated to "classical" approximation methods, because in common engineering practice, it is sometimes required to mainly have an idea of only a few of the first natural frequencies of a vibrating system. Only two approximate methods, viz., Rayleigh and Rayleigh–Ritz methods, are addressed for simple continuous system to have an overview of the method. In subsequent chapters, we will undertake these approximate methods in analyzing more complex problems such as vibration of plates.

2.5.1 Rayleigh's Method

Rayleigh's method may be applied to all continuous systems. This method requires expressions for maximum kinetic and potential energies of a system. The maximum kinetic and potential energies of the system must be equal since no energy is lost and no energy is fed into the system over one cycle of vibration. This gives us a quotient known as Rayleigh quotient. We will show the methodology for finding frequency by this method for a string.

If S and m denote the tensile force and mass per unit length of a fixed-fixed uniform string, then the potential energy V and the kinetic energy T for vibration of the string are, respectively, given as

$$V = \frac{S}{2} \int_0^l \left(\frac{dy}{dx}\right)^2 dx \qquad (2.182)$$

and

$$T = \frac{m}{2} \int_0^l (\dot{y})^2 dx \qquad (2.183)$$

If the maximum value of $y(x, t)$ is $Y(x)$, then the maximum value of potential energy is written as

$$V_{max} = \frac{S}{2} \int_0^l \left(\frac{dY(x)}{dx}\right)^2 dx \qquad (2.184)$$

The maximum kinetic energy can be obtained by assuming a harmonic function of the form
$y(x, t) = Y(x) \cos \omega t$ in Equation 2.183 and we get

$$T_{max} = \frac{\omega^2 m}{2} \int_0^l (Y(x))^2 dx \qquad (2.185)$$

Equating maximum kinetic and potential energies of the system, we obtain the Rayleigh quotient as

$$\omega^2 = \frac{\dfrac{S}{2} \displaystyle\int_0^l \left(\dfrac{dY(x)}{dx}\right)^2 dx}{\dfrac{m}{2} \displaystyle\int_0^l (Y(x))^2 dx} \qquad (2.186)$$

If the deflection function $Y(x)$ is known, then the natural frequency of the system may be computed. We will consider here the deflection function $Y(x)$ of the fixed-fixed uniform string as

$$Y(x) = x(l - x) \qquad (2.187)$$

So,

$$\frac{dY(x)}{dx} = l - 2x \qquad (2.188)$$

Substituting expressions from Equations 2.187 and 2.188 in Equation 2.186, the Rayleigh quotient becomes

$$\omega^2 = \frac{S}{m} \frac{\displaystyle\int_0^l (l - 2x)^2 \, dx}{\displaystyle\int_0^l (lx - x^2)^2 \, dx}$$

$$= \frac{S}{m} \frac{\left(l^3 + \dfrac{4}{3}l^3 - 2l^3\right)}{\left(\dfrac{l^5}{3} - \dfrac{l^5}{2} + \dfrac{l^5}{5}\right)}$$

$$= \frac{10S}{ml^2}$$

This gives the natural frequency of the vibration of string as

$$\omega = 3.162\sqrt{\frac{S}{ml^2}} \tag{2.189}$$

2.5.2 Rayleigh–Ritz Method

This method is an extension of Rayleigh's method. In Rayleigh's method, approximate value of the lowest frequency (first frequency) is obtained using a single function approximation. In Rayleigh–Ritz method, we will consider a linear combination of the several assumed functions satisfying some boundary condition with the hope to obtain a closer approximation to the exact values of the natural modes of vibration. However, approximate value of the frequencies does depend on the selection of the assumed functions. This book will cover an intelligent and systematic way of generating these assumed functions that in turn is used to have excellent results in a variety of vibration problems. As mentioned earlier, these will be discussed in later chapters of this book. In general, to have a rough idea, the number of frequencies to be computed is equal to the number of arbitrary functions used. Hence, we should consider a very large number of approximating functions (say n) to get accurate number of frequencies (say n_f), where N is generally very large than n_f, i.e., $N \gg n_f$.

Let us suppose that n assumed functions are selected for approximating a deflection function $Y(x)$ for vibration of a string as discussed in the previous section. We will then write the deflection expression as

$$Y(x) = \sum_{i=1}^{n} c_i y_i(x) = c_1 y_1(x) + c_2 y_2(x) + \cdots + c_n y_n(x) \tag{2.190}$$

where $c_1, c_2, \ldots, c_n$ are the constants to be determined and $y_1, y_2, \ldots, y_n$ are the known functions of the spatial coordinate that also satisfy some boundary condition of the problem. Now, putting Equation 2.190 in the Rayleigh quotient (Equation 2.186) and for stationarity of the natural frequencies, we will equate the first partial derivatives with respect to each of the constants c_i to zero. Accordingly, we can write

$$\frac{\partial \omega^2}{\partial c_i} = 0, \quad i = 1, 2, \dots, n \tag{2.191}$$

Equation 2.191 represents a set of n algebraic equations in n unknowns $c_1, c_2, \dots, c_n$, which can be solved for n natural frequencies and mode shapes. In the following paragraphs, the string vibration problem will be again solved using Rayleigh–Ritz method by taking first one term, and then first two terms in the deflection expression, Equation 2.190.

Using the first term in Equation 2.190 for the string vibration problem:

Let the deflection function be taken as $y_1(x) = x(l - x)$. Accordingly, Equation 2.190 may be written as

$$Y(x) = c_1 x(l - x) \tag{2.192}$$

and

$$\frac{dY}{dx} = c_1 l - 2c_1 x \tag{2.193}$$

Substituting expressions from Equations 2.192 and 2.193 in Equation 2.186, the Rayleigh quotient becomes

$$\omega^2 = \frac{S}{m} \frac{\displaystyle\int_0^l (c_1(l - 2x))^2 dx}{\displaystyle\int_0^l (c_1(lx - x^2))^2 dx} = \frac{E}{F} \text{ (say)} \tag{2.194}$$

Then $\frac{\partial \omega^2}{\partial c_1} = 0$ gives $\frac{F\frac{\partial E}{\partial c_1} - E\frac{\partial F}{\partial c_1}}{F^2} = 0$, which can be written as

$$\frac{\partial E}{\partial c_1} - \frac{E}{F}\frac{\partial F}{\partial c_1} = \frac{\partial E}{\partial c_1} - \omega^2 \frac{\partial F}{\partial c_1} = 0 \tag{2.195}$$

Putting the values of E and F from Equation 2.194 in 2.195, we have

$$c_1\left(\frac{Sl^3}{3} - \frac{m\omega^2 l^5}{30}\right) = 0 \tag{2.196}$$

In the above equation, c_1 cannot be zero and so the term in the bracket is zero, which directly gives the first frequency of the system

$$\omega = 3.162\sqrt{\frac{S}{ml^2}} \tag{2.197}$$

It is to be noted that by taking one term in the approximation, the method turns out to be Rayleigh's method. The above is shown for understanding of the methods only. Next, we will discuss the Rayleigh–Ritz method by considering two terms in the deflection expression, Equation 2.190.

Using the first two terms in Equation 2.190 for the string vibration problem:

Here, we will consider the deflection expression having two terms as

$$Y(x) = \sum_{i=1}^{2} c_i y_i(x) = c_1 y_1(x) + c_2 y_2(x) \tag{2.198}$$

where

$$y_1(x) = x(l - x), \quad y_2(x) = x^2(x - l)^2 \tag{2.199}$$

The Rayleigh quotient may be obtained as in Equation 2.194. There are two terms, which give two equations when the Rayleigh quotient is differentiated partially with respect to the two constants c_1 and c_2, respectively. This yields

$$\frac{\partial \omega^2}{\partial c_1} = 0, \quad \frac{\partial \omega^2}{\partial c_2} = 0 \tag{2.200}$$

Thus, the two equations corresponding to Equation 2.200 may be written as in Equation 2.195

$$\frac{\partial E}{\partial c_1} - \omega^2 \frac{\partial F}{\partial c_1} = 0 \tag{2.201}$$

and

$$\frac{\partial E}{\partial c_2} - \omega^2 \frac{\partial F}{\partial c_2} = 0 \tag{2.202}$$

Putting the values of E and F from Equation 2.194 in Equations 2.201 and 2.202, we have the two simultaneous equations in two unknowns as

$$c_1\left(\frac{Sl^3}{3} - \frac{m\omega^2 l^5}{30}\right) + c_2\left(\frac{Sl^5}{15} - \frac{m\omega^2 l^7}{140}\right) = 0 \tag{2.203}$$

$$c_1\left(\frac{Sl^5}{15} - \frac{m\omega^2 l^7}{140}\right) + c_2\left(\frac{2Sl^7}{105} - \frac{m\omega^2 l^9}{630}\right) = 0 \tag{2.204}$$

The above equations may be written in matrix form

$$
\begin{bmatrix}
\left(\frac{Sl^3}{3} - \frac{m\omega^2 l^5}{30}\right) & \left(\frac{Sl^5}{15} - \frac{m\omega^2 l^7}{140}\right) \\
\left(\frac{Sl^5}{15} - \frac{m\omega^2 l^7}{140}\right) & \left(\frac{2Sl^7}{105} - \frac{m\omega^2 l^9}{630}\right)
\end{bmatrix}
\begin{Bmatrix} c_1 \\ c_2 \end{Bmatrix} = \begin{Bmatrix} 0 \\ 0 \end{Bmatrix}
\tag{2.205}
$$

The matrix equation obtained in Equation 2.205 may be solved for the constants c_1 and c_2 by equating the determinant of the square matrix to zero, and the solution is written as

$$
\omega_1 = 3.142\sqrt{\frac{S}{ml^2}}
\tag{2.206}
$$

and

$$
\omega_2 = 10.120\sqrt{\frac{S}{ml^2}}
\tag{2.207}
$$

One may compare the results of the first frequency in Equations 2.197 and 2.206 by taking one term and two terms, respectively, in the deflection approximation. It is to be noted that if the number of terms is taken large in the deflection approximation, then frequency will converge to a constant value tending to the exact value of the frequencies of the system. These will all be addressed and discussed in subsequent chapters.

Further Reading

Bathe, K.J. 1996. *Finite Element Procedures*, Prentice Hall, Englewood Cliffs, New Jersey.

Bhat, R.B. and Chakraverty, S. 2007. *Numerical Analysis in Engineering*, Narosa Pub. House, New Delhi.

Chopra, A.K. 2004. *Dynamics of Structures—Theory and Applications to Earthquake Engineering*, 2nd ed., Prentice Hall of India, New Delhi.

Gantmacher, F.R. 1977. *The Theory of Matrices*, Vol. 1, Chelsea Publishing Company, New York.

Gorman, D.J. 1975. *Free Vibration Analysis of Beams and Shafts*, Wiley, New York.

Horn, R.A. and Johnson C.R. 1985. *Matrix Analysis*, Cambridge University Press Cambridge.

Meirovitch, L. 1997. *Principles and Techniques of Vibration*, Prentice Hall, Englewood Cliffs, New Jersey.

Newland, D.E. 1989. *Mechanical Vibration Analysis and Computation*, Longman Scientific and Technical, London.

Petyt, M. 1998. *Introduction to Finite Element Vibration Analysis*, Cambridge University Press, Cambridge.

Reddy, J.N. 1986. *Applied Functional Analysis and Variational Methods in Engineering*, McGraw-Hill, Singapore.

Timoshenko, S.P. and Young D.H. 1948. *Advanced Dynamics*, McGraw-Hill, New York.

3

Vibration Basics for Plates

Study of vibration of plates is an extremely important area owing to its wide variety of engineering applications such as in aeronautical, civil, and mechanical engineering. Since the members, viz., beams, plates, and shells, form integral parts of structures, it is essential for a design engineer to have a prior knowledge of the first few modes of vibration characteristics before finalizing the design of a given structure. In particular, plates with different shapes, boundary conditions at the edges, and various complicating effects have often found applications in different structures such as aerospace, machine design, telephone industry, nuclear reactor technology, naval structures, and earthquake-resistant structures. A plate may be defined as a solid body bounded by two parallel, flat surfaces having two dimensions far greater than the third.

The vibration of plates is an old topic in which a lot of work has already been done in the past decades. In earlier periods, results were computed for simple cases only where the analytical solution could be found. The lack of good computational facilities made it almost impossible to get reasonably accurate results even in these simple cases. This may be the cause for why in spite of a lot of theoretical developments, numerical results were available only for a few cases. With the invention of fast computers, there was a tremendous increase in the research work using approximate and numerical methods for simple as well as complex plate vibration problems. Now, we have some very fast and efficient algorithms that can solve these problems in a very short time and give comparatively accurate results. It is also worth mentioning that methods like finite element methods, boundary integral equation methods, finite difference methods, and the methods of weighted residuals have made handling any shape and any type of boundary conditions possible.

Different theories have been introduced to handle the vibration of plate problems. Correspondingly, many powerful new methods have also been developed to analyze these problems. In the following sections, an overview of basic equations, theories, stress–strain relations, etc., are addressed related to the vibration basics for plates.

3.1 Stress–Strain Relations

Generalized Hooke's law relates the stresses and strains in tensor notation as

$$
\begin{Bmatrix} \sigma_x \\ \sigma_y \\ \sigma_z \\ \sigma_{yz} \\ \sigma_{zx} \\ \sigma_{xy} \end{Bmatrix} = \begin{bmatrix} c_{11} & c_{12} & c_{13} & c_{14} & c_{15} & c_{16} \\ c_{21} & c_{22} & c_{23} & c_{24} & c_{25} & c_{26} \\ c_{31} & c_{32} & c_{33} & c_{34} & c_{35} & c_{36} \\ c_{41} & c_{42} & c_{43} & c_{44} & c_{45} & c_{46} \\ c_{51} & c_{52} & c_{53} & c_{54} & c_{55} & c_{56} \\ c_{61} & c_{62} & c_{63} & c_{64} & c_{65} & c_{66} \end{bmatrix} \begin{Bmatrix} \varepsilon_x \\ \varepsilon_y \\ \varepsilon_z \\ \varepsilon_{yz} \\ \varepsilon_{zx} \\ \varepsilon_{xy} \end{Bmatrix} \tag{3.1}
$$

where σ_x, σ_y, σ_z are the normal stresses (those perpendicular to the plane of the face as shown in Figure 3.1) and σ_{yz}, σ_{zx}, σ_{xy} are the shear stresses (those parallel to the plane of the face as shown in Figure 3.1). On the other hand, normal strains are denoted by ε_x, ε_y, ε_z and the shear strains by ε_{yz}, ε_{zx}, ε_{xy}, respectively. The above relations for stresses and strains may be written in contracted notation as

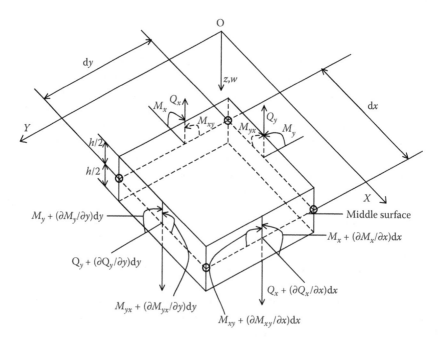

FIGURE 3.1
Moment and force resultants on plate.

$$\begin{Bmatrix} \sigma_1 \\ \sigma_2 \\ \sigma_3 \\ \sigma_4 \\ \sigma_5 \\ \sigma_6 \end{Bmatrix} = \begin{bmatrix} c_{11} & c_{12} & c_{13} & c_{14} & c_{15} & c_{16} \\ c_{21} & c_{22} & c_{23} & c_{24} & c_{25} & c_{26} \\ c_{31} & c_{32} & c_{33} & c_{34} & c_{35} & c_{36} \\ c_{41} & c_{42} & c_{43} & c_{44} & c_{45} & c_{46} \\ c_{51} & c_{52} & c_{53} & c_{54} & c_{55} & c_{56} \\ c_{61} & c_{62} & c_{63} & c_{64} & c_{65} & c_{66} \end{bmatrix} \begin{Bmatrix} \varepsilon_1 \\ \varepsilon_2 \\ \varepsilon_3 \\ \varepsilon_4 \\ \varepsilon_5 \\ \varepsilon_6 \end{Bmatrix} \tag{3.2}$$

where ε_4, ε_5, ε_6 are called the engineering shearing strains and are related to the tensor shearing strains, viz., ε_{yz}, ε_{zx}, ε_{xy}, by the following relation:

$$\begin{Bmatrix} \varepsilon_4 \\ \varepsilon_5 \\ \varepsilon_6 \end{Bmatrix} = 2 \begin{Bmatrix} \varepsilon_{yz} \\ \varepsilon_{zx} \\ \varepsilon_{xy} \end{Bmatrix} \tag{3.3}$$

Equation 3.2 may be written in the form

$$\{\sigma\} = [C]\{\varepsilon\} \tag{3.4}$$

Sometimes, it is also written in Cartesian tensor notation

$$\sigma_i = c_{ij}\varepsilon_j, \quad i, j = 1, 2, \ldots, 6 \tag{3.5}$$

Matrix $[C]$ is known as stiffness matrix. Then, the inverse of these stress–strain relations is written in the form

$$\{\varepsilon\} = [S]\{\sigma\} \tag{3.6}$$

It may also be written in Cartesian tensor notation

$$\varepsilon_i = s_{ij}\sigma_j, \quad i, j = 1, 2, \ldots, 6 \tag{3.7}$$

The matrix $[S]$ in this case is known as compliance matrix. There are 36 components in both the stiffness and compliance matrices and these are referred as elastic constants. But, this reduces to 21 because of the symmetry of the stress and strain tensors such as

$$c_{ij} = c_{ji}$$

and

$$s_{ij} = s_{ji}$$

Equations 3.1 and 3.2 are referred as anisotropic materials because there are no planes of symmetry for the material properties. If there is one plane of symmetry say $z = 0$, i.e., the xy-plane, then we have the stress–strain relation

$$\begin{Bmatrix} \sigma_1 \\ \sigma_2 \\ \sigma_3 \\ \sigma_4 \\ \sigma_5 \\ \sigma_6 \end{Bmatrix} = \begin{bmatrix} c_{11} & c_{12} & c_{13} & 0 & 0 & c_{16} \\ c_{12} & c_{22} & c_{23} & 0 & 0 & c_{26} \\ c_{13} & c_{23} & c_{33} & 0 & 0 & c_{36} \\ 0 & 0 & 0 & c_{44} & c_{45} & 0 \\ 0 & 0 & 0 & c_{45} & c_{55} & 0 \\ c_{16} & c_{26} & c_{36} & 0 & 0 & c_{66} \end{bmatrix} \begin{Bmatrix} \varepsilon_1 \\ \varepsilon_2 \\ \varepsilon_3 \\ \varepsilon_4 \\ \varepsilon_5 \\ \varepsilon_6 \end{Bmatrix} \tag{3.8}$$

There are 13 independent constants in this case because we have

$$c_{14} = c_{15} = c_{24} = c_{25} = c_{34} = c_{35} = c_{46} = c_{56} = 0 \tag{3.9}$$

When a material has elastic symmetry with respect to two mutually orthogonal planes, it will also have elastic symmetry with respect to a third plane that is orthogonal to the other two. Then, the stress–strain relation will have the form

$$\begin{Bmatrix} \sigma_1 \\ \sigma_2 \\ \sigma_3 \\ \sigma_4 \\ \sigma_5 \\ \sigma_6 \end{Bmatrix} = \begin{bmatrix} c_{11} & c_{12} & c_{13} & 0 & 0 & 0 \\ c_{12} & c_{22} & c_{23} & 0 & 0 & 0 \\ c_{13} & c_{23} & c_{33} & 0 & 0 & 0 \\ 0 & 0 & 0 & c_{44} & 0 & 0 \\ 0 & 0 & 0 & 0 & c_{55} & 0 \\ 0 & 0 & 0 & 0 & 0 & c_{66} \end{bmatrix} \begin{Bmatrix} \varepsilon_1 \\ \varepsilon_2 \\ \varepsilon_3 \\ \varepsilon_4 \\ \varepsilon_5 \\ \varepsilon_6 \end{Bmatrix} \tag{3.10}$$

In this case, the material is said to be orthotropic and there are now only nine independent constants in the stiffness matrix. From the above, we have

$$c_{16} = c_{26} = c_{36} = c_{45} = 0 \tag{3.11}$$

In tensor notation, the above equation is written as

$$\begin{Bmatrix} \sigma_x \\ \sigma_y \\ \sigma_z \\ \sigma_{yz} \\ \sigma_{zx} \\ \sigma_{xy} \end{Bmatrix} = \begin{bmatrix} c_{11} & c_{12} & c_{13} & 0 & 0 & 0 \\ c_{12} & c_{22} & c_{23} & 0 & 0 & 0 \\ c_{13} & c_{23} & c_{33} & 0 & 0 & 0 \\ 0 & 0 & 0 & c_{44} & 0 & 0 \\ 0 & 0 & 0 & 0 & c_{55} & 0 \\ 0 & 0 & 0 & 0 & 0 & c_{66} \end{bmatrix} \begin{Bmatrix} \varepsilon_x \\ \varepsilon_y \\ \varepsilon_z \\ \varepsilon_{yz} \\ \varepsilon_{zx} \\ \varepsilon_{xy} \end{Bmatrix} \tag{3.12}$$

It may be noted in this case that there is no interaction between normal stresses σ_x, σ_y, σ_z and shearing strains ε_{yz}, ε_{zx}, ε_{xy}. Moreover, there is also no interaction between shearing stresses and normal strains. Similarly, there is no interaction between shearing stresses and shearing strains in different planes.

If one of the three coordinate planes say xy-plane is isotropic, which means that the material properties in that plane are independent of direction, then we have the stress–strain relations as

$$\begin{Bmatrix} \sigma_1 \\ \sigma_2 \\ \sigma_3 \\ \sigma_4 \\ \sigma_5 \\ \sigma_6 \end{Bmatrix} = \begin{bmatrix} c_{11} & c_{12} & c_{13} & 0 & 0 & 0 \\ c_{12} & c_{11} & c_{13} & 0 & 0 & 0 \\ c_{13} & c_{13} & c_{33} & 0 & 0 & 0 \\ 0 & 0 & 0 & c_{44} & 0 & 0 \\ 0 & 0 & 0 & 0 & c_{44} & 0 \\ 0 & 0 & 0 & 0 & 0 & (c_{11} - c_{12})/2 \end{bmatrix} \begin{Bmatrix} \varepsilon_1 \\ \varepsilon_2 \\ \varepsilon_3 \\ \varepsilon_4 \\ \varepsilon_5 \\ \varepsilon_6 \end{Bmatrix} \quad (3.13)$$

As such, we have only five independent constants and then the material is said to be transversely isotropic. Here, one may see that

$$c_{22} = c_{11}, \quad c_{23} = c_{13}, \quad c_{55} = c_{44}, \quad \text{and} \quad c_{66} = (c_{11} - c_{12})/2 \quad (3.14)$$

Finally, for complete isotropy, i.e., if there are infinite number of planes of material property symmetry, then we will remain with only two independent constants. Accordingly, we will have

$$c_{13} = c_{12}, \quad c_{33} = c_{11}, \quad c_{44} = (c_{11} - c_{12})/2 \quad (3.15)$$

Thus the stress–strain relation in case of isotropic case is written in the form

$$\begin{Bmatrix} \sigma_1 \\ \sigma_2 \\ \sigma_3 \\ \sigma_4 \\ \sigma_5 \\ \sigma_6 \end{Bmatrix} = \begin{bmatrix} c_{11} & c_{12} & c_{12} & 0 & 0 & 0 \\ c_{12} & c_{11} & c_{12} & 0 & 0 & 0 \\ c_{12} & c_{12} & c_{11} & 0 & 0 & 0 \\ 0 & 0 & 0 & (c_{11} - c_{12})/2 & 0 & 0 \\ 0 & 0 & 0 & 0 & (c_{11} - c_{12})/2 & 0 \\ 0 & 0 & 0 & 0 & 0 & (c_{11} - c_{12})/2 \end{bmatrix} \begin{Bmatrix} \varepsilon_1 \\ \varepsilon_2 \\ \varepsilon_3 \\ \varepsilon_4 \\ \varepsilon_5 \\ \varepsilon_6 \end{Bmatrix}$$
$$(3.16)$$

We will now write the strain–stress relations for orthotropic, transversely isotropic, and isotropic cases. From the above discussions, the strain–stress relation for orthotropic material with nine independent constants is written using compliance matrix as

$$\begin{Bmatrix} \varepsilon_1 \\ \varepsilon_2 \\ \varepsilon_3 \\ \varepsilon_4 \\ \varepsilon_5 \\ \varepsilon_6 \end{Bmatrix} = \begin{bmatrix} s_{11} & s_{12} & s_{13} & 0 & 0 & 0 \\ s_{12} & s_{22} & s_{23} & 0 & 0 & 0 \\ s_{13} & s_{23} & s_{33} & 0 & 0 & 0 \\ 0 & 0 & 0 & s_{44} & 0 & 0 \\ 0 & 0 & 0 & 0 & s_{55} & 0 \\ 0 & 0 & 0 & 0 & 0 & s_{66} \end{bmatrix} \begin{Bmatrix} \sigma_1 \\ \sigma_2 \\ \sigma_3 \\ \sigma_4 \\ \sigma_5 \\ \sigma_6 \end{Bmatrix} \quad (3.17)$$

The strain–stress relation for transversely isotropic material with five independent constants is written using compliance matrix as

$$
\begin{Bmatrix} \varepsilon_1 \\ \varepsilon_2 \\ \varepsilon_3 \\ \varepsilon_4 \\ \varepsilon_5 \\ \varepsilon_6 \end{Bmatrix} = \begin{bmatrix} s_{11} & s_{12} & s_{13} & 0 & 0 & 0 \\ s_{12} & s_{11} & s_{13} & 0 & 0 & 0 \\ s_{13} & s_{13} & s_{33} & 0 & 0 & 0 \\ 0 & 0 & 0 & s_{44} & 0 & 0 \\ 0 & 0 & 0 & 0 & s_{44} & 0 \\ 0 & 0 & 0 & 0 & 0 & (s_{11} - s_{12})/2 \end{bmatrix} \begin{Bmatrix} \sigma_1 \\ \sigma_2 \\ \sigma_3 \\ \sigma_4 \\ \sigma_5 \\ \sigma_6 \end{Bmatrix}
\tag{3.18}
$$

Similarly, the strain–stress relation for isotropic material with two independent constants may be written in the form

$$
\begin{Bmatrix} \varepsilon_1 \\ \varepsilon_2 \\ \varepsilon_3 \\ \varepsilon_4 \\ \varepsilon_5 \\ \varepsilon_6 \end{Bmatrix} = \begin{bmatrix} s_{11} & s_{12} & s_{12} & 0 & 0 & 0 \\ s_{12} & s_{11} & s_{12} & 0 & 0 & 0 \\ s_{12} & s_{12} & s_{11} & 0 & 0 & 0 \\ 0 & 0 & 0 & (s_{11} - s_{12})/2 & 0 & 0 \\ 0 & 0 & 0 & 0 & (s_{11} - s_{12})/2 & 0 \\ 0 & 0 & 0 & 0 & 0 & (s_{11} - s_{12})/2 \end{bmatrix} \begin{Bmatrix} \sigma_1 \\ \sigma_2 \\ \sigma_3 \\ \sigma_4 \\ \sigma_5 \\ \sigma_6 \end{Bmatrix}
\tag{3.19}
$$

3.1.1 Engineering Constants

The common engineering constants include Young's modulus, Poisson's ratio, and the shear modulus. Moreover, we assume that E_i is Young's modulus in the ith direction where $i = 1, 2, 3$; v_{ij} is Poisson's ratio for transverse strain in jth direction when stressed in the ith direction, that is

$$
v_{ij} = -\frac{\varepsilon_j}{\varepsilon_i}
\tag{3.20}
$$

and G_{ij} is the shear modulus in i–j planes, then the general form of strain–stress relation for orthotropic material is given by

$$
\begin{Bmatrix} \varepsilon_1 \\ \varepsilon_2 \\ \varepsilon_3 \\ \varepsilon_4 \\ \varepsilon_5 \\ \varepsilon_6 \end{Bmatrix} = \begin{bmatrix} \frac{1}{E_1} & -\frac{v_{12}}{E_1} & \frac{v_{13}}{E_1} & 0 & 0 & 0 \\ -\frac{v_{21}}{E_2} & \frac{1}{E_2} & -\frac{v_{23}}{E_2} & 0 & 0 & 0 \\ -\frac{v_{31}}{E_3} & -\frac{v_{32}}{E_3} & \frac{1}{E_3} & 0 & 0 & 0 \\ 0 & 0 & 0 & \frac{1}{2G_{23}} & 0 & 0 \\ 0 & 0 & 0 & 0 & \frac{1}{2G_{13}} & 0 \\ 0 & 0 & 0 & 0 & 0 & \frac{1}{2G_{12}} \end{bmatrix} \begin{Bmatrix} \sigma_1 \\ \sigma_2 \\ \sigma_3 \\ \sigma_4 \\ \sigma_5 \\ \sigma_6 \end{Bmatrix}
\tag{3.21}
$$

Of the 12 engineering constants in Equation 3.21, we should have only nine independent constants. Accordingly, there are three reciprocal relations

$$
\frac{v_{12}}{E_1} = \frac{v_{21}}{E_2}; \quad \frac{v_{23}}{E_2} = \frac{v_{32}}{E_3}; \quad \frac{v_{13}}{E_1} = \frac{v_{31}}{E_3}
\tag{3.22}
$$

So, we have nine independent constants, viz., E_1, E_2, E_3; ν_{12}, ν_{13}, ν_{23}; G_{12}, G_{13}, G_{23}, for orthotropic materials.

For a transversely isotropic material, if we have 1–2 plane (i.e., x–y plane in Cartesian coordinate system) as the special plane of isotropy, then we have four more relations satisfying

$$E_1 = E_2; \quad G_{13} = G_{23}; \quad \nu_{13} = \nu_{23}; \quad G_{12} = \frac{E_1}{2(1 + \nu_{12})} \tag{3.23}$$

and therefore five constants are left as independent for transversely isotropic material.

Next, in the case of completely isotropic material, in which the mechanical and physical properties do not vary with orientation, we have two relations as

$$E_1 = E_2 = E_3 = E(\text{say}) \tag{3.24}$$

and

$$G_{12} = G_{13} = G_{23} = G = \frac{E}{2(1 + \nu)} \tag{3.25}$$

Finally, we have two independent constants for isotropic materials. By inverting the compliance matrix in Equation 3.21, the terms of stiffness matrix may easily be written for stress–strain relations also.

3.1.2 Plane Stress

If a lamina is in a plane 1–2 (i.e., say in x–y plane) as shown in Figure 3.2, then a plane stress is defined by taking $\sigma_z = 0$, $\sigma_{yz} = 0$, $\sigma_{zx} = 0$, i.e., $\sigma_3 = 0$, $\sigma_4 = 0$, $\sigma_5 = 0$. Thus, from Equation 3.21, the strain–stress relations with respect to compliance matrix for orthotropic material may be obtained as

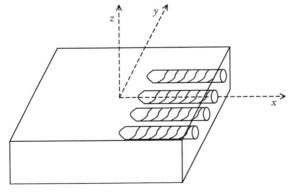

FIGURE 3.2
Plane stress (unidirectionally reinforced lamina).

$$\begin{Bmatrix} \varepsilon_1 \\ \varepsilon_2 \\ \varepsilon_6 \end{Bmatrix} = \begin{bmatrix} \frac{1}{E_1} & -\frac{\nu_{12}}{E_1} & 0 \\ -\frac{\nu_{21}}{E_2} & \frac{1}{E_2} & 0 \\ 0 & 0 & \frac{1}{2G_{12}} \end{bmatrix} \begin{Bmatrix} \sigma_1 \\ \sigma_2 \\ \sigma_6 \end{Bmatrix} \tag{3.26}$$

where we also have

$$\frac{\nu_{12}}{E_1} = \frac{\nu_{21}}{E_2} \tag{3.27}$$

By inverting Equation 3.26, the stress–strain relation in terms of stiffness matrix can be found. It may be noted here that four independent constants (or material properties), viz., E_1, E_2, ν_{12}, G_{12}, exist along with the reciprocal relation (Equation 3.27) for orthotropic material under plane stress.

Similarly, strain–stress relations for isotropic materials under plane stress are given by Equations 3.21 through 3.23 as

$$\begin{Bmatrix} \varepsilon_1 \\ \varepsilon_2 \\ \varepsilon_6 \end{Bmatrix} = \begin{bmatrix} \frac{1}{E} & -\frac{\nu}{E} & 0 \\ -\frac{\nu}{E} & \frac{1}{E} & 0 \\ 0 & 0 & \frac{2(1+\nu)}{E} \end{bmatrix} \begin{Bmatrix} \sigma_1 \\ \sigma_2 \\ \sigma_6 \end{Bmatrix} \tag{3.28}$$

Again, the stress–strain relations in terms of stiffness matrix may be written for isotropic materials under plane stress as

$$\begin{Bmatrix} \sigma_1 \\ \sigma_2 \\ \sigma_6 \end{Bmatrix} = \begin{bmatrix} \frac{E}{1-\nu^2} & \frac{\nu E}{1-\nu^2} & 0 \\ \frac{\nu E}{1-\nu^2} & \frac{E}{1-\nu^2} & 0 \\ 0 & 0 & \frac{E}{2(1+\nu)} \end{bmatrix} \begin{Bmatrix} \varepsilon_1 \\ \varepsilon_2 \\ \varepsilon_6 \end{Bmatrix} \tag{3.29}$$

Now considering the Cartesian coordinates, the strain–stress relations for an isotropic material in general are written from Equations 3.21 through 3.23 as

$$\varepsilon_x = \frac{1}{E} \left[\sigma_x - \nu(\sigma_y + \sigma_z) \right] \tag{3.30}$$

$$\varepsilon_y = \frac{1}{E} \left[\sigma_y - \nu(\sigma_x + \sigma_z) \right] \tag{3.31}$$

$$\varepsilon_z = \frac{1}{E} \left[\sigma_z - \nu(\sigma_x + \sigma_y) \right] \tag{3.32}$$

$$\varepsilon_{xy} = \frac{1}{2G} \sigma_{xy} \tag{3.33}$$

$$\varepsilon_{yz} = \frac{1}{2G} \sigma_{yz} \tag{3.34}$$

$$\varepsilon_{zx} = \frac{1}{2G} \sigma_{zx} \tag{3.35}$$

and

$$G = \frac{E}{2(1+\nu)} \tag{3.36}$$

For plane stress problem, we may similarly write the stress–strain in Cartesian coordinates from Equation 3.29 as

$$\sigma_x = \frac{E}{1-\nu^2}(\varepsilon_x + \nu\varepsilon_y) \tag{3.37}$$

$$\sigma_y = \frac{E}{1-\nu^2}(\varepsilon_y + \nu\varepsilon_x) \tag{3.38}$$

$$\sigma_{xy} = G\varepsilon_{xy} \tag{3.39}$$

3.2 Plate Theory

This book considers only the classical plate theory in the analysis along with the solution of plates of various shapes and complicating effects. Here, we will first address the classical plate theory, and then the Mindlin plate theory will be outlined for the sake of completeness.

Classical plate theory or Kirchhoff plate theory is based on the following assumptions:

1. Thickness of the plate is small when compared with other dimensions.
2. Normal stresses in the direction transverse to the plate are taken to be negligibly small.
3. Effect of rotatory inertia is negligible.
4. Normal to the undeformed middle surface remains straight and normal to the deformed middle surface and unstretched in length.

From the classical plate theory, it is to be noted that the plate equations are approximate. Refined equations, in general, are more accurate and applicable to higher modes than the equations of classical theory. The last assumption of the classical plate theory, viz., normal to the undeformed middle surface remains normal to the deformed middle surface tries to neglect the effect of transverse shear deformation. The transverse shear effects as well as the rotatory inertia effect are important when the plate is relatively thick or when higher-mode vibration characteristics are needed. The above theory was refined first by Timoshenko (1921) by including the effects of transverse shear and rotatory inertia in beam equations. Accordingly, transverse shear effect was then introduced in the plate equations by Reissner (1945).

Again, both transverse shear effect and rotatory inertia effect were included in the equation of motion of a plate by Mindlin (1951). Thus, the theory of the classical plate equation considering the effect of transverse shear (Reissner (1945)) relaxes the normality condition and the fourth assumption in the above classical theory will read as: "normal to the undeformed middle surface remains straight and unstretched in length but not necessarily normal to the deformed middle surface."

The above assumption implies a nonzero transverse shear strain giving an error to the formulation. Thereby, Mindlin (1951), as mentioned above, modified the third assumption of the classical plate theory too and it reads: "the effect of rotatory inertia is included," along with the above-modified assumption given by Reissner (1945). The final theory (modified third and fourth assumptions in the classical theory) is known as Mindlin plate theory.

3.3 Strain–Displacement Relations

Let $\bar{u}$, $\bar{v}$, and $\bar{w}$ denote displacements at a point (x, y, z) in a body or a plate, then the strain–displacement relations relating the displacements that result from the elastic body being strained due to the applied load are given by

$$\varepsilon_x = \frac{\partial \bar{u}}{\partial x} \tag{3.40}$$

$$\varepsilon_y = \frac{\partial \bar{v}}{\partial y} \tag{3.41}$$

$$\varepsilon_z = \frac{\partial \bar{w}}{\partial z} \tag{3.42}$$

$$\varepsilon_{xy} = \frac{1}{2} \left(\frac{\partial \bar{u}}{\partial y} + \frac{\partial \bar{v}}{\partial x} \right) \tag{3.43}$$

$$\varepsilon_{xz} = \frac{1}{2} \left(\frac{\partial \bar{u}}{\partial z} + \frac{\partial \bar{w}}{\partial x} \right) \tag{3.44}$$

$$\varepsilon_{yz} = \frac{1}{2} \left(\frac{\partial \bar{v}}{\partial z} + \frac{\partial \bar{w}}{\partial y} \right) \tag{3.45}$$

3.4 Compatibility Equations

Compatibility equations ensure that the displacements of an elastic body are continuous and single valued, though in the analysis of plate vibrations, the compatibility equations are not utilized, they require only the terms of displacements. But for completeness, these equations are incorporated and are written as

$$\frac{\partial^2 \varepsilon_x}{\partial y \partial z} = \frac{\partial}{\partial x}\left(\frac{\partial \varepsilon_{xz}}{\partial y} + \frac{\partial \varepsilon_{xy}}{\partial z} - \frac{\partial \varepsilon_{yz}}{\partial x}\right) \tag{3.46}$$

$$\frac{\partial^2 \varepsilon_y}{\partial z \partial x} = \frac{\partial}{\partial y}\left(\frac{\partial \varepsilon_{xy}}{\partial z} + \frac{\partial \varepsilon_{yz}}{\partial x} - \frac{\partial \varepsilon_{zx}}{\partial y}\right) \tag{3.47}$$

$$\frac{\partial^2 \varepsilon_z}{\partial x \partial y} = \frac{\partial}{\partial z}\left(\frac{\partial \varepsilon_{yz}}{\partial x} + \frac{\partial \varepsilon_{xz}}{\partial y} - \frac{\partial \varepsilon_{xy}}{\partial z}\right) \tag{3.48}$$

$$2\frac{\partial^2 \varepsilon_{xy}}{\partial x \partial y} = \frac{\partial^2 \varepsilon_x}{\partial y^2} + \frac{\partial^2 \varepsilon_y}{\partial x^2} \tag{3.49}$$

$$2\frac{\partial^2 \varepsilon_{yz}}{\partial y \partial z} = \frac{\partial^2 \varepsilon_y}{\partial z^2} + \frac{\partial^2 \varepsilon_z}{\partial y^2} \tag{3.50}$$

$$2\frac{\partial^2 \varepsilon_{zx}}{\partial z \partial x} = \frac{\partial^2 \varepsilon_z}{\partial x^2} + \frac{\partial^2 \varepsilon_x}{\partial z^2} \tag{3.51}$$

3.5 Kinematics of Deformation of Plates

Let us consider the deformation of mid-surface of a plate. A portion of a plate of thickness h and having inside region (domain) R is shown in Figure 3.3 in which xy-plane lies in the middle surface in undeformed position. A normal load distribution $q(x, y)$ is assumed to act on the top of the plate. Shear force Q and bending moment M act on the edge ∂R of the plate. We now consider the deformation parallel to the mid-surface of the plate. Accordingly, there will be two actions, first one due to the stretching and the second due to bending. The stretching actions are due to loads at the boundary of the plate. For these, let us suppose that u_s and v_s are the horizontal displacement components at any point (x, y, z) in the plate, which is identical in the middle surface of the plate. Thus, we may write

$$u_s(x, y, z) = u_s(x, y, 0) = u_s(x, y) \tag{3.52}$$

$$v_s(x, y, z) = v_s(x, y, 0) = v_s(x, y) \tag{3.53}$$

This is because the lines joining the surfaces of the plate and normal to the xy-plane in the undeformed geometry translate horizontally owing to the action of stretching.

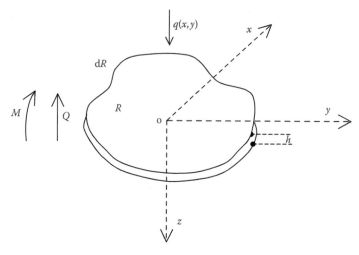

FIGURE 3.3
Middle surface of portion of a plate.

The other contribution, as mentioned earlier, is due to bending. Because of this and as per the classical plate theory, lines normal to the mid-surface in the undeformed geometry remain normal to this surface in the deformed geometry.

Now, the displacements in the x- and y-directions due to the bending action are denoted by u_b and v_b. Because of the bending actions, there will be a rotation as in rigid body elements and also vertical translation. Accordingly, the displacements in x- and y-directions for the bending action may be written as

$$u_b(x, y, z) = -z \frac{\partial w(x, y)}{\partial x} \tag{3.54}$$

$$v_b(x, y, z) = -z \frac{\partial w(x, y)}{\partial y} \tag{3.55}$$

Thus, considering both the horizontal (Equations 3.52 and 3.53) and vertical (Equations 3.54 and 3.55) displacements, the combining actions are written as

$$\bar{u} = u_s(x, y) - z \frac{\partial w(x, y)}{\partial x} \tag{3.56}$$

$$\bar{v} = v_s(x, y) - z \frac{\partial w(x, y)}{\partial y} \tag{3.57}$$

$$\bar{w} = w(x, y) \tag{3.58}$$

It is worth mentioning here that any displacement field $(\bar{u}, \bar{v}, \bar{w})$ may now be completely described by the displacement of the mid-plane (u_s, v_s, w). For the sake of convenience, we will write Equations 3.56 through 3.58 as follows:

$$\overline{u} = u - z\frac{\partial w}{\partial x} \tag{3.59}$$

$$\overline{v} = v - z\frac{\partial w}{\partial y} \tag{3.60}$$

$$\overline{w} = w \tag{3.61}$$

where (u, v, w) are the displacement of the middle surface of the plate at the point $(x, y, 0)$. Then, according to classical plate theory, the displacement functions $(\overline{u}, \overline{v}, \overline{w})$ of the plate at a point (x, y, z) are the approximations as given in Equations 3.59 through 3.61.

As such, now the strain fields in terms of the displacements are written as

$$\varepsilon_x = \frac{\partial \overline{u}}{\partial x} = \frac{\partial u}{\partial x} - z\frac{\partial^2 w}{\partial x^2} \tag{3.62}$$

$$\varepsilon_y = \frac{\partial \overline{v}}{\partial y} = \frac{\partial v}{\partial y} - z\frac{\partial^2 w}{\partial y^2} \tag{3.63}$$

$$\varepsilon_{xy} = \frac{1}{2}\left(\frac{\partial \overline{u}}{\partial y} + \frac{\partial \overline{v}}{\partial x}\right) = \frac{1}{2}\left(\frac{\partial u}{\partial y} + \frac{\partial v}{\partial x}\right) - z\frac{\partial^2 w}{\partial x \partial y} \tag{3.64}$$

All other strains are zero according to the classical plate theory.

3.6 Biharmonic Equation

An element of a plate with stresses at the mid-plane of the plate has been shown in Figure 3.4. These stresses vary in the z-direction over the thickness h of the plate. Then, the shear force intensities per unit length are defined as

$$Q_x = \int_{-h/2}^{h/2} \sigma_{xz} dz \tag{3.65}$$

$$Q_y = \int_{-h/2}^{h/2} \sigma_{yz} dz \tag{3.66}$$

Let us also define the bending moment intensities per unit length by

$$M_x = \int_{-h/2}^{h/2} \sigma_x z dz \tag{3.67}$$

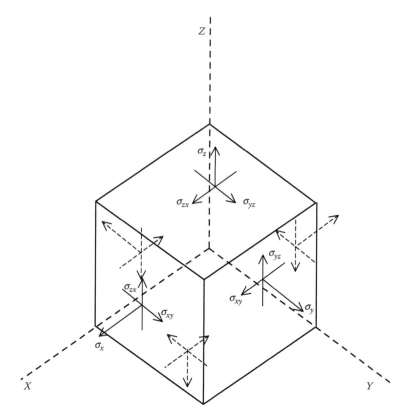

FIGURE 3.4
Plate element showing stresses.

$$M_y = \int\limits_{-h/2}^{h/2} \sigma_y z dz \qquad (3.68)$$

Finally, the twisting moment intensities per unit length are given by

$$M_{xy} = \int\limits_{-h/2}^{h/2} \sigma_{xy} z dz \qquad (3.69)$$

$$M_{yx} = \int\limits_{-h/2}^{h/2} \sigma_{yx} z dz \qquad (3.70)$$

We also have

$$M_{xy} = M_{yx} \tag{3.71}$$

Figure 3.1 depicts the shear force, bending, and twisting moment intensities on sections perpendicular to x- and y-axes.

Substituting the values of σ_x, σ_y, and σ_{xy} from Equations 3.37 through 3.39 in Equations 3.67 through 3.70, respectively, we obtain

$$M_x = \int_{-h/2}^{h/2} \frac{E}{(1-\nu^2)}(\varepsilon_x + \nu\varepsilon_y)z\,dz \tag{3.72}$$

$$M_y = \int_{-h/2}^{h/2} \frac{E}{(1-\nu^2)}(\varepsilon_y + \nu\varepsilon_x)z\,dz \tag{3.73}$$

$$M_{xy} = \int_{-h/2}^{h/2} 2G\varepsilon_{xy}z\,dz \tag{3.74}$$

Now, using strains from Equations 3.40 through 3.45 along with Equations 3.56 through 3.58 in Equations 3.72 through 3.74, respectively, and then integrating, one gets

$$M_x = -D\left(\frac{\partial^2 w}{\partial x^2} + \nu\frac{\partial^2 w}{\partial y^2}\right) \tag{3.75}$$

$$M_y = -D\left(\frac{\partial^2 w}{\partial y^2} + \nu\frac{\partial^2 w}{\partial x^2}\right) \tag{3.76}$$

and

$$M_{xy} = -D(1-\nu)\frac{\partial^2 w}{\partial x \partial y} \tag{3.77}$$

where

$$D = \frac{Eh^3}{12(1-\nu^2)} \tag{3.78}$$

is called the flexural rigidity of the plate material. The quantities Q_x, Q_y, M_x, M_y, and M_{xy} can be related by considering equilibrium of the plate element as shown in Figure 3.1. Thus, for equilibrium in the x-direction, in the absence of body forces, we have

$$\frac{\partial \sigma_x}{\partial x} + \frac{\partial \sigma_{xy}}{\partial y} + \frac{\partial \sigma_{xz}}{\partial z} = 0$$

Multiplying this equation by z and integrating over the thickness of the plate and noting that the operation $\partial/\partial x$ and $\partial/\partial y$ can be interchanged, one can get

$$Q_x = \frac{\partial M_x}{\partial x} + \frac{\partial M_{xy}}{\partial y} \qquad (3.79)$$

Similarly, we can integrate the equation of equilibrium in the y-direction to obtain

$$Q_y = \frac{\partial M_y}{\partial y} + \frac{\partial M_{xy}}{\partial x} \qquad (3.80)$$

Considering the integration over the thickness of the last equilibrium equation, viz.,

$$\frac{\partial \sigma_{xz}}{\partial x} + \frac{\partial \sigma_{yz}}{\partial y} + \frac{\partial \sigma_{zz}}{\partial z} = 0$$

we obtain

$$\frac{\partial Q_x}{\partial x} + \frac{\partial Q_y}{\partial y} + q(x,y) = 0 \qquad (3.81)$$

where $q(x,y)$ is a normal load distribution on the top face of the plate. Now putting Q_x and Q_y from Equations 3.79 and 3.80 in Equation 3.81, we get

$$\frac{\partial^2 M_x}{\partial x^2} + 2\frac{\partial^2 M_{xy}}{\partial x \partial y} + \frac{\partial^2 M_y}{\partial y^2} + q = 0 \qquad (3.82)$$

Finally, using Equations 3.75 through 3.78, we can write Equation 3.82 as

$$\frac{\partial^4 w}{\partial x^4} + 2\frac{\partial^4 w}{\partial x^2 \partial y^2} + \frac{\partial^4 w}{\partial y^4} + \frac{q}{D} = 0 \qquad (3.83)$$

This is usually written as

$$\nabla^4 w = \frac{q}{D} \qquad (3.84)$$

This is the nonhomogeneous biharmonic equation first obtained by Sophie Germain in 1815 and is the governing equation of the classical plate theory.

3.7 Minimum Total Potential Energy Approach for Biharmonic Equation

There are three energy principles used in structural mechanics, viz.,

1. Theorem of minimum potential energy
2. Theorem of minimum complimentary energy
3. Reissner's variational theorem

It is, however, to be noted that minimum complimentary energy is less useful for the type of problems addressed in this section. Moreover, Reissner's variational theorem is useful in solving problems that include transverse shear deformation. As such, only the minimum potential energy will be discussed. The theorem of minimum potential energy is defined as the following: Of all the displacements satisfying compatibility and the prescribed boundary conditions, those that satisfy equilibrium equations make the potential energy a minimum. To use the above theorem, it is first necessary to define the stress–strain relations to replace the stresses in the strain energy expression by strains and next it is needed to write the strain–displacement relations to place the strains in terms of displacements.

Here, a plate of domain R and edge C is considered subjected to a distributed load $q(x,y)$. For linear elastic analysis, the strain energy U of the plate may be evaluated by the integral

$$U = \frac{1}{2} \iint_R \left[\int_{-h/2}^{h/2} (\sigma_x \varepsilon_x + 2\sigma_{xy}\varepsilon_{xy} + \sigma_y \varepsilon_y)dz \right] dxdy \tag{3.85}$$

Substituting the values of σ_x, σ_{xy}, and σ_y from Equations 3.37 through 3.39 in Equation 3.85, we obtain

$$U = \frac{1}{2} \iint_R \left[\int_{-h/2}^{h/2} \left\{ \frac{E}{(1-\nu^2)}(\varepsilon_x + \nu\varepsilon_y)\varepsilon_x + 2\frac{E}{(1+\nu)}\varepsilon_{xy}\varepsilon_{xy} \right. \right.$$

$$\left. \left. + \frac{E}{(1-\nu^2)}(\varepsilon_y + \nu\varepsilon_y)\varepsilon_y \right\}dz \right] dxdy$$

$$= \frac{E}{2(1-\nu^2)} \int_{-h/2}^{h/2} \left[\varepsilon_x^2 + \varepsilon_y^2 + 2(1-\nu)\varepsilon_{xy}^2 + 2\nu\varepsilon_x\varepsilon_y \right] dzdxdy \tag{3.86}$$

If $q(x,y)$ is the load distribution acting on the mid-surface of the plate, then the potential energy V may be given by

$$V = -\iint_R q(x,y)w(x,y)dxdy \tag{3.87}$$

Expressing the strains by Equations 3.62 through 3.64 considering the stretching and bending terms in Equation 3.86, we can write the total potential energy as

$$S = \frac{E}{2(1-\nu^2)} \iint_R \int_{-h/2}^{h/2} \left[\left(\frac{\partial u_s}{\partial x} - z\frac{\partial^2 w}{\partial x^2} \right) + \left(\frac{\partial v_s}{\partial y} - z\frac{\partial^2 w}{\partial y^2} \right) \right.$$

$$+ 2(1-\nu)\left\{ \frac{1}{2}\left(\frac{\partial u_s}{\partial y} + \frac{\partial v_s}{\partial x} \right) - Z\frac{\partial^2 w}{\partial x \partial y} \right\}^2$$

$$\left. + 2\nu\left(\frac{\partial u_s}{\partial x} - z\frac{\partial^2 w}{\partial x^2} \right)\left(\frac{\partial v_s}{\partial x} - z\frac{\partial^2 w}{\partial y^2} \right) \right] dzdxdy - \iint_R qwdxdy \tag{3.88}$$

Integrating Equation 3.88 with respect to Z, we write the above in the form

$$S = \frac{D'}{2}\iint_R \left[\left(\frac{\partial u_s}{\partial x} \right)^2 + \left(\frac{\partial v_s}{\partial y} \right)^2 + 2\nu\frac{\partial u_s}{\partial x}\frac{\partial v_s}{\partial y} + \frac{1-\nu}{2}\left(\frac{\partial u_s}{\partial y} + \frac{\partial v_s}{\partial x} \right)^2 \right] dxdy$$

$$+ \frac{D}{2}\iint_R \left[\left(\frac{\partial^2 w}{\partial x^2} \right)^2 + \left(\frac{\partial^2 w}{\partial y^2} \right)^2 + 2\nu\frac{\partial^2 w}{\partial x^2}\cdot\frac{\partial^2 w}{\partial y^2} + 2(1-\nu)\left(\frac{\partial^2 w}{\partial x \partial y} \right)^2 \right] dxdy$$

$$- \iint_R qwdxdy \tag{3.89}$$

where the constants D and D' are given by

$$D = \frac{Eh^3}{12(1-\nu^2)} \tag{3.90}$$

and

$$D' = \frac{Eh}{(1-\nu^2)} \tag{3.91}$$

known as flexural rigidity and extensional stiffness, respectively.

It is now clear that the total potential energy functional includes three dependent variables, viz., the stretching components u_s and v_s and the vertical displacement w due to bending. Next, we consider only the bending effects because we deal with the plate subject to transverse loads. Accordingly, the components u_s and v_s due to stretching are neglected and so the total potential energy may be obtained from Equation 3.89 as

$$S = \frac{D}{2} \iint_R \left[\left(\frac{\partial^2 w}{\partial x^2}\right)^2 + \left(\frac{\partial^2 w}{\partial y^2}\right)^2 + 2\nu \frac{\partial^2 w}{\partial x^2} \cdot \frac{\partial^2 w}{\partial y^2} + 2(1-\nu)\left(\frac{\partial^2 w}{\partial x \partial y}\right)^2 \right] dxdy$$

$$- \iint_R qw \, dxdy \tag{3.92}$$

which may be written as

$$S = \frac{D}{2} \iint_R \left[(\nabla^2 w)^2 + 2(1-\nu)\left\{ \left(\frac{\partial^2 w}{\partial x \partial y}\right)^2 - \frac{\partial^2 w}{\partial x^2} \cdot \frac{\partial^2 w}{\partial y^2} \right\} \right] dxdy$$

$$- \iint_R qw \, dxdy \tag{3.93}$$

We will now extremize the above as

$$\delta S = 0$$

This gives

$$\frac{D}{2} \iint_R \left[2(\nabla^2 w)\left(\frac{\partial^2 \delta w}{\partial x^2} + \frac{\partial^2 \delta w}{\partial y^2}\right) + 2(1-\nu) \right.$$

$$\left. \times \left(2\frac{\partial^2 w}{\partial x \partial y} \cdot \frac{\partial^2 \delta w}{\partial x \partial y} + 2\frac{\partial^2 w}{\partial x \partial y} \cdot \frac{\partial^2 \delta w}{\partial y \partial x} - \frac{\partial^2 w}{\partial x^2} \cdot \frac{\partial^2 \delta w}{\partial y^2} - \frac{\partial^2 w}{\partial y^2} \cdot \frac{\partial^2 \delta w}{\partial x^2} \right) \right] dxdy$$

$$- \iint_R q \delta w \, dxdy = 0 \tag{3.94}$$

Applying Green's theorem, we may get each term of Equation 3.94 as

$$\iint_R (\nabla^2 w)\frac{\partial^2 \delta w}{\partial x^2} dxdy = \int_c (\nabla^2 w)\frac{\partial \delta w}{\partial x} dy - \left\{ \int_C \delta w \frac{\partial}{\partial x}(\nabla^2 w) dy - \iint_R \delta w \frac{\partial^2}{\partial x^2}(\nabla^2 w) dxdy \right\}$$

$$\tag{3.95}$$

$$\iint_R (\nabla^2 w)\frac{\partial^2 \delta w}{\partial y^2} dxdy = -\int_c (\nabla^2 w)\frac{\partial \delta w}{\partial y} dx - \left\{ -\int_C \delta w \frac{\partial}{\partial y}(\nabla^2 w) dx - \iint_R \delta w \frac{\partial^2}{\partial y^2}(\nabla^2 w) dxdy \right\}$$

$$\tag{3.96}$$

$$\iint_R \frac{\partial^2 w}{\partial x \partial y}\frac{\partial^2 \delta w}{\partial x \partial y} dxdy = \int_C \frac{\partial^2 w}{\partial x \partial y} \cdot \frac{\partial \delta w}{\partial y} dy - \left\{ -\int_C \frac{\partial^3 w}{\partial x^2 \partial y}\delta w \, dx - \iint_R \delta w \cdot \frac{\partial^4 w}{\partial x^2 \partial y^2} dxdy \right\}$$

$$\tag{3.97}$$

$$\iint_R \frac{\partial^2 w}{\partial y \partial x} \frac{\partial^2 \delta w}{\partial y \partial x} dxdy = -\int_C \frac{\partial^2 w}{\partial y \partial x} \cdot \frac{\partial \delta w}{\partial x} dx - \left\{ \int_C \frac{\partial^3 w}{\partial y^2 \partial x} \delta w dy - \iint_R \delta w \cdot \frac{\partial^4 w}{\partial y^2 \partial x^2} dxdy \right\}$$

(3.98)

$$\iint_R \frac{\partial^2 w}{\partial x^2} \frac{\partial^2 \delta w}{\partial y^2} dxdy = -\int_C \frac{\partial^2 w}{\partial x^2} \cdot \frac{\partial \delta w}{\partial y} dx - \left\{ -\int_C \delta w \cdot \frac{\partial^3 w}{\partial x^2 \partial y} dx - \iint_R \delta w \cdot \frac{\partial^4 w}{\partial x^2 \partial y^2} dydx \right\}$$

(3.99)

$$\iint_R \frac{\partial^2 w}{\partial y^2} \frac{\partial^2 \delta w}{\partial x^2} dxdy = \int_C \frac{\partial^2 w}{\partial y^2} \cdot \frac{\partial \delta w}{\partial x} dy - \left\{ \int_C \delta w \cdot \frac{\partial^3 w}{\partial x \partial y^2} dy - \iint_R \delta w \cdot \frac{\partial^4 w}{\partial x^2 \partial y^2} dxdy \right\}$$

(3.100)

Putting the expressions from Equations 3.95 through 3.100 in Equation 3.94, one may finally obtain

$$\iint_R (D\nabla^4 w - q)\delta w dxdy + D\int_C \left(\frac{\partial^2 w}{\partial x^2} + \nu \frac{\partial^2 w}{\partial y^2} \right) \frac{\partial \delta w}{\partial x} dy$$

$$- D\int_C \left(\frac{\partial^2 w}{\partial y^2} + \nu \frac{\partial^2 w}{\partial x^2} \right) \frac{\partial \delta w}{\partial y} dx + D\int_C (1-\nu) \frac{\partial^2 w}{\partial x \partial y} \frac{\partial \delta w}{\partial y} dy$$

$$- D\int_C (1-\nu) \frac{\partial^2 w}{\partial y \partial x} \frac{\partial \delta w}{\partial x} dx + D\int_C \left(\frac{\partial^3 w}{\partial y^3} + \nu \frac{\partial^3 w}{\partial x^2 \partial y} \right) \delta w dx$$

$$- D\int_C \left(\frac{\partial^3 w}{\partial x^3} + \nu \frac{\partial^3 w}{\partial x \partial y^2} \right) \delta w dy + D\int_C (1-\nu) \frac{\partial^3 w}{\partial x^2 \partial y} \delta w dx$$

$$- D\int_C (1-\nu) \frac{\partial^3 w}{\partial x \partial y^2} \delta w dy = 0$$

(3.101)

Now, writing the above in terms of M_x, M_y, and M_{xy} from Equations 3.75 through 3.77, we get

$$\iint_R (D\nabla^4 w - q)\delta w dxdy - \int_C M_x \frac{\partial \delta w}{\partial x} dy + \int_C M_y \frac{\partial \delta w}{\partial y} dx - \int_C M_{xy} \frac{\partial \delta w}{\partial y} dy$$

$$+ \int_C M_{xy} \frac{\partial \delta w}{\partial x} dx - \int_C \left(\frac{\partial M_y}{\partial y} + \frac{\partial M_{xy}}{\partial x} \right) \delta w dx + \int_C \left(\frac{\partial M_x}{\partial x} + \frac{\partial M_{xy}}{\partial y} \right) \delta w dy = 0$$

(3.102)

Putting again the values of Q_x and Q_y from Equations 3.79 and 3.80 in Equation 3.102 turns into the form

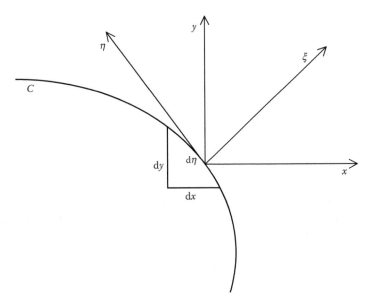

FIGURE 3.5
Portion of path C with normal and tangential.

$$\iint\limits_{R}[D\nabla^4 w - q]\delta w dxdy - \int\limits_{C} M_x \frac{\partial \delta w}{\partial x}dy + \int\limits_{C} M_y \frac{\partial \delta w}{\partial y}dx - \int\limits_{C} M_{xy} \frac{\partial \delta w}{\partial y}dy$$

$$+ \int\limits_{C} M_{xy}\frac{\partial \delta w}{\partial x}dx - \int\limits_{C} Q_y \delta w dx + \int\limits_{C} Q_x \delta w dy = 0 \qquad (3.103)$$

Examining a portion of the path C as shown in Figure 3.5 and considering ξ and η to be a rectangular set of coordinates at a point on the boundary, the final variation of the total potential energy may be obtained as

$$\iint\limits_{R}(D\nabla^4 w - q)\delta w dxdy - \int\limits_{C} M_\xi \delta\left(\frac{\partial w}{\partial \xi}\right)d\eta + \int\limits_{C}\left(Q_\xi + \frac{\partial M_{\xi\eta}}{\partial \eta}\right)\delta w d\eta = 0 \quad (3.104)$$

The Euler–Lagrange equation for this problem is

$$\nabla^4 w = \frac{q}{D} \qquad (3.105)$$

which is the biharmonic equation as also obtained in Equation 3.84 where

$$\nabla^4 w = \frac{\partial^4 w}{\partial x^4} + 2\frac{\partial^4 w}{\partial x^2 \partial y^2} + \frac{\partial^4 w}{\partial y^4}$$

and

$$M_\xi = M_x\left(\frac{dy}{d\eta}\right)^2 + M_y\left(\frac{dx}{d\eta}\right)^2 + 2M_{xy}\frac{dy}{d\eta}\frac{dx}{d\eta} \tag{3.106}$$

$$Q_\xi = Q_y\left(\frac{dx}{d\eta}\right) + Q_x\left(\frac{dy}{d\eta}\right) \tag{3.107}$$

$$M_{\xi\eta} = M_x\left(\frac{dy}{d\eta}\right)\left(\frac{dx}{d\eta}\right) + M_y\left(\frac{dy}{d\eta}\right)\left(\frac{dx}{d\eta}\right) + M_{xy}\left(\frac{dy}{d\eta}\right)^2 - M_{xy}\left(\frac{dx}{d\eta}\right)^2 \tag{3.108}$$

From Equation 3.104, we may have now the two sets of boundary conditions such as natural and kinematic boundary conditions for which one of the following conditions is required to be satisfied:

1. Either $M_\xi = 0$ or $\frac{\partial w}{\partial \xi}$ is prescribed.
2. Either $Q_\xi + \frac{\partial M_{\xi\eta}}{\partial \eta} = 0$ or w is prescribed.

3.8 Equation of Motion for Vibration of Plates by Hamilton's Principle

Kinetic energy T of a vibrating plate is given by

$$T = \frac{1}{2}\iint_R h\rho\dot{w}^2\,dxdy \tag{3.109}$$

where R denotes the transverse area of the plate, ρ denotes the density, and h is the plate thickness. Total potential energy has been derived in Equation 3.93 as

$$S = \frac{D}{2}\iint_R\left[\left(\nabla^2 w\right)^2 + 2(1-\nu)\left\{\left(\frac{\partial^2 w}{\partial x\partial y}\right)^2 - \frac{\partial^2 w}{\partial x^2}\cdot\frac{\partial^2 w}{\partial y^2}\right\}\right]dxdy - \iint_R qw\,dxdy \tag{3.110}$$

By Hamilton's principle, we have first variation

$$\delta(T - S) = 0 \tag{3.111}$$

This will give as before

$$\rho\left[\int_{t_1}^{t_2}\iint_R\left[\frac{h\rho}{2}\dot{w}^2-\frac{D}{2}\left\{(\nabla^2 w)^2+2(1-\nu)\left(\left(\frac{\partial^2 w}{\partial x\partial y}\right)^2-\frac{\partial^2 w}{\partial x^2}\cdot\frac{\partial^2 w}{\partial y^2}\right)\right\}-qw\right]dxdydt=0\right.$$

(3.112)

It may be noted here (due to dynamic analysis) that the integration includes time variable also. Now, taking the first variation by operator approach, we obtain

$$\int_{t_1}^{t_2}\iint_R\left[\rho h\frac{\partial w}{\partial t}\cdot\frac{\partial\delta w}{\partial t}-D\left\{(\nabla^2 w)(\nabla^2\delta w)+2(1-\nu)\right.\right.$$

$$\times\left(\frac{\partial^2 w}{\partial x\partial y}\frac{\partial^2\delta W}{\partial x\partial y}-\frac{1}{2}\frac{\partial^2 w}{\partial x^2}\cdot\frac{\partial^2\delta w}{\partial y^2}-\frac{1}{2}\frac{\partial^2 w}{\partial y^2}\cdot\frac{\partial^2\delta w}{\partial x^2}\right)\right\}-q\delta w\right]dxdydt=0 \quad (3.113)$$

The variations after first term in the above equation are found by Green's theorem as in Section 3.7. Here, the first term is given by

$$\iint_R\left[\int_{t_1}^{t_2}\rho h\frac{\partial w}{\partial t}\cdot\frac{\partial\delta w}{\partial t}dt\right]dxdy=\iint_R\rho h\left[\left(\frac{\partial w}{\partial t}\cdot\delta w\right)_{t_1}^{t_2}-\int_{t_1}^{t_2}\frac{\partial^2 w}{\partial t^2}\delta w dt\right]dxdy$$

which turns into

$$0-\int_{t_1}^{t_2}\iint_R h\rho\frac{\partial^2 w}{\partial t^2}\delta w dxdydt \qquad (3.114)$$

Using the above facts, we may write Equation 3.113 in the form

$$\int_{t_1}^{t_2}\iint_R\left[-h\rho\frac{\partial^2 w}{\partial t^2}\delta w-D\left\{\frac{\partial^2}{\partial x^2}(\nabla^2 w)+\frac{\partial^2}{\partial y^2}(\nabla^2 w)\right.\right.$$

$$+2(1-\nu)\left(\frac{\partial^4 w}{\partial x^2\partial y^2}-\frac{1}{2}\frac{\partial^2 w}{\partial x^2\partial y^2}-\frac{1}{2}\frac{\partial^4 w}{\partial x^2\partial y^2}\right)\right\}\delta w-q\delta w\right]dxdydt=0$$

(3.115)

The line integrals are similar to those obtained in Section 3.7 and those are not taken in the above equation because the dynamic term from kinetic energy gives no contribution to the line integrals.

So, the above equation becomes

$$\int_{t_1}^{t_2}\left[\iint_R\{-h\rho\ddot{w}-D\nabla^4w+q\}\delta w\,dxdy\right]dt=0 \qquad (3.116)$$

Thus, we have from Equation 3.116 the differential equation of motion of plate as

$$D\nabla^4w+\rho h\ddot{w}=q \qquad (3.117)$$

where $D=\frac{Eh^3}{12(1-\nu^2)}$, which is not a function of x and y, but a constant at this stage. In the next section, it will be considered a function of x and y and the corresponding equation of motion will be derived.

3.9　Differential Equation for Transverse Motion of Plates by Elastic Equilibrium

Consider an elemental parallelepiped cut out of the plate as shown in Figure 3.1, where we assign positive internal forces and moments to the near faces. To satisfy the equilibrium of the element, negative internal forces and moments must act on its far sides. Then, the equation of motion of the plate element in transverse direction is given as

$$\rho h\,dxdy\frac{\partial^2 w}{\partial t^2}=\frac{\partial Q_x}{\partial x}\,dxdy+\frac{\partial Q_y}{\partial y}\,dxdy \qquad (3.118)$$

where t is time variable; x, y are the space variables in the domain R occupied by the plate; and ρ and h are density and plate thickness respectively. Dividing the above equation by $dxdy$, we have

$$\rho h\frac{\partial^2 w}{\partial t^2}=\frac{\partial Q_x}{\partial x}+\frac{\partial Q_y}{\partial y} \qquad (3.119)$$

Taking moments of all the forces acting on the element about the line through the center of the element and parallel to y-axis, it turns out to be

$$\left(M_x+\frac{\partial M_x}{\partial x}dx\right)dy-M_x dy+\left(M_{yx}+\frac{\partial M_{yx}}{\partial y}dy\right)dx-M_{yx}dx$$
$$-\left(Q_x+\frac{\partial Q_x}{\partial x}dx\right)dy\frac{dx}{2}-Q_x\,dy.\frac{dx}{2}=0 \qquad (3.120)$$

Neglecting higher-order terms and simplifying, we get

$$\frac{\partial M_x}{\partial x} + \frac{\partial M_{yx}}{\partial y} = Q_x \tag{3.121}$$

Similarly, taking moments about the line through the center of the element and parallel to x-axis, the following is obtained:

$$\frac{\partial M_y}{\partial y} + \frac{\partial M_{xy}}{\partial x} = Q_y \tag{3.122}$$

Substituting Q_x and Q_y from Equations 3.121 and 3.122 in Equation 3.119 and using $M_{xy} = M_{yx}$ reduces to

$$\rho h \frac{\partial^2 w}{\partial t^2} = \frac{\partial^2 M_x}{\partial x^2} + 2 \frac{\partial^2 M_{xy}}{\partial x \partial y} + \frac{\partial^2 M_y}{\partial y^2} \tag{3.123}$$

Putting the values of M_x, M_{xy}, and M_y from Equations 3.75 through 3.77 and noting now that D is, in general, a function of x and y, we get

$$\nabla^2(D\nabla^2 w) - (1-\nu)\left(\frac{\partial^2 D}{\partial y^2}\frac{\partial^2 w}{\partial x^2} - 2\frac{\partial^2 D}{\partial x \partial y}\frac{\partial^2 w}{\partial x \partial y} + \frac{\partial^2 D}{\partial x^2}\cdot\frac{\partial^2 w}{\partial y^2}\right) + \rho h \frac{\partial^2 w}{\partial t^2} = 0 \tag{3.124}$$

where

$$\nabla^2 = \frac{\partial^2}{\partial x^2} + \frac{\partial^2}{\partial y^2} \tag{3.125}$$

Equation 3.124 is the differential equation governing the transverse motion of a plate with variable D, i.e., the flexural rigidity being a function of E, h, and ρ and these may not be constant. In particular, for a plate with variable thickness, Equation 3.124 is the governing equation of motion.

3.10 Boundary Conditions

Let C be the boundary of the plate as shown in Figure 3.5 (Equation 3.104). Then, as discussed in Section 3.7, we require one of the following conditions on C:

1. Either $M_\xi = 0$ or $\partial w/\partial \xi$ is prescribed where ξ denotes the normal to the plate boundary.
2. Either $Q_\xi + \frac{\partial M_{\xi\eta}}{\partial \eta} = 0$ or w is prescribed, where η denotes the tangent to the boundary.

From the above, we have the following boundary conditions of a plate (if all classes of elastically restrained edges are neglected).

1. Clamped boundary

$$w = 0$$

and

$$\frac{\partial w}{\partial \xi} = 0 \text{ on } C \qquad (3.126)$$

2. Simply supported boundary

$$w = 0$$

and

$$M_\xi = 0 \text{ on } C \qquad (3.127)$$

3. Completely free boundary

$$M_\xi = 0$$

and

$$Q_\xi + \frac{\partial M_{\xi\eta}}{\partial \eta} = 0 \text{ on } C \qquad (3.128)$$

3.11   Various Forms of Equation of Motion of a Plate in Cartesian Coordinates

In a nutshell, we may now write the various forms of the governing equation of motion for plates as the following:

1. Biharmonic equation for plates from Equation 3.105 with a load distribution $q(x, y)$ and constant thickness

$$D\nabla^2 w = q(x, y) \qquad (3.129)$$

2. Equation of motion for bending vibration of a plate is given by Equation 3.117, which is a modified form of Equation 3.105 as

$$D\nabla^4 \omega + ph\frac{\partial^2 \omega}{\partial t^2} = q(x,y) \qquad (3.130)$$

3. Equation of motion of vibration of a plate with variable thickness, $h = h(x,y)$ from Equation 3.124

$$\nabla^2(D\nabla^2 w) - (1-\nu)\Diamond^4(D,w) + ph\frac{\partial^2 w}{\partial t^2} = q(x,y) \qquad (3.131)$$

where, $\Diamond^4(D,w) = \dfrac{\partial^2 D}{\partial y^2}\cdot\dfrac{\partial^2 w}{\partial x^2} - 2\dfrac{\partial^2 D}{\partial x \partial y}\dfrac{\partial^2 w}{\partial x \partial y} + \dfrac{\partial^2 D}{\partial x^2}\cdot\dfrac{\partial^2 w}{\partial y^2} \qquad (3.132)$

   called the die operator.
4. Equation of motion of vibration of a plate with a linear foundation modulus k may similarly be written as

$$\nabla^2(D\nabla^2 w) - (1-\nu)\Diamond^4(D,w) + ph\frac{\partial^2 w}{\partial t^2} + kw = q(x,y) \qquad (3.133)$$

If we put $q(x,y) = 0$ in the above equations of motion, then one can get the governing equations for free vibration study.

3.12 Formulations in Polar and Elliptical Coordinates

The previous sections addressed the equations of motion of a plate in terms of Cartesian coordinates. Although these coordinates may very well be used for circular and elliptical geometries, sometimes the circular and elliptic peripheries are handled in terms of polar and elliptical coordinates in simple cases of vibration analysis. Accordingly, in the following section, these will be discussed in few details.

3.12.1 Polar Coordinates

These coordinates are shown in Figure 3.6, where the polar coordinates of a point A may be written as

$$x = r\cos\theta \qquad (3.134)$$

and

$$y = r\sin\theta \qquad (3.135)$$

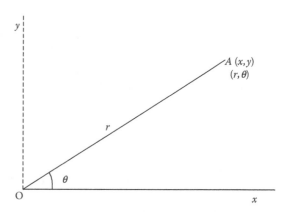

FIGURE 3.6
Polar coordinate system.

The bending and twisting moments are then given in polar coordinates similar to Cartesian coordinates as in Equations 3.75 through 3.77

$$M_r = -D\left[\frac{\partial^2 w}{\partial r^2} + \nu\left(\frac{1}{r}\frac{\partial w}{\partial r} + \frac{1}{r^2}\frac{\partial^2 w}{\partial \theta^2}\right)\right] \qquad (3.136)$$

$$M_\theta = -D\left[\frac{1}{2}\frac{\partial w}{\partial r} + \frac{1}{r^2}\frac{\partial^2 w}{\partial \theta^2} + \nu\frac{\partial^2 w}{\partial r^2}\right] \qquad (3.137)$$

$$M_{r\theta} = -D(1-\nu)\frac{\partial}{\partial r}\left(\frac{1}{r}\frac{\partial w}{\partial \theta}\right) \qquad (3.138)$$

The Laplacian operator in polar coordinates reduces to

$$\nabla^2(\,\cdot\,) = \frac{\partial^2(\,\cdot\,)}{\partial r^2} + \frac{1}{r}\frac{\partial(\,\cdot\,)}{\partial r} + \frac{1}{r^2}\frac{\partial^2(\,\cdot\,)}{\partial \theta^2} \qquad (3.139)$$

and the transverse shearing forces are given by

$$Q_r = -D\frac{\partial}{\partial r}(\nabla^2 w) \qquad (3.140)$$

$$Q_\theta = -D\frac{1}{r}\frac{\partial}{\partial \theta}(\nabla^2 w) \qquad (3.141)$$

Equation of motion in polar coordinates may be derived as given in Section 3.11, where the Laplacian operator is given by Equation 3.139 and the biharmonic operator is written as

$$\nabla^4(\,\cdot\,) = \nabla^2\nabla^2(\,\cdot\,) \qquad (3.142)$$

3.12.2 Elliptical Coordinates

These coordinates are denoted by (ξ, η) and Figure 3.7 shows this coordinate system along with the Cartesian system (x, y) and the relation between these systems may be written as

$$X = C \cosh \xi \cos \eta \tag{3.143}$$

$$Y = C \sinh \xi \sin \eta \tag{3.144}$$

where $2C$ is the interfocal distance.

Laplacian operator in elliptical coordinates may be derived as

$$\nabla^2 = \frac{\partial^2(\cdot)}{\partial x^2} + \frac{\partial^2(\cdot)}{\partial y^2} = \frac{2}{C^2(\cosh 2\xi - \cos 2\eta)} \left(\frac{\partial^2(\cdot)}{\partial \xi^2} + \frac{\partial^2(\cdot)}{\partial \eta^2} \right) \tag{3.145}$$

The bending moments, twisting moments, and the transverse shear can easily be derived from Cartesian equations given in earlier sections.

Again, the equation of motion in terms of elliptical coordinates may similarly be written, as given in Section 3.11 by using Equations 3.143 and 3.144 along with the Laplacian operator (Equation 3.145) and biharmonic operator (Equation 3.142).

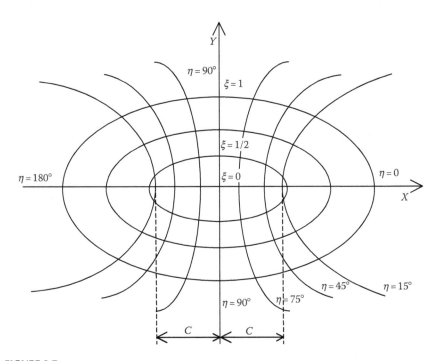

FIGURE 3.7
Elliptical coordinate system.

Further Reading

Jones, R.M. 1975. *Mechanics of Composite Materials*, Hemisphere Publishing Corporation, New York.

Liew, K.M., Wang, C.M., Xiang, Y. and Kitipornchai, S. 1998. *Vibration of Mindlin Plates*, Oxford, U.K., Elsevier Science Ltd.

Love, A.E.H. 1944. *A Treatise on the Mathematical Theory of Elasticity*, 4th ed., Dover Publications, New York.

Mindlin, R.D. 1951. Influence of rotatory inertia and shear on flexural motions of isotropic elastic plates. *Journal of Applied Mechanics*, 13: 31–38.

Petyt, M. 1998. *Introduction to Finite Element Vibration Analysis*, Cambridge University Press, Cambridge.

Reddy, J.N. 1997. *Mechanics of Laminated Composite Plates: Theory and Analysis*, CRC Press, Boca Raton, FL.

Reissner, E. 1945. The effect of transverse shear deformation on the bending of elastic plates. *Journal of Applied Mechanics*, 67: A-69–A-77.

Shames, I.H. and Dym, C.L. 1985. *Energy and Finite Element Methods in Structural Mechanics*, McGraw-Hill, New York.

Sokolnikoff, I.S. 1956. *Mathematical Theory of Elasticity*, 2nd ed., McGraw-Hill Book Company, New York.

Szilard, R. 1974. *Theory and Analysis of Plates, Classical and Numerical Methods*, Prentice-Hall, Englewood Cliffs, NJ.

Timoshenko, S. 1921. On the correction for shear of the differential equation for transverse vibrations of prismatic bars. *Philosophical Magazine*, 41: 744–746.

Timoshenko, S. and Goodier, J.N. 1970. *Theory of Elasticity*, McGraw-Hill Book Company, New York.

Timoshenko, S. and Woinowsky-Krieger, A. 1959. *Theory of Plates and Shells*, 2nd ed., McGraw-Hill Book Company, Inc., New York.

Vinson, J.R. 1989. *The Behavior of Thin Walled Structures, Beams, Plates and Shells*, Kluwer Academic Publishers, the Netherlands.

Yu, Y.Y. 1996. *Vibrations of Elastic Plates*, Inc., Springer-Verlag, New York.

Zienkiewicz, O.C. 1977. *The Finite Element Method*, 3rd ed., McGraw-Hill, London.

4

Exact, Series-Type, and Approximate Methods for Transverse Vibration of Plates

We have obtained the equation of motion for the transverse vibration of a plate in Chapter 3 as

$$DV^4w + \rho h \frac{\partial^2 w}{\partial t^2} = 0 \qquad (4.1)$$

where
w is the transverse displacement of the plate
ρ is the density of the material of the plate
h is the thickness of the plate
$D = \frac{Eh^3}{12(1-\nu^2)}$ is the flexural rigidity
ν is the Poisson's ratio
E is the Young's modulus of elasticity
V^4 is the biharmonic operator

$$V^4w = \frac{\partial^4 w}{\partial x^4} + 2 \frac{\partial^4 w}{\partial x^2 xy^2} + \frac{\partial^4 w}{\partial y^4} \qquad (4.2)$$

Also, $V^4w = V^2(V^2w)$ where V^2 is the Laplacian operator, and this has been defined in Chapter 3 in the cases of rectangular, polar, and elliptical coordinates. For free vibration with circular (natural) frequency ω, we can write the motion of the plate in polar coordinates as

$$w(r, \theta, t) = W(r, \theta) e^{i\omega t} \qquad (4.3)$$

and in Cartesian coordinates

$$w(x, y, t) = W(x, y) e^{i\omega t} \qquad (4.4)$$

In general, this will be written in the form

$$w = We^{i\omega t} \qquad (4.5)$$

By substituting Equation 4.5 in Equation 4.1, we get

$$(\nabla^4 - \beta^4)W = 0 \tag{4.6}$$

$$\beta^4 = \frac{\rho h \omega^2}{D} \tag{4.7}$$

Equation 4.6 is then written as

$$(\nabla^2 + \beta^2)(\nabla^2 - \beta^2)W = 0 \tag{4.8}$$

whose solution may be obtained in the form of two linear differential equations:

$$(\nabla^2 + \beta^2)W_1 = 0 \tag{4.9}$$

$$(\nabla^2 - \beta^2)W_2 = 0 \tag{4.10}$$

One can write the solution of Equation 4.8 as the superposition of the solutions of Equations 4.9 and 4.10. Let W_1 and W_2 be the corresponding solutions. Then, one may have the solution W of the original differential Equation 4.8 as

$$W = W_1 + W_2 \tag{4.11}$$

4.1 Method of Solution in Polar Coordinates

The polar coordinate system has already been discussed in Chapter 3, where the Laplacian operator ∇^2 is given by

$$\nabla^2(\,\cdot\,) = \frac{\partial^2(\,\cdot\,)}{\partial r^2} + \frac{1}{r}\frac{\partial(\,\cdot\,)}{\partial r} + \frac{1}{r^2}\frac{\partial^2(\,\cdot\,)}{\partial \theta^2} \tag{4.12}$$

Let us consider the previous two differential equations, viz., Equations 4.9 and 4.10; first we will put

$$W_1(r, \theta) = R_1(r)\Theta_1(\theta) \tag{4.13}$$

in Equation 4.9 for finding out the solution as separation of variables and obtain the following:

$$\Theta_1 \frac{d^2 R_1}{dr^2} + \frac{\Theta_1}{r}\frac{dR_1}{dr} + \frac{R_1}{r^2}\frac{d^2\Theta_1}{d\theta^2} + \lambda^2 R_1\Theta_1 = 0$$

The above is written by multiplying with $\frac{r^2}{R_1\Theta_1}$ as

$$r^2\left[\left(\frac{d^2R_1}{dr^2}+\frac{1}{r}\frac{dR_1}{dr}\right)\frac{1}{R_1}+\beta^2\right]=-\frac{1}{\Theta_1}\frac{d^2\Theta_2}{d\theta^2} \tag{4.14}$$

Equation 4.14 is satisfied only if each expression in the above is equal to a constant (say) k^2. Thus, we obtain two ordinary differential equations (ODEs) as

$$\frac{d^2\Theta_1}{d\theta^2}+k^2\Theta_1=0 \tag{4.15}$$

$$\frac{d^2R_1}{dr^2}+\frac{1}{r}\frac{dR_1}{dr}+\left(\beta^2-\frac{k^2}{r^2}\right)R_1=0 \tag{4.16}$$

Solution of Equation 4.15 will become

$$\Theta_1=G_1\cos k\theta+H_1\sin k\theta \tag{4.17}$$

where G_1 and H_1 are constants. Next, we will introduce a variable

$$\xi=\beta r \tag{4.18}$$

in Equation 4.16 and thus we get a Bessel equation of fractional order as

$$\frac{d^2R_1}{d\xi^2}+\frac{1}{\xi}\frac{dR_1}{d\xi}+\left(1-\frac{k^2}{\xi^2}\right)R_1=0 \tag{4.19}$$

The solution in terms of Bessel functions of the first and second kinds, viz., $J_k(\xi)$ and $Y_k(\xi)$ with $\xi=\beta r$, may be written as

$$R_1=\overline{A}J_k(\beta_\gamma)+\overline{B}Y_k(\beta_r) \tag{4.20}$$

Therefore, from Equations 4.13, 4.17, and 4.20, we obtain

$$W_1(r,\theta)=\overline{A}J_k(\beta r)+\overline{B}Y_k(\beta r)(G_1\cos k\theta+H_1\sin k\theta) \tag{4.21}$$

Similarly, Equation 4.10 may again be solved by separation of variables by assuming a solution of the form

$$W_2(r,\theta)=R_2(r)\Theta_2(\theta) \tag{4.22}$$

Replacing β by $i\beta$, Equation 4.10 is expressed as

$$\left[\nabla^2+(i\beta)^2\right]W_2=0 \tag{4.23}$$

Putting Equation 4.22 in Equation 4.23 and multiplying by

$$\frac{r^2}{R_2 \Theta_2}$$

we can obtain two ODEs as before:

$$\frac{d^2\Theta_2}{d\theta^2} + k^2\Theta_2 = 0 \tag{4.24}$$

and

$$\frac{d^2 R_2}{dr^2} + \frac{1}{r}\frac{dR_2}{dr} + \left[(i\beta)^2 - \frac{k^2}{r^2} \right] R_2 = 0 \tag{4.25}$$

Again, introducing a new variable

$$\eta = i\beta r \tag{4.26}$$

in Equation 4.25, the solution in this case may be written as

$$R_2 = \overline{C}I_k(\beta r) + \overline{D}K_k(\beta r) \tag{4.27}$$

where $I_k(\beta r)$ and $K_k(\beta r)$ are the modified Bessel functions of first and second kinds, respectively. As done earlier, the solution of differential equation (Equation 4.24) is written in the form

$$\Theta_2 = G_2 \cos k\theta + H_2 \sin k\theta \tag{4.28}$$

It is to be noted that k can be a fractional number, whereas plates that are closed in θ direction indicate that Θ must be a function of period 2π. So, in this case k must be an integer. Therefore, we will write

$$k = n \tag{4.29}$$

where $n = 0, 1, 2, 3, \ldots$

Putting Equations 4.27 and 4.28 in Equation 4.22, the solution of Equation 4.10 becomes

$$W_2(r, \theta) = \left[\overline{C}I_k(\beta r) + \overline{D}K_k(\beta r) \right] (G_2 \cos k\theta + H_2 \sin k\theta) \tag{4.30}$$

Combining Equations 4.21 and 4.30 and putting Equation 4.29 for $k=n$ in Equation 4.11, we have

$$
\begin{aligned}
W(r,\theta) &= (\overline{A}J_n(\beta r) + \overline{B}Y_n(\beta r))(G_1 \cos n\theta + H_1 \sin n\theta) \\
&\quad + (\overline{C}I_n(\beta r) + \overline{D}K_n(\beta r))(G_2 \cos n\theta + H_2 \sin n\theta) \\
&= (AJ_n(\beta r) + BY_n(\beta r) + CI_n(\beta r) + DK_n(\beta r)) \cos n\theta \\
&\quad + (A^*J_n(\beta r) + B^*Y_n(\beta r) + C^*I_n(\lambda r) + D^*K_n(\beta r)) \sin n\theta
\end{aligned}
\tag{4.31}
$$

where the constants

$$
\begin{aligned}
A &= \overline{A}G_1, \quad B = \overline{B}G_1, \quad C = \overline{C}G_2, \quad D = \overline{D}G_2 \\
A^* &= \overline{A}H_1, \quad B^* = \overline{B}H_1, \quad C^* = \overline{C}H_2, \quad D^* = \overline{D}H_2
\end{aligned}
$$

General solution may now be written from Equation 4.31 in the form

$$
\begin{aligned}
W_n(r,\theta) &= [A_n J_n(\beta r) + B_n Y_n(\beta r) + c_n I_n(\beta r) + D_n K_n(\beta r)] \cos n\theta \\
&\quad + [A_n^* J_n(\beta r) + B_n^* Y_n(\beta r) + c_n^* I_n(\beta r) + D_n^* K_n(\beta r)] \sin n\theta
\end{aligned}
\tag{4.32}
$$

where

$$
\beta^4 = \frac{\rho h \omega^2}{D}
$$

Both $Y_n(\beta r)$ and $K_n(\beta r)$ are singular at $\beta r = 0$; i.e., at $r = 0$. So, for a plate such as a circular one, we take constants B_n, B_n^* and D_n, D_n^* as zero. We may also note that for a circular plate with a concentric hole (i.e., annular circular plate) of radius $b < r$, these functions must be considered because there is no singularity at the origin. But in the present case, two additional boundary conditions are to be specified at $r = b$.

4.1.1 Circular Plate

Let us consider a circular plate as shown in Figure 4.1, where a is its radius and let the origin of polar coordinate system be taken as the center of the circular plate. As mentioned in Section 4.1, for a circular plate the constants B_n, B_n^* and D_n, D_n^* in Equation 4.32 are neglected. Moreover, if the boundary conditions possess symmetry with respect to one or more diameters of the circular plate, then we also discard the terms involving $\sin n\theta$. With these clarifications, Equation 4.32 is now written as

$$
W_n(r\theta) = [A_n J_n(\beta r) + C_n I_n(\beta r)] \cos n\theta
\tag{4.33}
$$

where the subscript n will denote the number of nodal diameters. The coefficients A_n and C_n determine the mode shapes and may be solved from

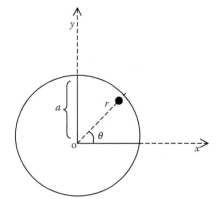

FIGURE 4.1
Circular plate.

the boundary conditions. As mentioned earlier, J_n and I_n are Bessel functions and modified Bessel functions of the first kind and

$$\beta^4 = \frac{\rho h \omega^2}{D} \tag{4.34}$$

In the next few sections, frequency equations for a circular plate with three types of boundary conditions, viz., clamped, simply supported, and completely free all around, will be addressed.

4.1.1.1 Circular Plate with Clamped Condition All Around

Let us consider a circular plate with radius a as shown in Figure 4.2. If the edge of the plate is clamped, then we have the boundary conditions

$$W(r, \theta)|_{r=a} = 0 \tag{4.35}$$

and

$$\frac{\partial W}{\partial r}(r, \theta)\bigg|_{r=a} = 0 \tag{4.36}$$

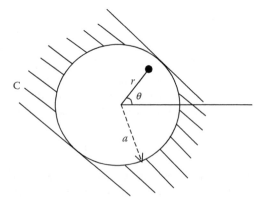

FIGURE 4.2
Clamped circular plate.

Putting Equation 4.33 in Equations 4.35 and 4.36, we have

$$A_n J_n(\beta a) + C_n I_n(\beta a) = 0 \tag{4.37}$$

$$A_n J_n'(\beta a) + C_n I_n'(\beta a) = 0 \tag{4.38}$$

where

$$J_n'(.) = \frac{dJ_n(.)}{dr} \quad \text{and} \quad I_n'(.) = \frac{dI_n(.)}{dr}$$

The above two equations, viz., Equations 4.37 and 4.38, are written in matrix form as

$$\begin{bmatrix} J_n(\lambda) & I_n(\lambda) \\ J_n'(\lambda) & I_n'(\lambda) \end{bmatrix} \begin{Bmatrix} A_n \\ C_n \end{Bmatrix} = \begin{Bmatrix} 0 \\ 0 \end{Bmatrix} \tag{4.39}$$

where

$$\lambda = \beta a \tag{4.40}$$

For nontrivial solution, we get the characteristic determinant as zero, i.e.,

$$\begin{vmatrix} J_n(\lambda) & I_n(\lambda) \\ J_n'(\lambda) & I_n'(\lambda) \end{vmatrix} = 0 \tag{4.41}$$

This gives the frequency equation

$$J_n(\lambda) I_n'(\lambda) - J_n'(\lambda) I_n(\lambda) = 0 \tag{4.42}$$

Utilizing the recursion relation of Bessel's function

$$J_n'(\lambda) = \frac{\eta}{\lambda} J_n(\lambda) - J_{\eta+1}(\lambda) \tag{4.43}$$

and

$$I_n'(\lambda) = \frac{n}{\lambda} I_n(\lambda) + I_{n+1}(\lambda) \tag{4.44}$$

in Equation 4.42, the frequency equation is written as

$$J_n(\lambda) I_{n+1}(\lambda) + I_n(\lambda) J_{n+1}(\lambda) = 0 \tag{4.45}$$

where the eigenvalue λ is the frequency parameter that is given by

$$\lambda^2 = \beta^2 a^2 = a^2 \omega \sqrt{\frac{ph}{D}} \tag{4.46}$$

and the corresponding natural frequencies are written as

$$\omega = \frac{\lambda^2}{a^2}\sqrt{\frac{D}{\rho h}} \qquad (4.47)$$

To find the mode shapes, we may use either of Equations 4.39. If we take the first one, then we may write

$$\frac{A_n}{C_n} = -\frac{I_n(\lambda)}{J_n(\lambda)} \qquad (4.48)$$

4.1.1.2 Circular Plate with Simply Supported Condition All Around

In this case (Figure 4.3), the boundary conditions are given by

$$W(r, \theta)|_{r=a} = 0 \qquad (4.49)$$

and

$$M_r(r, \theta)|_{r=a} = 0 \qquad (4.50)$$

This translates into

$$W(a) = 0 \qquad (4.51)$$

and

$$M_r(a) = 0 \qquad (4.52)$$

where from Chapter 3, we have the bending moment

$$M_r = -D\left[\frac{\partial^2 w}{\partial r^2} + \frac{\nu}{r}\frac{\partial w}{\partial r}\right] = 0 \qquad (4.53)$$

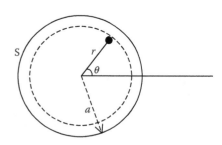

FIGURE 4.3
Simply supported circular plate.

Substituting Equation 4.33 in Equation 4.51, we get

$$A_n J_n(\beta a) + C_n I_n(\beta a) = 0 \tag{4.54}$$

Again putting Equation 4.33 with Equation 4.53 in Equation 4.52, we have

$$A_n \left\{ J_n''(\beta a) + \frac{\nu}{\beta a} J_n'(\beta a) \right\} + C_n \left\{ I_n''(\beta a) + \frac{\nu}{\beta a} I_n'(\beta a) \right\} = 0 \tag{4.55}$$

Equations 4.54 and 4.55 are now written as

$$\begin{bmatrix} J_n(\lambda) & I_n(\lambda) \\ J_n''(\lambda) + \frac{\nu}{\lambda} J_n'(\lambda) & I_n''(\lambda) + \frac{\nu}{\lambda} I_n'(\lambda) \end{bmatrix} \begin{Bmatrix} A_n \\ C_n \end{Bmatrix} = \begin{Bmatrix} 0 \\ 0 \end{Bmatrix} \tag{4.56}$$

where $\lambda = \beta a$.

For nontrivial solution, the characteristic determinant is set to zero. Thus, we get the frequency equation as

$$\begin{vmatrix} J_n(\lambda) & I_n(\lambda) \\ J_n''(\lambda) + \frac{\nu}{\lambda} J_n'(\lambda) & I_n''(\lambda) + \frac{\nu}{\lambda} I_n'(\lambda) \end{vmatrix} = 0 \tag{4.57}$$

Differentiating Equation 4.43 and using the identity

$$J_{n+2} = \frac{2}{\lambda}(n+1)J_{n+1} - J_n \tag{4.58}$$

we may obtain

$$J_n'' = \left[\frac{n(n-1)}{\lambda^2} - 1 \right] J_n + \frac{1}{\lambda} J_{n+1} \tag{4.59}$$

and then we have

$$J_n''(\lambda) + \frac{\nu}{\lambda} J_n'(\lambda) = \left[\frac{n(n-1)}{\lambda^2} - 1 + \frac{n\nu}{\lambda^2} \right] J_n + \frac{1}{\lambda}(1 - \nu)J_{n+1} \tag{4.60}$$

Again, differentiating the identity (Equation 4.44) and using the relation

$$I_{n+2} = I_n - \frac{2}{\lambda}(n+1)I_{n+1} \tag{4.61}$$

we get

$$I_n'' = \left[\frac{n(n-1)}{\lambda^2} + 1 \right] I_n - \frac{1}{\lambda} I_{n+1} \tag{4.62}$$

and obtain

$$I_n''(\lambda) + \frac{\nu}{\lambda}I_n'(\lambda) = \left[\frac{n(n-1)}{\lambda^2} + 1 + \frac{n\nu}{\lambda^2}\right]I_n + \frac{1}{\lambda}(\nu - 1)I_{n+1} \tag{4.63}$$

From Equation 4.56, by expanding the determinant and utilizing Equations 4.60 and 4.61, the frequency equation becomes

$$\frac{I_{n+1}(\lambda)}{I_n(\lambda)} + \frac{J_{n+1}(\lambda)}{J_n(\lambda)} = \frac{2\lambda}{1 - \nu} \tag{4.64}$$

Solution of Equation 4.64 gives the frequency parameter λ. Corresponding mode shapes for simply supported circular plate are determined from the first of Equation 4.57, i.e.,

$$\frac{A_n}{C_n} = -\frac{I_n(\lambda)}{J_n(\lambda)} \tag{4.65}$$

4.1.1.3 Circular Plate with Completely Free Condition All Around

We have the boundary conditions for completely free case (Figure 4.4) as

$$M_r(r, \theta)|_{r=a} = 0 \tag{4.66}$$

$$V_r(r, \theta)|_{r=a} = 0 \tag{4.67}$$

which translates to

$$M_r(a) = 0 \tag{4.68}$$

$$\text{and} \quad V_r(a) = 0 \tag{4.69}$$

For the first boundary condition (Equation 4.68), we arrive similarly as in Equation 4.55

$$A_n\left[J_n''(\beta a) + \frac{\nu}{\beta a}J_n'(\beta a)\right] + C_n\left[I_n''(\beta a) + \frac{\nu}{\beta a}I_n'(\beta a)\right] = 0 \tag{4.70}$$

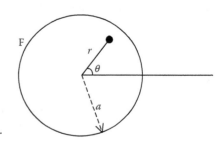

FIGURE 4.4
Completely free circular plate.

For the second boundary condition, viz., Equation 4.69, we will write

$$V_r = Q_r + \frac{1}{r}\frac{\partial Mr\theta}{\partial\theta}$$

$$= -D\frac{\partial}{\partial r}(\nabla^2 w)$$

So, we have the condition as

$$\frac{\partial}{\partial r}\left(\frac{\partial^2 w}{\partial r^2} + \frac{1}{r}\frac{\partial w}{\partial r}\right)\bigg|_{r=a} = 0 \qquad (4.71)$$

By substituting Equation 4.33 in Equation 4.71, we can get

$$A_n\left[J_n''' + \frac{1}{\lambda}J_n'' - \frac{1}{\lambda^2}J_n'\right] + C_n\left[I_n''' + \frac{1}{\lambda}I_n'' - \frac{1}{\lambda^2}I_n'\right] = 0 \qquad (4.72)$$

Here, we will not give the details regarding the derivation of the frequency equation, but the same may be obtained from Equations 4.70 and 4.72 as

$$\frac{\lambda^2 J_n(\lambda) + (1-\nu)\left[\lambda J_n'(\lambda) - n^2 J_n(\lambda)\right]}{\lambda^2 I_n(\lambda) - (1-\nu)\left[\lambda I_n'(\lambda) - n^2 I_n(\lambda)\right]} = \frac{\lambda^3 J_n'(\lambda) + (1-n)n^2\left[\lambda J_n'(\lambda) - J_n(\lambda)\right]}{\lambda^3 I_n'(\lambda) - (1-\nu)n^2\left[\lambda I_n'(\lambda) - I_n(\lambda)\right]}$$

$$(4.73)$$

Solution of Equation 4.73 gives the frequency parameter λ.

Next, we will study circular plates with concentric circular holes. These are generally called annular plates.

4.1.2 Annular Plates

As mentioned, these plates have circular outer boundary and concentric circular inner boundary. It has already been seen that three types of boundary conditions exist in a circular plate, viz., clamped, simply supported, and completely free. Because of the circular hole in an annular plate, this (hole) will also have three boundary conditions: clamped, simply supported, and completely free. As such, there exist nine possible combinations of simple boundary conditions for the two boundaries (Figure 4.5). Various authors (given in references) have introduced the frequency equations for nine combinations of the boundary conditions for axisymmetric, one diametral node, and two diametral nodes. In this section, we will only incorporate the frequency determinant for axisymmetric case, which can be used for getting the exact results, though in later chapters, these annular plates will be considered in greater detail for obtaining the first few modes of vibrations. Accordingly, these nine cases as per the axisymmetric modes will be addressed. We denote the boundary conditions clamped, simply supported,

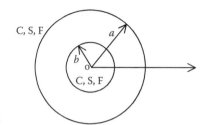

FIGURE 4.5
Annular circular plate.

and free as C, S, and F, respectively. In the annular region, the first symbol designates outer boundary and the second denotes inner boundary.

4.1.2.1 Circular Annular Plate with Outer and Inner Both Clamped (C–C)

Here, the first symbol "C" and the second symbol "C" denote, respectively, the clamped outer and clamped inner boundaries. Let us consider an annular circular plate with outer radius a and inner radius b as shown in Figures 4.6a with the said boundary conditions.

Now, we will substitute the solution (Equation 4.32) with the $\cos n\theta$ term only into the clamped boundary conditions at $r = a$ and $r = b$. So, we will have

$$W(r, \theta)|_{r=a} = W(r, \theta)|_{r=b} = 0 \qquad (4.74a)$$

and

$$\left.\frac{dW}{dr}(r, \theta)\right|_{r=a} = \left.\frac{dW}{dr}(r, \theta)\right|_{r=b} = 0 \qquad (4.74b)$$

This will give four homogeneous equations in four unknowns A_n, B_n, C_n, and D_n. Similar to the complete circular plates in the previous section for a nontrivial solution, the determinant of coefficient will be zero. The frequency

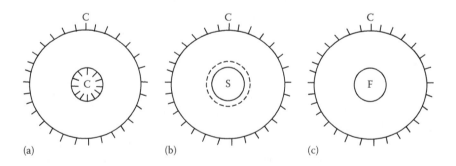

(a)   (b)   (c)

FIGURE 4.6
Annular circular plate with outer as clamped and inner (a) clamped, (b) simply supported, and (c) completely free.

determinant will consist of Bessel function of higher orders. These are reduced to first and zeroth order by using various identities of Bessel's functions. Here, the details in this regard will not be included. Only the frequency determinant for axisymmetric case (i.e., $n = 0$) is given below.

$$\begin{vmatrix} J_0(\lambda) & Y_0(\lambda) & I_0(\lambda) & K_0(\lambda) \\ J_1(\lambda) & Y_1(\lambda) & -I_1(\lambda) & K_1(\lambda) \\ J_0(\alpha\lambda) & Y_0(\alpha\lambda) & I_0(\alpha\lambda) & K_0(\alpha\lambda) \\ J_1(\alpha\lambda) & Y_1(\alpha\lambda) & -I_1(\alpha\lambda) & K_1(\alpha\lambda) \end{vmatrix} \tag{4.75}$$

where $\alpha = b/a = m$, which is the ratio of inner radius to outer radius of the annular plate. Solution of the determinant (Equation 4.75) will give the frequency parameter for various m.

4.1.2.2 Circular Annular Plate with Outer Clamped and Inner Simply Supported (C–S)

Again, C–S denotes outer clamped and inner simply supported. Here, the boundary condition at $r = a$ is clamped and at $r = b$ is simply supported (Figure 4.6b). Accordingly, we will have

$$W(r,\theta)|_{r=a} = 0, \quad \frac{\partial W}{\partial r}(r,\theta)\bigg|_{r=a} = 0 \tag{4.76a}$$

$$W(r,\theta)|_{r=b} = 0 \quad \text{and} \quad M_r(r,\theta)|_{r=b} = 0 \tag{4.76b}$$

Putting again the solution (Equation 4.32) with $\cos n\theta$ term into Equations 4.76a and 4.76b and simplifying, we have the frequency determinant for axisymmetric case ($n = 0$) as

$$\begin{vmatrix} J_0(\lambda) & Y_0(\lambda) & I_0(\lambda) & K_0(\lambda) \\ J_1(\lambda) & Y_1(\lambda) & -I_1(\lambda) & K_1(\lambda) \\ J_0(\alpha\lambda) & Y_0(\alpha\lambda) & I_0(\alpha\lambda) & K_0(\alpha\lambda) \\ J_1(\alpha\lambda) & Y_1(\alpha\lambda) & \rho\alpha I_0(\alpha\lambda) - I_1(\alpha\lambda) & P\alpha K_0(\alpha\lambda) + K_1(\alpha\lambda) \end{vmatrix} = 0 \tag{4.77}$$

where

$$P = \frac{2\lambda}{1-\nu}$$

$$\alpha = b/a = m$$

Frequency parameters may be obtained by the solution of Equation 4.77 for various $m = b/a$.

In the remaining seven boundary conditions, the frequency determinants for $n = 0$ (axisymmetric) case are written in the following subsections, as the reader may now be clear with the boundary conditions that are to be satisfied at each specified boundaries of the annular plate.

4.1.2.3 Circular Annular Plate with Outer Clamped and Inner Free (C–F)

Frequency determinant for $n=0$ in this case is obtained as (Figure 4.6c).

$$
\begin{vmatrix}
J_0(\lambda) & Y_0(\lambda) & I_0(\lambda) & K_0(\lambda) \\
J_1(\lambda) & Y_1(\lambda) & -I_1(\lambda) & K_1(\lambda) \\
J_0(\alpha\lambda) & Y_0(\alpha\lambda) & -I_0(\alpha\lambda)+QI_1(\alpha\lambda) & -K_0(\alpha\lambda)-QK_1(\alpha\lambda) \\
J_1(\alpha\lambda) & Y_1(\alpha\lambda) & I_1(\alpha\lambda) & -K_1(\alpha\lambda)
\end{vmatrix} = 0 \qquad (4.78)
$$

where

$$
\alpha = b/a = m
$$

$$
Q = \frac{2(1-\nu)}{\alpha\lambda}
$$

4.1.2.4 Circular Annular Plate with Outer Simply Supported and Inner Clamped (S–C)

Here, the frequency determinant is found to be (Figure 4.7a)

$$
\begin{vmatrix}
J_0(\lambda) & Y_0(\lambda) & I_0(\lambda) & K_0(\lambda) \\
J_1(\lambda) & Y_1(\lambda) & PI_0(\lambda)-I_1(\lambda) & Pk_0(\lambda)+K_1(\lambda) \\
J_0(\alpha\lambda) & Y_0(\alpha\lambda) & I_0(\alpha\lambda) & K_0(\alpha\lambda) \\
J_1(\alpha\lambda) & Y_1(\alpha\lambda) & -I_1(\alpha\lambda) & K_1(\alpha\lambda)
\end{vmatrix} = 0 \qquad (4.79)
$$

where

$$
\alpha = b/a = m
$$

$$
P = \frac{2\lambda}{1-\nu}
$$

4.1.2.5 Circular Annular Plate with Outer and Inner Both Simply Supported (S–S)

For this boundary condition, the frequency determinant is (Figure 4.7b)

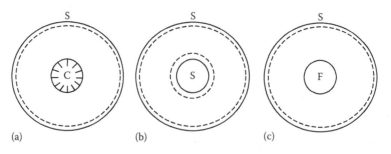

(a)   (b)   (c)

FIGURE 4.7
Annular circular plate with outer as simply supported and inner (a) clamped, (b) simply supported, and (c) completely free.

$$\begin{vmatrix} J_o(\lambda) & Y_o(\lambda) & I_o(\lambda) & K_o(\lambda) \\ J_1(\lambda) & Y_1(\lambda) & PI_o(\lambda) - I_1(\lambda) & Pk_o(\lambda) + K_1(\lambda) \\ J_o(\alpha\lambda) & Y_o(\alpha\lambda) & I_o(\alpha\lambda) & K_o(\alpha\lambda) \\ J_1(\alpha\lambda) & Y_1(\alpha\lambda) & P\alpha I_o(\alpha\lambda) - I_1(\alpha\lambda) & P\alpha K_o(\alpha\lambda) + K_1(\alpha\lambda) \end{vmatrix} = 0 \quad (4.80)$$

where α and P are as given above.

4.1.2.6 Circular Annular Plate with Outer Simply Supported and Inner Free (S–F)

Frequency determinant here may be given by (Figure 4.7c)

$$\begin{vmatrix} J_o(\lambda) & Y_o(\lambda) & I_o(\lambda) & K_o(\lambda) \\ J_1(\lambda) & Y_1(\lambda) & PI_o(\lambda) - I_1(\lambda) & PK_o(\lambda) + k_1(\lambda) \\ J_o(\alpha\lambda) & Y_o(\alpha\lambda) & -I_o(\alpha\lambda) + QI_1(\alpha\lambda) & -K_o(\alpha\lambda) - QK_1(\alpha\lambda) \\ J_1(\alpha\lambda) & Y_1(\alpha\lambda) & I_1(\alpha\lambda) & -K_1(\alpha\lambda) \end{vmatrix} = 0 \quad (4.81)$$

where P and α are defined earlier.

4.1.2.7 Circular Annular Plate with Outer Free and Inner Clamped (F–C)

The frequency determinant is obtained as (Figure 4.8a)

$$\begin{vmatrix} J_o(\lambda) & Y_o(\lambda) & -I_o(\lambda) + RI_1(\lambda) & -k_o(\lambda) - RK_1(\lambda) \\ J_1(\lambda) & Y_1(\lambda) & I_1(\lambda) & -K_1(\lambda) \\ J_o(\alpha\lambda) & Y_o(\alpha\lambda) & I_o(\alpha\lambda) & K_o(\alpha\lambda) \\ J_1(\alpha\lambda) & Y_1(\alpha\lambda) & -I_1(\alpha\lambda) & K_1(\alpha\lambda) \end{vmatrix} = 0 \quad (4.82)$$

where $R = \dfrac{2(1 - \nu)}{\lambda}$.

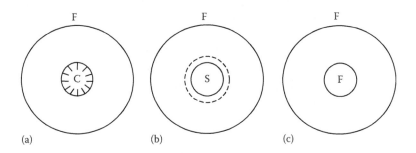

FIGURE 4.8
Annular circular plate with outer as completely free and inner (a) clamped, (b) simply supported, and (c) completely free.

4.1.2.8 Circular Annular Plate with Outer Free and Inner Simply Supported (F–S)

For $n=0$, the frequency determinant is written in the form (Figure 4.8b)

$$\begin{vmatrix} J_0(\lambda) & Y_0(\lambda) & -I_0(\lambda)+RI_1(\lambda) & -K_0(\lambda)-RK_1(\lambda) \\ J_1(\lambda) & Y_1(\lambda) & I_1(\lambda) & -K_1(\lambda) \\ J_0(\alpha\lambda) & Y_0(\alpha\lambda) & I_0(\alpha\lambda) & K_0(\alpha\lambda) \\ J_1(\alpha\lambda) & Y_1(\alpha\lambda) & P\alpha I_0(\alpha\lambda)-I_1(\alpha\lambda) & P\alpha K_0(\alpha\lambda)+K_1(\alpha\lambda) \end{vmatrix} = 0 \qquad (4.83)$$

where

$$R = \frac{2(1-\nu)}{\lambda}$$

$$P = \frac{2\lambda}{1-\nu}$$

4.1.2.9 Circular Annular Plate with Outer and Inner Both Free (F–F)

In this case for ($n=0$), the frequency determinant will become (Figure 4.8c)

$$\begin{vmatrix} J_0(\lambda) & Y_0(\lambda) & -I_0(\lambda)+RI_1(\lambda) & -K_0(\lambda)-RK_1(\lambda) \\ J_1(\lambda) & Y_1(\lambda) & I_1(\lambda) & -K_1(\lambda) \\ J_0(\alpha\lambda) & Y_0(\alpha\lambda) & -I_0(\alpha\lambda)+QI_1(\alpha\lambda) & -K_0(\alpha\lambda)-QK_1(\alpha\lambda) \\ J_1(\alpha\lambda) & Y_1(\alpha\lambda) & I_1(\alpha\lambda) & -K_1(\alpha\lambda) \end{vmatrix} = 0 \qquad (4.84)$$

where

$$R = \frac{2(1-\nu)}{\lambda}$$

$$Q = \frac{2(1-\nu)}{\alpha\lambda}$$

It is to be noted here that the lowest roots of $n=0$ and $n=1$ are rigid body translation and rotation modes in case of the annular plate when both outer and inner boundaries are free.

So, we have to obtain the frequency determinant for $n=2$ for finding the lowest root. Accordingly, the frequency determinant with $n=2$ for this boundary condition may be given as

$$\begin{vmatrix} J_0(\lambda) & Y_0(\lambda) & AI_0(\lambda)-BI_1(\lambda) & AK_0(\lambda)+BK_1(\lambda) \\ J_1(\lambda) & Y_1(\lambda) & CI_0(\lambda)-DI_1(\lambda) & CK_0(\lambda)+DK_1(\lambda) \\ J_0(\alpha\lambda) & Y_0(\alpha\lambda) & \overline{A}I_0(\alpha\lambda)-\overline{B}I_1(\alpha\lambda) & \overline{A}K_0(\alpha\lambda)+\overline{B}K_1(\alpha\lambda) \\ J_1(\alpha\lambda) & Y_1(\alpha\lambda) & \overline{C}I_1(\alpha\lambda)-\overline{D}I_1(\alpha\lambda) & \overline{C}K_0(\alpha\lambda)+\overline{D}K_1(\alpha\lambda) \end{vmatrix} = 0 \qquad (4.85)$$

where

$$A = 1 - \left(\frac{\lambda}{4}-\frac{3+\nu}{2\lambda}\right)C$$

$$B = \frac{\lambda}{4}+\frac{3+\nu}{2\lambda}-\left(\frac{\lambda}{4}-\frac{3+\nu}{2\lambda}\right)D$$

$$C = \frac{48(1-\nu)\lambda}{12(1-\nu^2) - \lambda^4}$$

$$D = \frac{12(1-\nu)(7+\nu+\lambda^2) - \lambda^4}{12(1-\nu)^2 - \lambda^4}$$

$$\overline{A} = 1 - \left(\frac{\alpha\lambda}{4} - \frac{3+\nu}{2\alpha\lambda}\right)\overline{C}$$

$$\overline{B} = \frac{\alpha\lambda}{4} + \frac{3+\nu}{2\alpha\lambda} - \left(\frac{\alpha\lambda}{4} - \frac{3+\nu}{2\alpha\lambda}\right)\overline{D}$$

$$\overline{C} = \frac{48(1-\nu)\alpha\lambda}{12(1-\nu^2) - (\alpha\lambda)^4}$$

and

$$\overline{D} = \frac{12(1-\nu)[7+\nu+(\alpha\lambda)^2] - (\alpha\lambda)^4}{12(1-\nu)^2 - (\alpha\lambda)^4}$$

Equations 4.75 and 4.77 through 4.84 give the form of frequency determinants that may be obtained for the possible nine simple boundary conditions for annular circular plates. From these determinants, the frequency parameters for axisymmetric case may be computed for desired value of m. Next, we will provide the two rows in each case of the boundary conditions that may be placed in the determinant to generate the frequency determinants for any of the above boundary conditions.

(i) Clamped boundary

$$\left\{ \begin{array}{cccc} J_0(x) & Y_0(x) & I_0(x) & K_0(x) \\ J_1(x) & Y_1(x) & -I_1(x) & K_1(x) \end{array} \right\}$$

(ii) Simply supported boundary

$$\left\{ \begin{array}{cccc} J_0(x) & Y_0(x) & I_0(x) & K_0(x) \\ J_1(x) & Y_1(x) & \overline{P}I_0(x) - I_1(x) & \overline{P}K_0(x) + K_1(x) \end{array} \right\}$$

(iii) Completely free boundary

$$\left\{ \begin{array}{cccc} J_0(x) & Y_0(x) & -I_0(x) + \overline{Q}I_1(x) & -K_0(x) - \overline{Q}K_1(x) \\ J_1(x) & Y_1(x) & I_1(x) & -K_1(x) \end{array} \right\}$$

One has to put

$$\left. \begin{array}{l} x = \lambda \\ \overline{P} = \dfrac{2\lambda}{1-\nu} \\ \text{and} \\ \overline{Q} = \dfrac{2(1-\nu)}{\lambda} \end{array} \right\}$$

in two rows of the above cases (i) through (iii) to have each of the above-mentioned boundary conditions for the outer edges. Similarly, for getting each of the boundary conditions in the inner boundary of the annular circular plate, we have to substitute

$$\left.\begin{array}{l} x = \alpha\lambda \\[4pt] \overline{P} = \dfrac{2\alpha\lambda}{1 - \nu} \\[4pt] \text{and} \\[4pt] \overline{Q} = \dfrac{2(1 - \nu)}{\alpha\lambda} \end{array}\right\}$$

in the two rows of the cases (i) through (iii).

4.2   Method of Solution in Elliptical Coordinate System

For exact analysis of vibration problem in elliptical coordinates, such as an elliptic plate, the flexural vibration displacement W of the plate in Cartesian system (x, y) is considered first. We had the differential equation of motion for a plate from Equation 4.6 as

$$(\nabla^4 - \beta^4)W = 0 \tag{4.86}$$

where

$$\beta^4 = \frac{\rho h \omega^2}{D} \tag{4.87}$$

Solution of Equation 4.86 was written in terms of two linear differential equations:

$$(\nabla^2 + \beta^2)W_1 = 0 \tag{4.88}$$

and

$$(\nabla^2 - \beta^2)W_2 = 0 \tag{4.89}$$

The solution W of the original differential equation is

$$W = W_1 + W_2 \tag{4.90}$$

Equations 4.88 and 4.89 are written in the form

$$\frac{\partial^2 W_1}{\partial x^2} + \frac{\partial^2 W_1}{\partial y^2} + \beta^2 W_1 = 0 \tag{4.91}$$

and

$$\frac{\partial^2 W_2}{\partial x^2} + \frac{\partial^2 W_2}{\partial y^2} - \beta^2 W_2 = 0 \tag{4.92}$$

Let us now introduce the elliptical coordinates (ξ, η) as shown in Figure 3.7 such that

$$\left. \begin{array}{l} x = C \cosh \xi \cos \eta \\ y = C \sinh \xi \sin \eta \end{array} \right\} \tag{4.93}$$

where C is the semifocal length of the ellipse. As given in Chapter 3, the Laplacian operator in elliptical coordinate is given by

$$\begin{aligned} \nabla^2(\cdot) &= \frac{\partial^2}{\partial x^2}(\cdot) + \frac{\partial^2}{\partial y^2}(\cdot) \\ &= \frac{2}{C^2(\cosh 2\xi - \cos 2\eta)} \left(\frac{\partial^2}{\partial \xi^2} + \frac{\partial^2}{\partial \eta^2} \right) \end{aligned} \tag{4.94}$$

Putting the Laplacian operator in elliptical coordinates from Equation 4.94 into the differential equation (Equation 4.91), we arrive at

$$\frac{\partial^2 W_1}{\partial \xi^2} + \frac{\partial^2 W_1}{\partial \eta^2} + \frac{\beta^2 C^2}{2}(\cosh 2\xi - \cos 2\eta)W_1 = 0$$

Putting

$$BC = 2k \tag{4.95}$$

we get

$$\frac{\partial^2 W_1}{\partial \xi^2} + \frac{\partial^2 W_1}{\partial \eta^2} + 2k^2(\cosh 2\xi - \cos 2\eta)W_1 = 0 \tag{4.96}$$

Similarly, we can have from the differential equation (Equation 4.92)

$$\frac{\partial^2 W_2}{\partial \xi^2} + \frac{\partial^2 W_2}{\partial \eta^2} - 2k^2(\cosh 2\xi - \cos 2\eta)W_2 = 0 \tag{4.97}$$

The solution of the differential equation of motion of an elliptic plate may be obtained from Equations 4.96 and 4.97 in terms of Mathieu functions.

Here, we will write the above two equations in terms of Mathieu's differential equation only. For variable separation, we put

$$W_1(\xi, \eta) = U(\xi)V(\eta) \tag{4.98}$$

in the first Equation 4.96 and then it leads to

$$V\frac{\partial^2 U}{\partial \xi^2} + U\frac{\partial^2 V}{\partial \eta^2} + 2k^2(\cosh 2\xi - \cos 2\eta)UV = 0$$

Dividing both sides by UV again, two ODEs are obtained:

$$\frac{1}{U}\frac{\partial^2 U}{\partial \xi^2} + 2k^2 \cosh 2\xi = a \tag{4.99}$$

and

$$\frac{1}{V}\frac{\partial^2 V}{\partial \eta^2} - 2k^2 \cos 2\eta = -a \tag{4.100}$$

where a is the arbitrary separation constant.

Equations 4.99 and 4.100 are now written as Mathieu's differential equations:

$$\frac{\partial^2 U}{\partial \xi^2} - (a - 2q \cos h2\xi)U = 0 \tag{4.101}$$

and

$$\frac{\partial^2 V}{\partial \eta^2} + (a - 2q \cos 2\eta)V = 0 \tag{4.102}$$

where

$$q = k^2 \tag{4.103}$$

Equation 4.102 is called Mathieu's differential equation and Equation 4.101 is known as Mathieu's modified differential equation.

Similarly, in Equation 4.97, we put

$$W_2(\xi, \eta) = \aleph(\xi)\psi(\eta) \tag{4.104}$$

which then will become

$$\psi\frac{\partial^2 \aleph}{\partial \xi^2} + \aleph\frac{\partial^2 \psi}{\partial \eta^2} - 2k^2(\cos h2\xi - \cos 2\eta)\aleph\psi = 0$$

Dividing both sides by $\aleph\psi$, we obtain two ODEs:

$$\frac{1}{\aleph}\frac{\partial^2 \aleph}{\partial \xi^2} - 2k^2 \cosh 2\xi = b \tag{4.105}$$

and

$$\frac{1}{\psi}\frac{\partial^2 \psi}{\partial \eta^2} + 2k^2 \cos 2\eta = -b \qquad (4.106)$$

where b is the arbitrary separation constant. As before, the Equations 4.105 and 4.106 may be written as Mathieu's differential equations

$$\frac{\partial^2 \aleph}{\partial \xi^2} - [b - 2(-q)\cosh 2\xi]\aleph = 0 \qquad (4.107)$$

and

$$\frac{\partial^2 \psi}{\partial \eta^2} + [b - 2(-q)\cos 2\eta]\psi = 0 \qquad (4.108)$$

where $q = K^2$.

The solution of Mathieu's differential equations (Equations 4.101 and 4.102) for U and V and Equations 4.107 and 4.108 for $\aleph$ and ψ are substituted in Equations 4.98 and 4.104 for W_1 and W_2, respectively. Thus, W may be obtained from Equation 4.90.

We will incorporate here directly the solutions for W_1 and W_2, which are

$$W_1 = \sum_{m=0}^{\infty} \left[A_m Ce_m(\xi, q) + B_m Fey_{ym}(\xi, q)\right] ce_m(\eta, q)$$

$$+ \sum_{m=1}^{\infty} [C_m Se_m(\xi, q) + D_m Gey_m(\xi, q)]se_m(\eta, q) \qquad (4.109)$$

and

$$W_2 = \sum_{m=0}^{\infty} \left[\overline{A}_m Ce_m(\xi, -q) + \overline{B}_m Fek_m(\xi, -q)\right] ce_m(\eta, -q)$$

$$+ \sum_{m=1}^{\infty} \left[\overline{C}_m Se_m(\xi, -q) + \overline{D}_m Gek_m(\xi, -q)\right] se_m(\eta, -q) \qquad (4.110)$$

where Ce_m, ce_m, Se_m, se_m, Fey_m, Fek_m, Gey_m, and Gek_m are ordinary and modified Mathieu functions of order m.

A_m, B_m, C_m, D_m, $\overline{A}_m$, $\overline{B}_m$, $\overline{C}_m$, and $\overline{D}_m$ are the constants. As in circular plate when we consider the problem of elliptical plate (Figure 4.9) without hole, we may write the final solution from Equations 4.109 and 4.110 as

$$W_1 = \sum_{m=0}^{\infty} \left[A_m Ce_m(\xi, q)ce_m(\eta, q) + \overline{A}_m Ce_m(\xi, -q)ce_m(\eta, -q)\right]$$

$$+ \sum_{m=1}^{\infty} \left[C_m Se_m(\xi, q) + se_m(\eta, q) + \overline{C}_m Se_m(\xi, -q)se_m(\eta, -q)\right] \qquad (4.111)$$

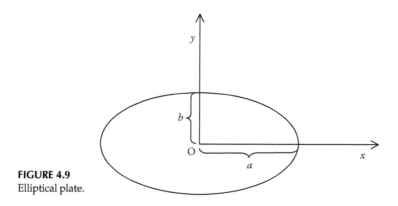

FIGURE 4.9
Elliptical plate.

In Figure 4.9, a and b denote, respectively, the semimajor and semiminor axes of the elliptic plate. For symmetric vibration (i.e., mode shapes having symmetry with respect to both the axes x and y), Equation 4.111 will become

$$W = \sum_{m=0}^{\infty} \left[A_m Ce_m(\xi, q)ce_m(\eta, q) + \bar{A}_m Ce_m(\xi, -q)ce_m(\eta, -q) \right] \qquad (4.112)$$

4.3 Method of Solution in Rectangular Coordinate System

Rectangular coordinate system will be used here for free vibration of rectangular plates. Let us consider a uniform rectangular plate in the domain of xy-plane as shown in Figure 4.10, where a and b are the two sides of the rectangular region. The equation of motion for free vibration of plate is given in Equation 4.1. It has been also shown that this equation may be written in the form of Equation 4.8, i.e.,

$$(\nabla^2 + \beta^2)(\nabla^2 - \beta^2)W = 0 \qquad (4.113)$$

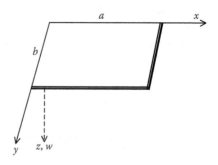

FIGURE 4.10
Rectangular plate showing the rectangular coordinate system.

where

$$\beta^4 = \frac{\rho h \omega^2}{D} \tag{4.114}$$

and ∇^2 in this case has been shown to be

$$\nabla^2(\cdot) = \frac{\partial^2(\cdot)}{\partial x^2} + \frac{\partial^2(\cdot)}{\partial y^2} \tag{4.115}$$

Equation 4.113 is again written as

$$(\nabla^2 + \beta^2)(\nabla^2 + (i\beta)^2)W = 0$$

The above may now yield two linear ODEs:

$$(\nabla^2 + \beta^2)W_1 = 0 \tag{4.116a}$$

and

$$(\nabla^2 + (i\beta)^2)W_2 = 0 \tag{4.116b}$$

Then, the final solution is expressed in the form

$$W = W_1 + W_2 = 0 \tag{4.117}$$

due to linearity of the problem. For the solution of Equation 4.116, we will put

$$W_1(x, y) = X_1(x)Y_1(y) \tag{4.118}$$

and obtain the ODE

$$Y_1 \frac{d^2 X_1}{dx^2} + X_1 \frac{d^2 Y_1}{dy^2} + X_1 Y_1 \beta^2 = 0$$

This differential equation is now written in the form of two equations:

$$\frac{d^2 X_1}{dx^2} + \alpha_1^2 X_1 = 0 \tag{4.119}$$

and

$$\frac{d^2 Y_1}{dy^2} + \alpha_2^2 Y_1 = 0 \tag{4.120}$$

where

$$\beta^2 = \alpha_1^2 + \alpha_2^2 \tag{4.121}$$

Solutions of the second-order differential equations (Equations 4.119 and 4.120) are given by

$$X_1(x) = \bar{A}\sin\alpha_1 x + \bar{B}\cos\alpha_1 x \tag{4.122}$$

$$Y_1(y) = \bar{C}\sin\alpha_2 y + \bar{D}\cos\alpha_2 y \tag{4.123}$$

and the solution of Equation 4.118 is obtained as

$$\begin{aligned}
W_1(x,y) &= (\bar{A}\sin\alpha_1 x + \bar{B}\cos\alpha_1 x)(\bar{C}\sin\alpha_2 y + \bar{D}\cos\alpha_2 y) \\
&= A\sin\alpha_1 x\sin\alpha_2 y + B\sin\alpha_1 x\cos\alpha_2 y + C\cos\alpha_1 x\sin\alpha_2 y \\
&\quad + D\cos\alpha_1 x\cos\alpha_2 y
\end{aligned} \tag{4.124}$$

where $A = \bar{A}\bar{C}$, $B = \bar{A}\bar{D}$, $C = \bar{B}\bar{C}$, and $D = \bar{B}\bar{D}$.

Similarly, if we put

$$W_2(x,y) = X_2(x)Y_2(y) \tag{4.125}$$

in Equation 4.117, then we can derive the solution as

$$X_2(x) = \bar{E}\sinh\alpha_1 x + \bar{F}\cosh\alpha_1 x \tag{4.126}$$

$$Y_2(y) = \bar{G}\sinh\alpha_2 y + \bar{H}\cosh\alpha_2 y \tag{4.127}$$

Using Equations 4.126 and 4.127 in Equation 4.125, $W_2(x,y)$ turns out to be

$$\begin{aligned}
W_2(x,y) &= E\sinh\alpha_1 x\sinh\alpha_2 y + F\sinh\alpha_1 x\cosh\alpha_2 y \\
&\quad + G\cosh\alpha_1 x\sinh\alpha_2 y + H\cosh\alpha_1 x\cosh\alpha_2 y
\end{aligned} \tag{4.128}$$

where the constants are

$$E = \bar{E}\bar{G}, \quad F = \bar{E}\bar{H}, \quad G = \bar{F}\bar{G}, \quad \text{and} \quad H = \bar{F}\bar{H}$$

Combining Equations 4.124 and 4.128, the complete solution for $W(x,y)$ is obtained as

$$\begin{aligned}
W(x,y) &= A\sin\alpha_1 x\sin\alpha_2 y + B\sin\alpha_1 x\cos\alpha_2 y \\
&\quad + C\cos\alpha_1 x\sin\alpha_2 y + D\cos\alpha_1 x\cos\alpha_2 y \\
&\quad + E\sinh\alpha_1 x\sinh\alpha_2 y + F\sinh\alpha_1 x\cosh\alpha_2 y \\
&\quad + G\cosh\alpha_1 x\sinh\alpha_2 y + H\cos\alpha_1 x\cosh\alpha_2 y
\end{aligned} \tag{4.129}$$

The constants A, B, C, D, E, F, G, and H depend on the boundary conditions of the plate. It is worth mentioning now that there exist altogether 21 combinations of classical boundary conditions, viz., clamped (C), simply supported (S), and completely free (F), for a rectangular plate. Gorman (1982) and Leissa (1969) give the solution and other details regarding most of the boundary conditions. In this chapter, we only study one case of the boundary condition of a rectangular plate. The results and solutions related to all boundary conditions will be covered in the later chapters of this book. The three types of classical boundary condition equations for rectangular plates are addressed next.

(i) Clamped (C) boundary condition in rectangular plate
   Along $x=0$ and $x=a$, the condition would be

$$W = 0 \quad \text{and} \quad \frac{\partial W}{\partial x} = 0 \tag{4.130}$$

The conditions along $y=0$ and $y=b$ for clamped edge are

$$W = 0 \quad \text{and} \quad \frac{\partial W}{\partial y} = 0 \tag{4.131}$$

(ii) Simply supported (S) boundary condition in rectangular plate
   Along $x=0$ and $x=a$, the conditions are given by

$$W = 0 \quad \text{and} \quad M_x = 0 \tag{4.132}$$

Conditions along $y=0$ and $y=b$ are written as

$$W = 0 \quad \text{and} \quad M_y = 0 \tag{4.133}$$

Equation 4.132 gives the conditions
along $x=0$ and $x=a$:

$$W = 0 \quad \text{and} \quad M_x = -D\left[\frac{\partial^2 W}{\partial x^2} + \nu\frac{\partial^2 W}{\partial y^2}\right] = 0$$

Then, we obtain from above

$$W = 0 \quad \text{and} \quad \frac{\partial^2 W}{\partial x^2} = 0 \tag{4.134}$$

because x is constant along the two directions.

Similarly, Equation 4.133 gives the conditions along $y=0$ and $y=b$ as

$$W = 0 \quad \text{and} \quad M_y = -D\left[\frac{\partial^2 W}{\partial y^2} + v\frac{\partial^2 W}{\partial x^2}\right] = 0$$

from which we obtain

$$W = 0 \quad \text{and} \quad \frac{\partial^2 W}{\partial y^2} = 0 \tag{4.135}$$

because y is constant along the above two directions.

(iii) Free (F) boundary condition in rectangular plate

Along $x=0$ and $x=a$, the boundary conditions are

$$M_x = -D\left[\frac{\partial^2 W}{\partial x^2} + v\frac{\partial^2 W}{\partial y^2}\right] = 0$$

and

$$Q_x - \frac{\partial M_{xy}}{\partial y} = -D\frac{\partial}{\partial x}\left[\frac{\partial^2 W}{\partial x^2} + (2-v)\frac{\partial^2 W}{\partial y^2}\right] = 0$$

Combining the above two gives the boundary conditions along $x=0$ and $x=a$ as

$$\frac{\partial^2 W}{\partial x^2} = 0 \quad \text{and} \quad \frac{\partial^2 W}{\partial x^3} = 0 \tag{4.136}$$

The free conditions along $y=0$ and $y=b$ will become

$$M_y = -D\left[\frac{\partial^2 W}{\partial y^2} + v\frac{\partial^2 W}{\partial x^2}\right] = 0$$

$$Q_y + \frac{\partial M_{yx}}{\partial x} = -D\frac{\partial}{\partial y}\left[\frac{\partial^2 W}{\partial y^2} + (2-v)\frac{\partial^2 W}{\partial x^2}\right] = 0$$

Again combining the above two, we obtain the boundary conditions along $y=0$ and $y=b$ as

$$\frac{\partial^2 W}{\partial y^2} = 0 \quad \text{and} \quad \frac{\partial^3 W}{\partial y^3} = 0 \tag{4.137}$$

Now, we will consider some detailed analysis for a simple case of a boundary condition, viz., when all sides of a rectangular plate are simply supported. The corresponding boundary conditions (Equations 4.134 and 4.135) are

$$\text{at } x = 0 \quad \text{and} \quad x = a, \quad W = \frac{\partial^2 W}{\partial x^2} = 0 \tag{4.138}$$

$$\text{at } y = 0 \quad \text{and} \quad y = b, \quad W = \frac{\partial^2 W}{\partial y^2} = 0 \tag{4.139}$$

First, we write the solution for x-coordinate, i.e., $\overline{W}(x)$ from Equations 4.122 and 4.126 as

$$\begin{aligned} \overline{W}(x) &= X_1(x) + X_2(x) \\ &= \overline{A} \sin \alpha_1 x + \overline{B} \cos \alpha_1 x + \overline{E} \sinh \alpha_1 x + \overline{F} \cosh \alpha_1 x \end{aligned} \tag{4.140}$$

$$\overline{W}''(x) = \frac{\partial^2 W}{\partial x^2} = -\overline{A}\alpha_1^2 \sin \alpha_1 x - \overline{B}\alpha_1^2 \cos \alpha_1 x + \overline{E}\alpha_1^2 \sinh \alpha_1 x + \overline{F}\alpha_1^2 \cosh \alpha_1 x \tag{4.141}$$

Imposing the boundary condition $\overline{W}(0) = 0$ in Equation 4.140, we have

$$\overline{B} + \overline{F} = 0 \tag{4.142}$$

Using the boundary condition $\overline{W}(a) = 0$ in Equation 4.140, we get

$$\overline{A} \sin \alpha_1 a + \overline{B} \cos \alpha_1 a + \overline{E} \sinh \alpha_1 a + \overline{F} \cosh \alpha_1 a = 0 \tag{4.143}$$

Putting Equation 4.141 in boundary conditions $W''(0) = W''(a) = 0$, respectively, yields

$$-\alpha_1^2 \overline{B} + \alpha_1^2 \overline{F} = 0 \tag{4.144}$$

and

$$-\overline{A}\alpha_1^2 \sin \alpha_1 a - \overline{B}\alpha_1^2 \cos \alpha_1 a + \overline{E}\alpha_1^2 \sinh \alpha_1 a + \overline{F}\alpha_1^2 \cosh \alpha_1 a = 0 \tag{4.145}$$

From Equations 4.142 and 4.144, we may conclude that $\overline{B} = \overline{F} = 0$, since $\alpha_1^2 \neq 0$.

Therefore, Equations 4.143 and 4.145 are written as

$$\overline{A} \sin \alpha_1 a + \overline{E} \sinh \alpha_1 a = 0 \tag{4.146}$$

and

$$-\alpha_1^2 \overline{A} \sin \alpha_1 a + \alpha_1^2 \overline{E} \sinh \alpha_1 a = 0 \tag{4.147}$$

Equation 4.146 gives

$$\bar{E} \sinh \alpha_1 a = -\bar{A} \sin \alpha_1 a \qquad (4.148)$$

Putting Equation 4.148 in Equation 4.147, the following is obtained:

$$\bar{A} \sin \alpha_1 a = 0. \qquad (4.149)$$

From Equations 4.148 and 4.149, one may arrive at $\bar{E} = 0$ since $\sinh \alpha_1 a \neq 0$. So, the solution for $\bar{W}(x)$ becomes

$$\bar{W}(x) = \bar{A} \sin \alpha_1 x \qquad (4.150)$$

Then, Equation 4.149 yields either $\bar{A} = 0$ or $\sin \alpha_1 a = 0$. But $\bar{A}$ cannot be zero, so we are left with

$$\sin \alpha_1 a = 0 \qquad (4.151)$$

which leads to

$$\alpha_1 = \frac{m\pi}{a} \qquad (4.152)$$

where $m = 1, 2, 3, \ldots$

One may directly obtain Equation 4.152 if the Equations 4.142 through 4.145 are written in matrix form as

$$\begin{bmatrix} 0 & 1 & 0 & 1 \\ \sin \alpha_1 a & \cos \alpha_1 a & \sinh \alpha_1 a & \cosh \alpha_1 a \\ 0 & -\alpha_1^2 & 0 & \alpha_1^2 \\ -\alpha_1^2 \sin \alpha_1 a & -\alpha_1^2 \cos \alpha_1 a & \alpha_1^2 \sinh \alpha_1 a & \alpha_1^2 \cosh \alpha_1 a \end{bmatrix} \begin{Bmatrix} \bar{A} \\ \bar{B} \\ \bar{E} \\ \bar{F} \end{Bmatrix} = \begin{Bmatrix} 0 \\ 0 \\ 0 \\ 0 \end{Bmatrix} \qquad (4.153)$$

and the above matrix equation is satisfied only if the determinant is zero. Expanding the determinant, we may have directly

$$\sin \alpha_1 a \sinh \alpha_1 a = 0 \qquad (4.154)$$

Since $\sinh \alpha_1 a$ cannot be zero, so

$\sin \alpha_1 a = 0$, which again gives $\alpha_1 = \frac{m\pi}{a}$, which is the same as Equation 4.152.

In a similar fashion, if we take the y-coordinate solution, i.e., $\bar{W}(y)$ from Equations 4.123 and 4.127 and we write $\bar{W}(y)$ as

$$\bar{W}(y) = Y_1(y) + Y_2(y)$$
$$= \bar{C} \sin \alpha_2 y + \bar{D} \cos \alpha_2 y + \bar{G} \sinh \alpha_2 y + \bar{H} \cosh \alpha_2 y \qquad (4.155)$$

then one may arrive at

$$\sin \alpha_2 b = 0 \qquad (4.156a)$$

So, only the constant $\bar{C}$ will be nonzero. Accordingly, we have

$$\overline{W}(y) = \bar{C} \sin \alpha_2 y \qquad (4.156\text{b})$$

Therefore, the solution $W(x, y)$ is found from Equations 4.150 and 4.156b as

$$W(x, y) = A \sin \alpha_1 x \sin \alpha_2 y \qquad (4.157)$$

where $A = \bar{A}\bar{C}$

Moreover, similar to Equation 4.151, we will have for y-coordinate solution

$$\alpha_2 = \frac{n\pi}{b} \qquad (4.158)$$

where $n = 1, 2, 3, \ldots$

Substituting values of α_1 and α_2 from Equations 4.152 and 4.158, in Equations 4.121 and 4.114, we obtain

$$\beta^2 = \omega \sqrt{\frac{\rho h}{D}} = \pi^2 \left[\left(\frac{m}{a} \right)^2 + \left(\frac{n}{b} \right)^2 \right] \qquad (4.159)$$

and therefore the frequency ω is written with suffix m and n as

$$\omega_{mn} = \pi^2 \left[\left(\frac{m}{a} \right)^2 + \left(\frac{n}{b} \right)^2 \right] \sqrt{\frac{D}{\rho h}} \qquad (4.160)$$

$m, n = 1, 2, 3, \ldots$

Solution (Equation 4.157) may also be written as

$$W_{mn}(x, y) = A_{mn} \sin \frac{m\pi x}{a} \cdot \sin \frac{n\pi y}{b} \qquad (4.161)$$

which are also known as the corresponding eigenfunctions for a rectangular plate with sides a and b. In this case, the fundamental natural frequency occurs for $m = n = 1$.

4.4 Approximate Solution Methods

In the previous sections, either exact solution or series-type solutions are introduced for some standard problems. In vibration problems, sometimes it is not possible to get the exact or series solutions in particular to plate problems. The complexity due to the domain occupied by the plate and the specified boundary conditions make the problem more complex. Thereby, we need to use approximate methods as discussed in Chapter 2. There exist various approximate methods to handle these problems. Next, the approximate methods, viz., Rayleigh and Rayleigh–Ritz methods, will be discussed to handle the vibration of plate problems, because only these methods will be mainly used in the rest of the chapters of this book by using a newly developed solution methodology.

Energy formulations are sometimes an alternative to the differential equations that govern the equations of motion. Energy principles are the basis for many powerful numerical methods for solving complex problems such as plate vibration. There exists a direct connection between the energy and the differential equation approaches through the principle of virtual displacements as well as through different extremum principles. These will not be discussed here as it is beyond the scope of this book.

4.4.1 Rayleigh's Method for Plates

As discussed in Chapter 2, only the first, i.e., the fundamental frequency, may be approximated by this method. Let us assume that a plate is vibrating freely with natural frequency ω and let $\omega = \omega_1$ be its fundamental frequency. The kinetic energy T of the plate is written as

$$T = \frac{1}{2} \iint_R h\rho \dot{w}(x, y, t) dx dy \tag{4.162}$$

A solution of the form is assumed as

$$\omega(x, y, t) = W_1(x, y) \cos \omega_1 t \tag{4.163}$$

Substituting Equation 4.163 in Equation 4.162 and taking the maximum value of T as T_{max}, we write

$$T_{max} = \frac{1}{2} \omega^2 \iint_R h\rho W_1^2(x, y) dx dy \tag{4.164}$$

where for the maximum value, we have $\sin^2 \omega_1 t = 1$.

Also, the maximum strain energy from Equation 3.93 is written as

$$U = \frac{1}{2} \iint_R D \left[(\nabla^2 w)^2 + 2(1 - \nu) \left\{ \left(\frac{\partial^2 w}{\partial x \partial y} \right)^2 - \frac{\partial^2 w}{\partial x^2} \frac{\partial^2 w}{\partial y^2} \right\} \right] dx dy \tag{4.165}$$

where q is taken as zero in Equation 3.93. Again, putting Equation 4.163 in Equation 4.165 and taking the maximum value of U as U_{max}, one can obtain

$$U_{max} = \frac{1}{2} \iint_R D \left[(\nabla^2 W_1)^2 + 2(1 - \nu) \left\{ \left(\frac{\partial^2 W_1}{\partial x \partial y} \right)^2 - \frac{\partial^2 W_1}{\partial x^2} \cdot \frac{\partial^2 W_1}{\partial y^2} \right\} \right] dx dy \tag{4.166}$$

We now equate the maximum kinetic and strain energies of the system, as there is no dissipation. Accordingly, we have

$$T_{max} = U_{max} \tag{4.167}$$

Putting the values of T_{max} and U_{max} from Equations 4.164 and 4.166 in Equation 4.167, we obtain

$$\omega_1^2 = \frac{\iint\limits_R D\left[(\nabla^2 W_1)^2 + 2(1-\nu)\left\{W_{1,xy}^2 - W_{1,xx}W_{1,yy}\right\}\right]dxdy}{\iint\limits_R h\rho W_1^2 dxdy} \qquad (4.168)$$

where

$$W_{1,xy} = \frac{\partial^2 W_1}{\partial x \partial y}, \quad W_{1,xx} = \frac{\partial^2 W_1}{\partial x^2}, \quad \text{and} \quad W_{1,yy} = \frac{\partial^2 W_1}{\partial y^2}$$

The above is known as Rayleigh's quotient, which gives the first frequency directly if the integrals in Equation 4.168 are evaluated. For a constant thickness and homogeneous plate, Rayleigh's quotient in Rayleigh's method may easily be written from Equation 4.168 as

$$\omega_1^2 = \frac{D\iint\limits_R \left[(\nabla^2 W_1)^2 + 2(1-\nu)\left\{W_{1,xy}^2 - W_{1,xx}W_{1,yy}\right\}\right]dxdy}{h\rho \iint\limits_R W_1^2 dxdy} \qquad (4.169)$$

because in that case D, h, and ρ will be constant and so those may be kept out of the integrals. As discussed in Chapter 2, if W_1 as assumed in Equation 4.163 also satisfies boundary conditions of the problem, as well as if it happens to be a good approximation to the first (fundamental) mode shape, then the first frequency as given by Rayleigh's method will give even a better approximation of the fundamental frequency.

4.4.2 Rayleigh–Ritz Method for Plates

This method can yield higher modes along with a better approximation of the fundamental mode. The method requires a linear combination of assumed deflection shapes of structures in free harmonic vibration that satisfies at least the geometrical boundary conditions of the vibrating structure. For the plate vibrating freely with frequency ω, we assume again a solution of the form

$$w = W(x,y)\cos \omega t \qquad (4.170)$$

We can again easily obtain the maximum kinetic energy and maximum strain energy as in the Section 4.4.1. Therefore, we would have

$$T_{max} = \frac{1}{2}\omega^2 \iint\limits_R \rho h W^2 dxdy \qquad (4.171)$$

$$U_{\max} = \frac{1}{2}\iint_R D\left[(\nabla^2 W)^2 + 2(1-\nu)\left\{W_{xy}^2 - W_{xx}W_{yy}\right\}\right]dxdy \qquad (4.172)$$

As there is no dissipation, maximum kinetic energy and the maximum strain energy of the system are equated. Accordingly, the Rayleigh quotient in this case is obtained as

$$\omega^2 = \frac{E}{12\rho(1-\nu^2)}\frac{\displaystyle\iint_R h^3\left[(\nabla^2 W)^2 + 2(1-\nu)\left\{(W_{xy})^2 - W_{xx}W_{yy}\right\}\right]dxdy}{\displaystyle\iint_R hW^2 dxdy} \qquad (4.173)$$

The natural frequencies are then determined by minimizing the above quotient after putting suitable expressions for W that satisfy the prescribed boundary conditions of the system.

In this method, we assume an approximate solution involving some unknown constants. Rayleigh's quotient is then extremized as a function of these constants. This leads to a system of homogeneous linear equations. Equating to zero the determinant of the coefficient matrix, we get a polynomial equation in the frequency parameter. Solving this, the natural frequencies are obtained. The associated set of the unknown constants for a given frequency gives the mode shapes for that frequency. By varying the number of terms in the approximation, one can build up a sequence of solutions with the hope that these converge to the exact solution.

To fix the above idea, let us now assume the N-term approximation

$$W(x, y) = \sum_{j=1}^{N} C_j\phi_j(x, y)$$
$$= C_1\phi_1(x, y) + C_2\phi_2(x, y) + \cdots + C_n\phi_n(x, y) \qquad (4.174)$$

where $\phi_j(x, y)$ satisfies the boundary conditions and C_j are the constants to be determined. Before proceeding further, we rewrite the Rayleigh quotient by expanding the Laplacian operator and obtain

$$\rho\omega^2 = \frac{\displaystyle\iint_R h^3\left[W_{xx}^2 + W_{yy}^2 + 2\nu W_{xx}W_{yy} + 2(1-\nu)W_{xy}^2\right]dxdy}{\displaystyle\iint_R hW^2 dxdy} \qquad (4.175)$$

where

$$W_{xx} = \frac{\partial^2 W}{\partial x^2}, \quad W_{yy} = \frac{\partial^2 W}{\partial y^2}, \quad \text{and} \quad W_{xy} = \frac{\partial^2 W}{\partial x \partial y}$$

Substituting the *N*-term approximation (Equation 4.174) in Equation 4.175, one may get

$$
p\omega^2 = \frac{\iint_R \left[\left(\sum_{j=1}^N C_j \phi_j^{xx} \right)^2 + \left(\sum_{j=1}^N C_j \phi_j^{yy} \right)^2 + 2\nu \left(\sum_{j=1}^N C_j \phi_j^{xx} \right) \left(\sum_{j=1}^N C_j \phi_j^{yy} \right) + 2(1-\nu) \left(\sum_{j=1}^N C_j \phi_j^{xy} \right)^2 \right] dxdy}{\iint_R \left(\sum_{j=1}^N C_j \phi_j \right)^2 dxdy}
$$

(4.176)

where

$$
\phi_j^{xx} = \frac{\partial^2 \phi_j}{\partial x^2}
$$

$$
\phi_j^{yy} = \frac{\partial^2 \phi_j}{\partial y^2}
$$

$$
\phi_j^{xy} = \frac{\partial^2 \phi_j}{\partial x \partial y}
$$

and

$$
p = \frac{12\rho(1-\nu^2)}{E}
$$

(4.177)

Minimizing Equation 4.176 as a function of the coefficients $C_1, C_2, \ldots, C_N$, we write

$$
\frac{\partial(p\omega^2)}{\partial C_i} = 0
$$

(4.178)

which gives

$$
\iint_R \left(\sum_{j=1}^N C_j \phi_j \right)^2 dxdy \left[\iint_R \left\{ \phi_i^{xx} \sum_{j=1}^N C_j \phi_j^{xx} + \phi_i^{yy} \sum_{j=1}^N C_j \phi_j^{yy} + \nu \phi_i^{xx} \sum_{j=1}^N C_j \phi_j^{yy} \right. \right.
$$

$$
\left. + \nu \phi_i^{yy} \sum_{j=1}^N C_j \phi_j^{xx} + 2(1-\nu) \phi_i^{xy} \sum_{j=1}^N C_j \phi_j^{xy} \right\} dxdy \Bigg]
$$

$$
- \iint_R \phi_i \left(\sum_{j=1}^N C_j \phi_j dxdy \right) \left[\iint_R \left\{ \left(\sum_{j=1}^N C_j \phi_j^{xx} \right)^2 + \left(\sum_{j=1}^N C_j \phi_j^{yy} \right)^2 \right. \right.
$$

$$
\left. + 2\nu \left(\sum_{j=1}^N C_j \phi_j^{xx} \right) \left(\sum_{j=1}^N C_j \phi_j^{yy} \right) + 2(1-\nu) \left(\sum_{j=1}^N C_j \phi_j^{xy} \right)^2 \right\} dxdy \Bigg] = 0
$$

Using Equation 4.176, the above is written in the form

$$
\iint_R \left(\sum_{j=1}^N C_j \phi_j \right)^2 dxdy \left[\iint_R \left\{ \phi_i^{xx} \sum_{j=1}^N C_j \phi_j^{xx} + \phi_i^{yy} \sum_{j=1}^N C_j \phi_j^{yy} + \nu \phi_i^{xx} \sum_{j=1}^N C_j \phi_j^{yy} \right. \right.
$$
$$
\left. \left. + \nu \phi_i^{yy} \sum_{j=1}^N C_j \phi_j^{xx} + 2(1-\nu)\phi_i^{xy} \sum_{j=1}^N C_j \phi_j^{xy} \right\} dxdy \right]
$$
$$
- \left(\sum_{j=1}^N C_j \iint_R \phi_i \phi_j dxdy \right) \left[p\omega^2 \iint_R \left(\sum_{j=1}^N C_j \phi_j \right)^2 dxdy \right] = 0 \tag{4.179}
$$

Now since $\left(\sum_{j=1}^N C_j \phi_j \right)^2 \neq 0$, Equation 4.179 leads to

$$
\sum_{j=1}^N C_j \iint_R \left\{ \phi_i^{xx}\phi_j^{xx} + \phi_i^{yy}\phi_j^{yy} + \nu\left(\phi_i^{xx}\phi_j^{yy} + \phi_i^{yy}\phi_j^{xx} \right) + 2(1-\nu)\phi_i^{xy}\phi_j^{xy} \right\} dxdy
$$
$$
- p\omega^2 \sum_{j=1}^N C_j \iint \phi_i\phi_j dxdy = 0 \tag{4.180}
$$

If the following nondimensional forms

$$
X = \frac{x}{a}, \quad Y = \frac{y}{a}, \quad \text{and} \quad H = \frac{h}{h_o} \tag{4.181}
$$

are used in Equation 4.180, we have

$$
\sum_{j=1}^N C_j \iint_{R'} \frac{H^3}{a^4} \left[\phi_i^{XX}\phi_j^{XX} + \phi_i^{YY}\phi_j^{YY} + \nu\left(\phi_i^{XX}\phi_j^{YY} + \phi_i^{YY}\phi_j^{XX} \right) \right.
$$
$$
\left. + 2(1-\nu)\phi_i^{XY}\phi_j^{XY} \right] dXdY - p\omega^2 \sum_{j=1}^N C_j \iint_{R'} H\phi_i\phi_j dXdY \tag{4.182}
$$

where
　　a is some characteristic length of the problem
　　h_o is the thickness at some standard point, which we have taken as the
　　origin

Here, R' is the new domain in nondimensional form.
　　Equation 4.182 may now be written as

$$
\sum_{j=1}^N (a_{ij} - a^4\omega^2 pb_{ij})C_j = 0
$$

The above is rewritten in the form

$$\sum_{j=1}^{N} (a_{ij} - \lambda^2 b_{ij})C_j = 0 \qquad (4.183)$$

where $\lambda^2 = a^4 \omega^2 p$. Putting the value of p from Equation 4.177, we obtain

$$\lambda^2 = \frac{12\rho a^4 (1 - \nu^2)\omega^2}{Eh_o^2} \qquad (4.184)$$

known as the frequency parameter. The expressions a_{ij} and b_{ij} are, respectively,

$$a_{ij} = \iint_{R'} H^3 \left[\phi_i^{XX} \phi_j^{XX} + \phi_i^{YY} \phi_j^{YY} + \nu \left(\phi_j^{XX} \phi_j^{YY} + \phi_i^{YY} \phi_j^{XX} \right) + 2(1-\nu)\phi_i^{XY} \phi_j^{XY} \right] dXdY$$

$$(4.185)$$

and

$$b_{ij} = \iint_{R^1} H \phi_i \phi_j dXdY \qquad (4.186)$$

Equation 4.183 is known as the generalized eigenvalue problem, which can be solved for frequencies and mode shapes for free vibration of plates of various shapes and with different boundary conditions.

It is now important to note here from Equation 4.186 that if the functions ϕ_j considered in Equation 4.174 are orthonormal with respect to the weight function H, then

$$b_{ij} = \begin{cases} 0 & \text{for } i \neq j \\ 1 & \text{for } i = j \end{cases}$$

This will turn the generalized eigenvalue problem into a standard eigenvalue problem. However, this requires some assumption and analysis before choosing or generating the orthogonal polynomials that satisfy the boundary conditions. This procedure of applying the orthogonal polynomials in Rayleigh–Ritz method is applied recently throughout the globe to study vibration problems. Present book is mainly devoted to the study of vibration of plates using this newly emerging, intelligent, and powerful method. The benefit and the other usefulness will all be discussed in subsequent chapters for different types of vibration of plate problems. We discuss in Chapters 5 and 6 about generating and using the one- and two-dimensional orthogonal polynomials. Before switching over to the discussion of these polynomials in the next chapters, the following are some discussions related to strain energy procedures.

As per the above discussion, it is now understood that the strain energy stored in a plate due to deformation is very useful. In particular, to obtain approximate solution by Rayleigh–Ritz method, this energy expression is important to know. Accordingly, here the strain energy stored in elastic body in different cases is given.

Strain energy in polar coordinates:

$$U = \frac{D}{2} \iint_R \left[\left(\frac{\partial^2 w}{\partial r^2} + \frac{1}{r} \frac{\partial w}{\partial r} + \frac{1}{r^2} \frac{\partial^2 w}{\partial \theta^2} \right)^2 \right.$$

$$+ 2(1 - \nu) \left(\left\{ \frac{\partial}{\partial r} \left(\frac{1}{r} \frac{\partial w}{\partial \theta} \right) \right\}^2 - \frac{\partial^2 w}{\partial r^2} \left\{ \frac{1}{r} \frac{\partial w}{\partial r} + \frac{1}{r^2} \frac{\partial^2 w}{\partial \theta^2} \right\} \right) \Big] r \, dr \, d\theta \qquad (4.187)$$

Strain energy for rectangular orthotropic plate:

$$U = \frac{1}{2} \iint_R \left[D_x \left(\frac{\partial^2 w}{\partial x^2} \right)^2 + 2\nu_x D_y \frac{\partial^2 w}{\partial x^2} \cdot \frac{\partial^2 w}{\partial y^2} + D_y \left(\frac{\partial^2 w}{\partial y^2} \right)^2 + 4 D_{xy} \left(\frac{\partial^2 w}{\partial_x \partial_y} \right)^2 \right] dy \, dx$$

$$(4.188)$$

The D coefficients are bending rigidities defined by

$$D_x = \frac{E_x h^3}{12(1 - \nu_y \nu_x)}$$

$$D_y = \frac{E_y h^3}{12(1 - \nu_y \nu_x)}$$

$$D_y \nu_x = D_x \nu_y$$

and

$$D_{xy} = \frac{G_{xy} h^3}{12}$$

where E_x and E_y are Young's moduli, ν_x and ν_y are Poisson's ratios in x, y directions, and G_{xy} is shear modulus.

Strain energy for polar orthotropic plates:

$$U = \frac{1}{2} \iint_R \left[D_r \left(\frac{\partial^2 w}{\partial r^2} \right)^2 + 2\nu_{\theta r} D_r \frac{\partial^2 w}{\partial r^2} \left(\frac{1}{r} \frac{\partial w}{\partial r} + \frac{1}{r^2} \frac{\partial^2 w}{\partial \theta^2} \right) \right.$$

$$+ D_\theta \left(\frac{1}{r} \frac{\partial w}{\partial r} + \frac{1}{r^2} \frac{\partial^2 w}{\partial \theta^2} \right)^2 + 4 D_{r\theta} \left\{ \frac{\partial}{\partial r} \left(\frac{1}{r} \frac{\partial w}{\partial \theta} \right) \right\}^2 \Big] r \, d\theta \, dr \qquad (4.189)$$

where

$$D_r = \frac{E_r h^3}{12(1 - \nu_{r\theta}\nu_{\theta r})}$$

$$D_\theta = \frac{D_r E_\theta}{E_r}$$

and

$$D_{r\theta} = G_{r\theta}\frac{h^3}{12}$$

in which E_r and E_θ are Young's moduli in r and θ directions, respectively, $G_{r\theta}$ is the shear modulus, and $\nu_{r\theta}$, $\nu_{\theta r}$ are the Poisson ratios.

Bibliography

Airey, J. 1911. The vibration of circular plates and their relation to Bessel functions. *Proceedings of the Physical Society of London*, 23: 225–232.

Chakraverty, S. 1992. Numerical solution of vibration of plates, PhD Thesis. Indian Institute of Technology, Roorkee, India.

Gontkevich, V.S. 1964. *Natural Vibrations of Plates and Shells*. A.P. Filippov (Ed.), Nauk. Dumka (Kiev) (Transl.) By Lockheed Missiles & Space Co. (Sunnyvale, California.)

Gorman, D.J. 1982. *Free Vibration Analysis of Rectangular Plates*, Elsevier, Amsterdam, the Netherlands.

Joga-Rao, C.V. and Pickett, G. 1961. Vibrations of plates of irregular shapes and plates with holes. *Journal of the Aeronautical Society of India*, 13(3): 83–88.

Leissa, A.W. 1969. *Vibration of Plates*, NASA, Washington D.C.

McLachlan, N. 1947. *Theory and Application of Mathieu Functions*, Oxford University Press, London.

McLachlan, N. 1948. *Bessel Functions for Engineers*, The Oxford Engineering Science Series, Ser., Oxford University Press, London.

McLachlan, N.W. 1947. Vibrational problems in elliptical coordinates. *Quarterly of Applied Mathematics*, 5(3): 289–297.

Prescott, T. 1961. *Applied Elasticity*, Dover Pub. Inc., (originally published Longmans, Green & Co. 1924), New York.

Raju, P.N. 1962. Vibrations of annular plates. *Journal of the Aeronautical Society of India*, 14(2): 37–52.

Shames, I.H. and Dym, C.L. 1985. *Energy and Finite Element Methods in Structural Mechanics*, Hemisphere Publishing Corporation, McGraw-Hill Book Company, Washington, New York.

Soedel, W. 1993. *Vibrations of Shells and Plates*, 2nd ed., Marcel Dekker Inc., New York.

Vinson, J.R., *The Behaviour of Thin Walled Structures: Beams, Plates and Shells*, Kluwer Academic Publishers, the Netherlands.

5

Development of Characteristic Orthogonal Polynomials (COPs) in Vibration Problems

Vibration analysis of different shaped structures has been of interest to several engineering disciplines over several decades. Dynamic behavior of the structures is strongly dependent on the boundary conditions, geometrical shapes, material properties, different theories, and various complicating effects. Closed-form solutions are possible only for a limited set of simple boundary conditions and geometries. For the analysis of arbitrarily shaped structures, a variety of numerical methods such as finite element method, finite difference method, and boundary element method are usually applied. Although such discretization methods provide a general framework for general structures, they invariably result in problems with a large number of degrees of freedom. This deficiency is overcome by using the well-known Rayleigh–Ritz method.

Recently, tremendous work has been done by using a newly developed method of orthogonal polynomials in the Rayleigh–Ritz method first proposed in 1985. This method provides better accuracy of the results and is much efficient, simple, and easy for computer implementation. The importance of orthogonal polynomials is well known in the problems of numerical approximation. For generating such polynomials in one variable, we can start with a linearly independent set such as $\{1, x, x^2, x^3, \ldots\}$ and apply the well-known Gram–Schmidt process to generate them over the desired interval with an appropriate weight function. Fortunately, in this case the whole exercise is greatly simplified because of the existence of a three-term recurrence relation. This naturally saves a lot of computation. Perhaps, this is the reason why problems that are one-dimensional can easily be treated by using such polynomials. When certain boundary conditions are to be satisfied, one can start with the set $f(x)$ $\{1, x, x^2, x^3, \ldots\}$, where $f(x)$ satisfies the condition of the problem. Before going into the details of the characteristic orthogonal polynomials (COPs), we will discuss few terms and theories related to orthogonal polynomials that will be used in the contents in this book.

5.1 Preliminary Definitions

n-Tuple: An ordered set $\bar{x} = (x_1, x_2, \ldots, x_n)$ of n elements is called an n-tuple, which is also called a point or vector in n-dimensional space. The scalars $x_1, x_2, \ldots x_n$ are said to be the coordinates of the vector $\bar{x}$.

Equality of two vectors: Two vectors $\bar{x}$ and $\bar{y}$ where $\bar{x} = (x_1, x_2, \ldots, x_n)$ and $\bar{y} = (y_1, y_2, \ldots, y_n)$ are termed as equal iff (i.e., if and only if) $x_i = y_i$ for each $i = 1, 2, \ldots, n$. We define the zero vector denoted by $\bar{0}$ as the vector $(0, 0, \ldots, 0)$.

Addition of vectors: Addition of two vectors $\bar{x}$ and $\bar{y}$ is defined as

$$\bar{x} + \bar{y} = (x_1 + y_1, x_2 + y_2, \ldots, x_n + y_n)$$

Scalar multiplication: Let α be a scalar. Then, scalar multiplication is defined as

$$\alpha \bar{x} = (\alpha x_1, \alpha x_2, \ldots, \alpha x_n)$$

Linear vector space: A collection of vectors $\bar{a}, \bar{b}, \bar{c}, \ldots$ is known as a linear vector space or vector space or linear space V_n over the real number field R if the following rules for vector addition and scalar multiplication are satisfied:

Vector addition: For every pair of vectors $\bar{a}$ and $\bar{b}$, there corresponds a unique vector $(\bar{a} + \bar{b})$ in V_n such that:

(1) It is commutative: $\bar{a} + \bar{b} = \bar{b} + \bar{a}$
(2) It is associative: $(\bar{a} + \bar{b}) + \bar{c} = \bar{a} + (\bar{b} + \bar{c})$
(3) Existence of additive identity element: there exists a unique vector $\bar{0}$, independent of $\bar{a}$, such that

$$\bar{a} + \bar{0} = \bar{a} = \bar{0} + \bar{a}$$

(4) Existence of the additive inverse element: for every $\bar{a}$, there exists a unique vector (depending on $\bar{a}$) denoted by $-\bar{a}$ such that $\bar{a} + (-\bar{a}) = \bar{0}$

Scalar multiplication: for every vector $\bar{a}$ and every real number $\alpha \in R$, there corresponds a unique vector $\alpha \bar{a}$ (product) in V_n with the properties:

(1) Associativity: $\alpha(\beta \bar{a}) = (\alpha \beta) \bar{a}$, $\alpha, \beta \in R$
(2) Distributive with respect to scalar addition: $(\alpha + \beta) \bar{a} = \alpha \bar{a} + \beta \bar{a}$
(3) Distributive with respect to vector addition: $\alpha(\bar{a} + \bar{b}) = \alpha \bar{a} + \alpha \bar{b}$
(4) $1.\bar{a} = \bar{a}.1 = \bar{a}$

In short, one may prove a set of vectors to be a vector space in V_n in the following way:

(1) One should first define the rules of vector addition and scalar multiplication of a vector over the set.
(2) Closure property must be verified, i.e., if $\bar{a}, \bar{b} \in V_n$, then $\alpha \bar{a} + \beta \bar{b} \in V_n$ for all scalars $\alpha, \beta \in R$.

Linear combination of vectors: Let $\bar{x}^{(1)}, \bar{x}^{(2)}, \ldots, \bar{x}^{(n)}$ be n vectors, then any vector $\bar{x} = \alpha_1 \bar{x}^{(1)} + \alpha_2 \bar{x}^{(2)} + \cdots + \alpha_n \bar{x}^{(n)}$ where $\alpha_1, \alpha_2, \ldots, \alpha_n$ are scalars is called a linear combination of the vectors $\bar{x}^{(1)}, \bar{x}^{(2)}, \ldots, \bar{x}^{(n)}$.

5.1.1 Linear Dependence and Linear Independence of Vectors

Linear dependence:
If V_n is a linear vector space, then a finite nonempty set $\{\bar{x}^{(1)}, \bar{x}^{(2)}, \ldots, \bar{x}^{(m)}\}$ of vectors of V_n is said to be linearly dependent if there exists scalars $\alpha_1, \alpha_2, \ldots, \alpha_m$ not all of them zero (some of them may be zero) such that

$$\alpha_1 \bar{x}^{(1)} + \alpha_2 \bar{x}^{(2)} + \cdots + \alpha_m \bar{x}^{(m)} = 0 \tag{5.1}$$

Linear independence:
If V_n is a linear vector space, then a finite nonempty set $\{\bar{x}^{(1)}, \bar{x}^{(2)}, \ldots, \bar{x}^{(m)}\}$ of vectors of V_n is said to be linearly independent if every relation of the form

$$\alpha_1 \bar{x}^{(1)} + \alpha_2 \bar{x}^{(2)} + \cdots + \alpha_m \bar{x}^{(m)} = 0$$

implies that $\alpha_i = 0$ for each $1 \le i \le m$.

It is to be noted here that if $\alpha_m \ne 0$ (and other scalars are zero), then from Equation 5.1, we have

$$\bar{x}^{(m)} = \sum_{j=1}^{m-1} \beta_j \bar{x}^{(j)}, \quad \beta_j = -\frac{\alpha_j}{\alpha_m}, \quad j = 1, 2, \ldots, (m-1) \tag{5.2}$$

So, we may define that the given vectors are linearly dependent if and only if one of them is a linear combination of the other vectors.

Inner product and norm:
We now introduce norm to measure the length of a vector or the difference between two vectors of a linear vector space and to measure the angle between two vectors, the concept of an inner product is introduced. Inner product is analogous to the scalar product of geometric vectors.

Norm of a vector:
If we consider V as a linear vector space over the real number field R, then a norm on vector space V is a function that transforms every element $x \in V$ into a real number denoted by $\|x\|$ such that $\|x\|$ satisfies the following conditions:

(1) $\|x\| \geq 0$ and $\|x\| = 0$ if $x = 0$

(2) $\|\alpha x\| = |\alpha| \|x\|$, $\alpha \in R$

(3) $\|x + y\| \leq \|x\| + \|y\|$, $x, y \in V$

Moreover, a linear vector space on which a norm can be defined is called a normed linear space.

Inner product:
Let us consider that F is a field of real numbers. Then, an inner product on a linear vector space V is defined as a mapping from $V \times V$ into F, which assigns to each ordered pair of vectors x, y in V, a scalar $<x, y>$ in F in such a way that

(1) $< x,y >=< y,x >$

(2) $< \alpha x + \beta y, z >= \alpha < x, z > + \beta < y, z >$

(3) $< x, x > \geq 0$ and $< x, x > = 0$ iff $x = 0$

for any $x, y, z \in V$ and $\alpha, \beta \in F$.

The linear space V is then said to be an inner product space with respect to that specified inner product defined on it and a norm is defined with every inner product by $\|x\| = \sqrt{< x, x >}$

Orthogonal systems of vectors:
Two vectors $\bar{x}$ and $\bar{y}$ in an inner product space I_n are called orthogonal if their inner product is zero, i.e., if $<\bar{x}, \bar{y}> = 0$. The orthogonality, in general, tells us about perpendicularity. If the vectors are nonzero, then orthogonality implies that the angle between the two vectors is $\pi/2$. A set of nonzero vectors $\bar{x}^{(1)}$, $\bar{x}^{(2)}, \ldots, \bar{x}^{(k)}$ or in short $\{\bar{x}^{(i)}\}$ is called an orthogonal set if any two vectors of the set are orthogonal to each other, i.e.,

$$< \bar{x}^{(i)}, \bar{x}^{(j)} >= 0 \quad \text{for} \quad i \neq j \tag{5.3}$$

A set of vectors $\bar{x}^{(1)}, \bar{x}^{(2)}, \ldots, \bar{x}^{(k)}$ is said to be orthonormal

$$\text{if} < \bar{x}^{(i)}, \bar{x}^{(j)} >= \delta_{ij} \tag{5.4}$$

where δ_{ij} is called the Kronecker delta (i.e., $\delta_{ij} = 0$ for $i \neq j$ and $\delta_{ij} = 1$ for $i = j$).

Accordingly, any nonzero set of orthogonal vectors may be converted into an orthonormal set by replacing each vector $\bar{x}^{(i)}$ with $\bar{x}^{(i)}/\|\bar{x}^{(i)}\|$.

Orthogonal system of functions:
A set of functions $\{\phi_i(x)\}$ is said to be orthogonal over a set of points $\{x_i\}$ with respect to a weight function $W(x)$ if

$$< \phi_j(x), \phi_k(x) >= 0, \quad j \neq k \tag{5.5}$$

where the inner product may be defined as

$$< \phi_j(x), \phi_k(x) >= \sum_{i=0}^{N} W(x_i)\phi_j(x_i)\phi_k(x_i)$$

On a closed interval [a,b], the set of functions $\{\phi_i(x)\}$ is said to be orthogonal with respect to the weight function $W(x)$ if Equation 5.4 holds, and in this case the inner product is defined as

$$< \phi_j(x), \phi_k(x) >= \int_a^b W(x)\phi_j(x)\phi_k(x)dx \tag{5.6}$$

The orthogonal functions $\{\phi_i(x)\}$ may now be orthonormalized by the relation

$$\frac{\phi_i(x)}{\| \phi_i(x) \|}$$ and is denoted by $\hat{\phi}_i(x)$ and so $< \hat{\phi}_i(x), \hat{\phi}_j(x) >= \delta_{ij}$.

Sequence of orthogonal polynomials:
A sequence of polynomials, viz., $P_0(x), P_1(x), P_2(x),\ldots$ (finite or infinite) is orthogonal if $P_i(x)$ are all orthogonal to each other and each $P_i(x)$ is a polynomial of exact *i*th degree. In other words, this can be written as:
   Sequence of polynomials $P_i(x)$, $i=0, 1, 2,\ldots$ are orthogonal if

$$(1) < P_i(x), P_j(x) >= 0 \quad \text{for} \quad i \neq j$$
$$(2) \ P_i(x) = C_i x^i + P_{i-1} \text{ for each } i \text{ with } C_i \neq 0$$

where P_{i-1} is a polynomial of degree less than *i*.
   Let us consider now an example with the functions $P_0(x)=1$, $P_1(x)=(x-1)$, and $P_2(x)=(x^2 - 2x +2/3)$, which form a sequence of orthogonal polynomials and these are orthogonal on [0,2] with respect to the weight function $W(x)=1$.

$$< P_0,P_1 >= \int_0^2 1.(x-1)dx = 0$$

$$< P_0,P_2 >= \int_0^2 1.(x^2 - 2x + 2/3)dx = 0$$

and

$$< P_1,P_2 >= \int_0^2 (x-1)(x^2 - 2x + 2/3)dx = 0$$

Therefore $\qquad\qquad < P_i(x), P_j(x) >= 0 \quad \text{for} \quad i \neq j$

Hence, P_0, P_1, P_2 are said to be orthogonal polynomials.

It may easily be proved that the set of functions $\{\sin\frac{n\pi x}{l}\}$, $n = 1, 2, \ldots$ is orthogonal in $[0, l]$ and we can find the corresponding orthonormal set.

Here, the norm of ϕ_i may be written as

$$= \left\|\sin\frac{i\pi x}{l}\right\|$$

$$= \left\{\int_0^l \sin^2\left(\frac{i\pi x}{l}\right)dx\right\}^{1/2}$$

$$= \left[\frac{1}{2}\int_0^l\left(1 - \cos\frac{2i\pi x}{l}\right)dx\right]^{1/2}$$

$$= \left[\frac{1}{2}\left(x - \frac{l}{2i\pi}\sin\frac{2i\pi x}{l}\right)_0^l\right]^{1/2} = \left(\frac{l}{2}\right)^{1/2} = \sqrt{\frac{l}{2}}$$

Therefore, orthonormal set $\hat{\phi}_i(x)$ is written as

$$\{\phi_i(x)\} = \left\{\frac{\phi_i}{\|\phi_i\|}\right\} = \left\{\sqrt{\frac{2}{l}}\sin\frac{i\pi x}{l}\right\}$$

5.2 Construction of Orthogonal Polynomials

It has been already pointed out that in one-dimensional case, there exits a three-term recurrence relation for generating the orthogonal polynomials. But for higher dimensions, we have to use the Gram–Schmidt method to generate them. In the following discussions, the three-term recurrence relation and the Gram–Schmidt orthogonalization procedure are given for one-dimensional case.

5.2.1 Three-Term Recurrence Relation

It is possible to construct a sequence of orthogonal polynomials using the following three-term recurrence relation:

$$\phi_{k+1}(x) = (x - e_k)\phi_k(x) - p_k\phi_{k-1}(x), \quad k = 0, 1, 2, \ldots \tag{5.7}$$

and

$$\phi_{-1} = 0$$

where
$$e_k = \frac{<x\phi_k, \phi_k>}{<\phi_k, \phi_k>} \quad \text{and} \quad p_k = \frac{<x\phi_k, \phi_{k-1}>}{<\phi_{k-1}, \phi_{k-1}>} \tag{5.8}$$

Here, it is worth noting that we get the sequence of orthogonal polynomials with leading coefficients as 1. If the leading coefficient of the sequence of orthogonal polynomials is required to be other than unity, then in general we can write the above three-term recurrence relation as

$$\phi_{k+1}(x) = (d_k x - e_k)\phi_k(x) - p_k\phi_{k-1}(x), \quad k = 0, 1, 2, \dots \tag{5.9}$$

where d_k, e_k, and p_k can be obtained from the orthogonality property.

5.2.2 Gram–Schmidt Orthogonalization Procedure

The above recurrence relation is valid only for polynomials. Now, let us suppose that we are given a set of functions $(f_i(x), i = 1, 2, \dots)$ in $[a,b]$. From this set of functions, we can construct appropriate orthogonal functions by using a well-known procedure known as the Gram–Schmidt orthogonalization process as follows:

$$\phi_1 = f_1$$
$$\phi_2 = f_2 - \alpha_{21}\phi_1$$
$$\phi_3 = f_3 - \alpha_{31}\phi_1 - \alpha_{32}\phi_2 \tag{5.10}$$
$$\dots\dots\dots\dots\dots\dots$$
$$\dots\dots\dots\dots\dots\dots$$

where

$$\alpha_{21} = \frac{<f_2, \phi_1>}{<\phi_1, \phi_1>}, \quad \alpha_{31} = \frac{<f_3, \phi_1>}{<\phi_1, \phi_1>}, \quad \alpha_{32} = \frac{<f_3, \phi_2>}{<\phi_2, \phi_2>} \quad \text{etc.} \tag{5.11}$$

In compact form, we can write the above procedure as

$$\phi_1 = f_1$$
$$\phi_i = f_i - \sum_{j=1}^{i-1} \alpha_{ij}\phi_j \tag{5.12}$$

and

$$\alpha_{ij} = \frac{<f_i,\phi_j>}{<\phi_j,\phi_j>} = \frac{\int_a^b W(x)f_i(x)\phi_j(x)dx}{\int_a^b W(x)\phi_j(x)\phi_j(x)dx} \tag{5.13}$$

The above procedure is valid only when the inner product exists for the interval $[a,b]$ with respect to the weight function $W(x)$. Now, we will show that the functions $f_1(x)=1$, $f_2(x)=x$, and $f_3(x)=x^2$ in the interval [0,2] can be made orthogonal by Gram–Schmidt process where the weight function $W(x)=1$.

For this, we will write as per the Gram–Schmidt procedure

$$\phi_1(x) = f_1(x) = 1$$

$$\phi_2(x) = f_2 - \alpha_{21}\phi_1 = x - \alpha_{21}\phi_1$$

$$\phi_3(x) = f_3 - \alpha_{31}\phi_1 - \alpha_{32}\phi_2 = x^2 - \alpha_{31}\phi_1 - \alpha_{32}\phi_2$$

The constant α_{21} may be found from Equation 5.11 as

$$\alpha_{21} = \frac{<x,\phi_1>}{<\phi_1,\phi_1>} = \frac{\int_0^2 x dx}{\int_0^2 dx} = 1$$

So, we may write the second orthogonal polynomial from above as

$$\phi_2(x) = (x-1)$$

Then the constants α_{31} and α_{32} may also be found similarly (using Equation 5.11) in the following way:

$$\alpha_{31} = \frac{<x^2,\phi_1>}{<\phi_1,\phi_1>} = \frac{\int_0^2 x^2 dx}{\int_0^2 dx} = \frac{4}{3}$$

and

$$\alpha_{32} = \frac{<x^2,\phi_2>}{<\phi_2,\phi_2>} = \frac{\int_0^2 x^2(x-1)dx}{\int_0^2 (x-1)^2 dx} = 2$$

Accordingly, the third orthogonal polynomial may be written as

$$\phi_3(x) = x^2 - \frac{4}{3}.(1) - 2(x-1) = \left(x^2 - 2x + \frac{2}{3}\right)$$

Thus, one can generate the orthogonal polynomials $\{\phi_i\}$ from the set of functions $\{f_i\}$.

5.2.3 Standard Orthogonal Polynomials

Some well-known orthogonal polynomials, viz., Legendre and Chebyshev polynomials, do exist and those have been used also in vibration problems, but these are mainly suitable for one-dimensional problems or to problems that can be solved in terms of one-dimensional coordinates. It is to be noted here that although the Gram–Schmidt orthogonalization procedure may be applied to find a set of orthogonal polynomials, we often use the well-known orthogonal polynomials also. This is out of the scope of this text. However, we just give here first few of these polynomials.

Legendre polynomials:
The few orthogonal polynomials described here are known as Legendre polynomials, which are defined in the interval $[-1,1]$ only:

$$\phi_0 = 1$$

$$\phi_1 = x$$

$$\phi_2 = x^2 - 1/3$$

$$\phi_3 = x^3 - \frac{3}{5}x, \text{ etc.}$$

Chebyshev polynomials:
These polynomials are also defined in the interval $[-1,1]$ and the following are few orthogonal polynomials known as Chebyshev polynomials with the weight function $= \frac{1}{\sqrt{1-x^2}}$

$$\phi_0 = 1$$
$$\phi_1 = x$$

$$\phi_2 = x^2 - \frac{1}{2}$$

5.3 Characteristic Orthogonal Polynomials

Although it is advantageous to apply the well-known method of Rayleigh–Ritz in various engineering problems, it is often difficult to obtain the meaningful deflection shape functions in the said method. So, a class of characteristic orthogonal polynomial (COPs) can be constructed using Gram–Schmidt process and then these polynomials are employed as deflection functions in the Rayleigh–Ritz method. The orthogonal nature of the polynomials makes the analysis simple and straightforward. Moreover, ill-conditions of the problem may also be avoided.

It is well understood now that the Rayleigh–Ritz method is a very powerful technique that can be used to predict the natural frequencies and mode shapes of vibrating structures. The method requires a linear combination of assumed deflection shapes of structures in free harmonic vibration that satisfy at least the geometrical boundary conditions of the vibrating structure. Expressions for the maximum kinetic and potential energies are obtained in terms of the arbitrary constants in the deflection expression. By equating the maximum potential and kinetic energies, it is possible to obtain an expression for the natural frequency of the structure. Applying the condition of stationarity of the natural frequencies at the natural modes, the variation of natural frequencies with respect to the arbitrary constants is equated to zero to obtain an eigenvalue problem (Meirovitch, 1967). Solution of this eigenvalue problem provides the natural frequencies and mode shapes of the system. The assumed deflection shapes were normally formulated by inspection and sometimes by trial and error until Bhat (1985a,b) proposed a systematic method of constructing such functions in the form of COPs. The restrictions on the series are the following:

1. They satisfy the geometrical boundary conditions.
2. They are complete.
3. They do not inherently violate the natural boundary conditions.

When the above conditions are met, the numerical solutions converge to the exact solution and it depends also on the number of terms taken in the admissible series. Different series types, viz., trigonometric, hyperbolic, polynomial, give different results for the same number of terms in the series and the efficiency of the solution will depend to some extent on the type of series chosen (Brown and Stone (1997)). Bhat (1985a) used the Gram–Schmidt orthogonalization procedure to generate the COPs for one dimension and showed that the orthogonal polynomials offered improved convergence and better results.

To use COPs in the Rayleigh–Ritz method for the study of vibration problems, first the orthogonal polynomials are generated over the domain of the structural member, satisfying the appropriate boundary conditions. After the first member function is constructed as the simplest polynomial that satisfies the boundary conditions, the higher members are constructed using the well-known Gram–Schmidt procedure (Wendroff (1961); Szego (1967); Freud (1971); Askey (1975); Chihara (1978)). The developed COPs are then used in the Rayleigh–Ritz method for the extraction of the vibration characteristics. Next, we discuss the one- and two-dimensional COPs.

5.4 Characteristic Orthogonal Polynomials in One Dimension

Here, the first member of the orthogonal polynomial set $\phi_1(x)$, which is function of a single variable say x, is chosen as the simplest polynomial of the least order that satisfies both the geometrical and the natural boundary conditions. In general, the first member is written in the form

$$\phi_1(x) = c_0 + c_1 x + c_2 x^2 + c_3 x^3 + \cdots \tag{5.14}$$

where the constants c_i can be found by applying the boundary condition of the problem. It is to be mentioned here that this function should satisfy at least the geometrical boundary condition. But, if the function also satisfies the natural boundary condition, then the resulting solution will be far better.

The other members of the orthogonal set in the interval $a \leq x \leq b$ are generated using Gram–Schmidt process as follows (Bhat (1985a)):

$$\begin{aligned} \phi_2(x) &= (x - B_2)\phi_1(x) \\ \phi_k(x) &= (x - B_k)\phi_{k-1}(x) - C_k\phi_{k-2}(x) \end{aligned} \tag{5.15}$$

where

$$\begin{aligned} B_k &= \frac{\int_a^b x[\phi_{k-1}(x)]^2 W(x)dx}{\int_a^b [\phi_{k-1}(x)]^2 W(x)dx} \\ C_k &= \frac{\int_a^b x\phi_{k-1}(x)\phi_{k-2}(x)W(x)dx}{\int_a^b [\phi_{k-2}(x)]^2 W(x)dx} \end{aligned} \tag{5.16}$$

and $W(x)$ is the weight function. The polynomials $\phi_k(x)$ satisfy the orthogonality condition.

$$\int_a^b W(x)\phi_k(x)\phi_l(x)dx = 0 \quad \text{if} \quad k \neq l \tag{5.17}$$

$$\neq 0 \quad \text{if} \quad k = l$$

It is to be noted that even though $\phi_1(x)$ satisfies all the boundary conditions, both geometric and natural, the other members of the orthogonal set satisfy only the geometric boundary conditions.

Dickinson and Blasio (1986) and Kim and Dickinson (1987) further modified the method of Bhat (1985a) for the generation of COPs as

$$\phi_{k+1}(x) = \{f(x) - B_k\}\phi_k(x) - C_k\phi_{k-1}(x), \quad k = 1, 2, \ldots \tag{5.18}$$

where $f(x)$ is a generating function chosen to ensure that the higher-order orthogonal functions also satisfy all the boundary conditions.

Similarly, the vibration of structures such as rectangular plates whose deflection can be assumed in the form of product of one-dimensional COPs may be analyzed using one-dimensional COPs. Various studies related to the vibration problems of plates, beams, and with other complicating effects are handled using COPs. Those are surveyed in a recent paper by Chakraverty et al. (1999). Some details of those studies will be incorporated in later chapters.

5.4.1 Beam in [0,1] with Both Ends Clamped

Let us consider a beam in [0,1] with both ends clamped as shown in Figure 5.1.

Then, the boundary conditions are given by

$$\phi_1(0) = \phi_1'(0) = \phi_1(1) = \phi_1'(1) = 0 \tag{5.19}$$

Now, consider the deflection function

$$\phi_1(x) = a_0 + a_1 x + a_2 x^2 + a_3 x^3 + a_4 x^4 \tag{5.20}$$

Substituting the boundary conditions in the above equation, the coefficients a_i are determined to yield

$$\phi_1(x) = a_4(x^2 - 2x^3 + x^4) \tag{5.21}$$

The coefficient a_4 can be appropriately chosen so as to normalize $\phi_1(x)$ such that

$$\int_0^1 (\phi_1(x))^2 dx = 1 \tag{5.22}$$

Then, the Rayleigh–Ritz method may be applied as discussed earlier.

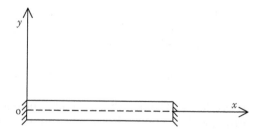

FIGURE 5.1
Beam with both ends clamped.

5.4.2 Condensation with COPs in the Vibration Problems

This section gives refinement in the above method by introducing condensation in the Rayleigh–Ritz method using the characteristic orthonormal polynomials for the deflection function. The method has been discussed here by taking a simple problem of vibration of a cantilever beam [0,1], i.e., of unit length where the left end $(x=0)$ is clamped and the right end $(x=1)$ is free. This can be solved by using one-dimensional COPs and taking only two terms for the deflection function, the efficiency and reliability of the method, is shown below.

For the present problem, let us choose two functions

$$f_1 = x^2 \quad \text{and} \quad f_2 = x^3 \tag{5.23}$$

It can easily be seen that the geometrical boundary conditions for the problem are satisfied by the above two functions. The COPs for the above functions in [0,1] can be constructed by the procedure as mentioned above and let us denote these by ϕ_1 and ϕ_2. For the generation of these orthogonal polynomials, the Gram–Schmidt orthogonalizing procedure is used and ϕ_1 and ϕ_2 are written as

$$\phi_1 = f_1$$
$$\phi_2 = f_2 - \alpha_{21}\phi_1$$

$$\alpha_{21} = \frac{\langle f_2, \phi_1 \rangle}{\langle \phi_1, \phi_1 \rangle}$$

where

$$\langle \phi_i, \phi_i \rangle = \int_0^1 (\phi_i)^2 dx$$

and is known as inner product as defined in the previous sections.

Using the above definition of the inner product with the orthogonalizing procedure, the orthogonal polynomials ϕ_1 and ϕ_2 may be written as

$$\phi_1 = x^2 \quad \text{and} \quad \phi_2 = 5x^2 - 6x^3 \tag{5.24}$$

The corresponding orthonormal polynomials may be generated by dividing each ϕ_i by its norm, i.e.

$$\hat{\phi}_i = \frac{\phi_i}{\|\phi_i\|} = \frac{\phi_i}{\langle \phi_i, \phi_i \rangle^{1/2}}$$

Then, the orthonormal polynomials can be constructed by the above procedure as

$$\hat{\phi}_1 = \sqrt{5}x^2 \quad \text{and} \quad \hat{\phi}_2 = \sqrt{7}[x^2(5 - 6x)] \tag{5.25}$$

Using these orthonormal polynomials, one can easily write down the mass "M" and stiffness "K" matrices for the cantilever beam as follows, in which the mass matrix will be an identity matrix due to the orthonormality of the shape functions (Meirovitch (1980) and Singh and Chakraverty (1994)):

$$M = \rho A \begin{bmatrix} 1 & 0 \\ 0 & 1 \end{bmatrix} \quad \text{and} \quad K = EI \begin{bmatrix} 20 & -16\sqrt{35} \\ -16\sqrt{35} & 1204 \end{bmatrix} \tag{5.26}$$

where ρ, A, E, I are material density, cross-sectional area, Young's modulus, and moment of inertia, respectively. Then, the equation of motion (in terms of Rayleigh–Ritz method) may be written as

$$([K] - \lambda[M])_{2\times2}\{q\} = \{0\} \tag{5.27}$$

where the frequency parameter is given by

$$\omega = \sqrt{\frac{\lambda EI}{\rho A}}$$

The orthonormal functions will render the problem (Equation 5.27) into a standard eigenvalue problem rather than a generalized one. This conversion to standard eigenvalue problem has been first discussed in Chakraverty (1992) and Singh and Chakraverty (1994) and they first time named their orthogonal polynomials as the BCOPs. However, details of those will be addressed in the next chapter. The straightforward solution of the above eigenvalue problem will give $\omega_1 = 3.5327$ and $\omega_2 = 34.8069$.

Now, the problem may be solved by reducing the order of the above eigenvalue problem by condensation (Meirovitch (1980)).

To reduce the order of the general eigenvalue problem of order $n \times n$

$$([K] - \lambda[M])_{n\times n}\{q\} = \{0\} \tag{5.28}$$

in the Rayleigh–Ritz method, it is possible to express the mass and stiffness matrices to lower order (as required) say $p \times p$, $p < n$ as

$$[M]_{n\times n} = \begin{bmatrix} M11_{p\times p} & M12_{p\times(n-p)} \\ M21_{(n-p)\times p} & M22_{(n-p)\times(n-p)} \end{bmatrix}$$

$$[K]_{n\times n} = \begin{bmatrix} K11_{p\times p} & K12_{p\times(n-p)} \\ K21_{(n-p)\times p} & K22_{(n-p)\times(n-p)} \end{bmatrix}$$

and

$$\{q\} = \left\{ \begin{matrix} q_p \\ q_{(n-p)} \end{matrix} \right\} = \left\{ \begin{matrix} q_1 \\ q_2 \end{matrix} \right\} \tag{5.29}$$

The condensation is done from Equation 5.28, using above mass and stiffness matrices and Equation 5.29 by writing as follows:

$$\begin{bmatrix} (K11_{p \times p} - \lambda M11_{p \times p}) & (K12_{p \times (n-p)} - \lambda M12_{p \times (n-p)}) \\ (K21_{(n-p) \times p} - \lambda M21_{(n-p) \times p}) & (K22_{(n-p) \times (n-p)} - \lambda M22_{(n-p) \times (n-p)}) \end{bmatrix} \left\{ \begin{matrix} q_1 \\ q_2 \end{matrix} \right\} = \left\{ \begin{matrix} 0 \\ 0 \end{matrix} \right\}$$

Ignoring the inertia terms in the second row of equations, we have

$$\{q_2\} = \left[-(K22_{(n-p) \times (n-p)})^{-1} (K21_{(n-p) \times p}) \right] \{q_1\}$$

Thus, $\{q\}$ may be written as

$$\{q\} = \left\{ \begin{matrix} 1 \\ -(K22_{(n-p) \times (n-p)})^{-1} (K21_{(n-p) \times p}) \end{matrix} \right\} \{q_1\}$$

Finally, the $p \times p$ condensed eigenvalue problem can be written as

$$\left\{ \begin{matrix} 1 \\ -(K22_{(n-p) \times (n-p)})^{-1} (K21_{(n-p) \times p}) \end{matrix} \right\}^T [K - \lambda M]$$

$$\times \left\{ \begin{matrix} 1 \\ -(K22_{(n-p) \times (n-p)})^{-1} (K21_{(n-p) \times p}) \end{matrix} \right\} = \{0\} \tag{5.30}$$

The condensed eigenvalue problem for the present example is then given by

$$\left\{ \begin{matrix} 1 \\ -K22^{-1}K21 \end{matrix} \right\}^{-1} \begin{bmatrix} K11 - \lambda & K12 \\ K21 & K22 - \lambda \end{bmatrix} \left\{ \begin{matrix} 1 \\ -K22^{-1}K21 \end{matrix} \right\} = \{0\} \tag{5.31}$$

Putting the values of K_{11}, K_{12}, K_{21}, and K_{22} from Equation 5.26 in the present condensation scheme given in Equation 5.31 and reducing it to a single coordinate and solving, the first frequency is obtained as $\omega_1 = 3.5328$. The exact solution of the problem may be found out and the first frequency is given by 3.5156.

If the orthonormal polynomials as developed are not used, i.e., if the polynomials mentioned in Equation 5.23 are used, then the mass and stiffness matrices will be given by

$$M = \rho A \begin{bmatrix} 1/5 & 1/6 \\ 1/6 & 1/7 \end{bmatrix} \quad \text{and} \quad K = EI \begin{bmatrix} 4 & 6 \\ 6 & 12 \end{bmatrix} \tag{5.32}$$

TABLE 5.1

Comparison and Evaluation for the Proposed Solutions

Frequency	Exact Solution	Non-Orthogonal Polynomial Solution	Orthonormal Polynomial Solution	Condensed Non-Orthonormal Polynomial Solution	Condensed Orthonormal Polynomial Solution
1	3.5156	3.5329	3.5327	3.80528	3.5328
2	22.0336	34.8881	34.8069	—	—

and then use of Equation 5.27 gives the frequencies as $\omega_1 = 3.5329$ and $\omega_2 = 34.8881$. Moreover, if we use the condensation as discussed above with the non-orthonormal polynomials, this gives the first frequency ω_1 as 3.80528. Thus, it is seen that the exact frequency is more closer by using the characteristic orthonormal polynomials with the present condensation method. In this regard, Table 5.1 gives easy comparison and evaluation for the above solutions:

It is to be noted from Table 5.1 that the orthonormal polynomial solution gives almost the same accuracy for the first frequency with less computation in comparison with the non-orthogonal/orthonormal polynomial solution. However, for the second frequency, the orthogonal/orthonormal polynomial solution has been obtained by taking only two terms of the series. So, the agreement for the second frequency is not good. The accuracy for this frequency and for the higher frequencies may be obtained by taking more terms in the orthonormal polynomial series as obtained in plate problems with noncondensed orthogonal polynomials by Chakraverty (1992) and Singh and Chakraverty (1994). The proposed condensed orthonormal polynomial solution gives a higher accuracy for the first frequency when compared with the condensed non-orthogonal/orthonormal polynomial solution.

The powerfulness and reliability of the COPs when used with condensation in the Rayleigh–Ritz method are now very well understood. Here, only the first two orthonormal polynomials have been taken and then it is shown how the condensation with the COPs may reduce the labor when one is interested in smaller number of vibration characteristics than in solving the whole eigenvalue problem. The orthonormal polynomials convert the mass matrix to an Identity matrix as shown in Chakraverty (1992) and Singh and Chakravery (1994). This also makes the condensation method more easy to handle and comparatively more efficient as discussed in Chapter 6.

5.4.3 Free Flexural Vibration of Rectangular Plate Using One-Dimensional Characteristic Orthogonal Polynomials

The vibration of structures such as rectangular plates whose deflection can be assumed in the form of product of one-dimensional COPs may be analyzed using one-dimensional COPs. Various studies related to the vibration problems of plates, beams, and with other complicating effects are handled using COPs. Those are surveyed in a recent paper by Chakraverty et al. (1999). Some details of those studies will be incorporated in later chapters.

However, here the problem of rectangular plates studied by Bhat (1985a) will be discussed for clarity. Accordingly, let us assume a rectangular plate with plate dimensions a and b, whose deflection due to free vibration may be expressed in terms of the COPs along x and y directions as

$$W(x,y) = \sum_m \sum_n A_{mn} \phi_m(x) \psi_n(y) \tag{5.33}$$

where $x = \xi/a$ and $y = \eta/b$, with ξ and η the coordinates along the two sides of the rectangular plate. Corresponding maximum kinetic energy T_{max} and maximum strain energy U_{max} then may, respectively, be written as

$$T_{max} = \frac{1}{2}\rho hab\omega^2 \int_0^1 \int_0^1 W^2(x,y)dxdy \tag{5.34}$$

$$U_{max} = \frac{1}{2}Dab \int_0^1 \int_0^1 \left[W_{xx}^2 + \beta^4 W_{yy}^2 + 2\nu\beta^2 W_{xx} W_{yy} + 2(1-\nu)\beta^2 W_{xy}^2 dxdy \right] \tag{5.35}$$

where
  $\beta = a/b$ is the side ratio
  ρ is the density of the material of the plate
  ν is the Poisson's ratio
  D is the flexural rigidity of the plate

Putting the deflection function (Equation 5.33) into the above maximum kinetic and strain energies and minimizing the Rayleigh quotient with respect to the constants A_{ij}, we get the eigenvalue problem

$$\sum_m \sum_n \left[C_{mnij} - \lambda E_{mi}^{(0,0)} F_{nj}^{(0,0)} \right] A_{mn} = 0 \tag{5.36}$$

where

$$C_{mnij} = E_{mi}^{(2,2)} F_{nj}^{(0,0)} + \beta^4 E_{mi}^{(0,0)} F_{nj}^{(2,2)} + \nu\beta^2 \left[E_{mi}^{(0,2)} F_{nj}^{(2,0)} + E_{mi}^{(2,0)} F_{nj}^{(0,2)} \right]$$
$$+ 2(1-\nu)\beta^2 E_{mi}^{(1,1)} F_{nj}^{(1,1)} \tag{5.37}$$

$$E_{mi}^{(p,q)} = \int_0^1 \frac{d^p\phi_m}{dx^p} \frac{d^q\phi_i}{dx^q} dx \tag{5.38}$$

$$F_{nj}^{(p,q)} = \int_0^1 \frac{d^p \psi_n}{dy^p} \frac{d^q \psi_j}{dy^q} dy \tag{5.39}$$

$$m, n, i, j = 1, 2, 3, \dots, p, q = 0, 1, 2$$

and

$$\lambda = \frac{\rho h \omega^2 a^4}{D} \tag{5.40}$$

Solution of the eigenvalue Equation 5.36 will give the vibration characteristics. It may be noted that the COPs as proposed yield superior solution as given in Bhat (1985a). The detailed results using orthogonal polynomials for various geometries will be included in later chapters.

In the above, a two-dimensional problem, viz., rectangular plate, has been analyzed using one-dimensional COPs. As such, studies related to one-dimensional problem, viz., beams and with various complicating effects, have been carried out by different researchers using one-dimensional orthogonal polynomials. On the other hand, two-dimensional and three-dimensional problems with regular geometry with and without complicating effects such as cutouts, anisotropic, laminated, and plates with different theories have been handled using one-dimensional orthogonal polynomials also by various investigators. As mentioned earlier, Chakraverty et al. (1999) reviewed the detailed studies in this respect.

5.5 Characteristic Orthogonal Polynomials in Two Dimensions

In view of the above, one may conclude that simple geometries, in particular one-dimensional problems and higher-dimensional problems whose deflection can be assumed as the product of one-dimensional functions, may be investigated by one-dimensional COPs. But, plates with bit complex shapes such as triangular and polygonal do require two-dimensional assumed functions. Accordingly, Bhat (1986b, 1987) proposed two-dimensional characteristic orthogonal polynomials (2DCOPs) to study vibration behavior of non-rectangular plates in the Rayleigh–Ritz method. Polynomials are generated by using again the Gram–Schmidt orthogonalization procedure in two variables (Askey (1975)). Polynomials in two dimensions may be generated by using simple monomials in a specific order

$$1, x, y, x^2, xy, y^2, \dots, x^i, x^{i-1}y, \dots, x^{i-k}y^k \tag{5.41}$$

The first member of the orthogonal polynomial set is so constructed that it satisfies at least the geometrical boundary condition of the plate. As in one-dimensional case, here also as proposed by Bhat (1987), the first member, $\phi_1(x, y)$ is taken to be of the form

$$\phi_1(x,y) = c_0 + c_1 x + c_2 y + c_3 x^2 + c_4 xy + c_5 y^2 + \cdots \tag{5.42}$$

where the constants c_i are to be determined by the specified boundary conditions of the plate. The steps about how the higher members of the orthogonal set and the first polynomial are actually generated will be shown in the following sections for an example of triangular plate as given in Bhat (1986b, 1987).

Accordingly, for a triangular plate, there will be six geometrical boundary conditions and so the first member of the orthogonal set may be written in the form

$$\phi_1(x, y) = c_0 + c_1 x + c_2 y + c_3 x^2 + c_4 xy + c_5 y^2 + c_6 x^3 \tag{5.43}$$

to satisfy all the six geometrical boundary conditions and the constants c_i can be found using those boundary conditions. After getting the first orthogonal polynomial, we will write as in Bhat (1987) the second orthogonal polynomial, i.e.,

$$\phi_2(x, y) = x^2 y + a_{21} \phi_1(x, y) \tag{5.44}$$

where $x^2 y$ is the next term of the monomials given in Equation 5.41 after the last monomial used in $\phi_1(x, y)$, i.e., the monomial $x^2 y$ appears after x^3 in the sequence. Here, a_{21} is the constant to be determined by orthogonal property. The orthogonality condition between the polynomials ϕ_1 and ϕ_2 requires that

$$\iint_R \eta(x, y) \phi_1(x, y) \phi_2(x, y) dx dy = \begin{cases} 0 & \text{if } i \neq j \\ \mu_i & \text{if } i = j \end{cases} \tag{5.45}$$

where $\eta(x, y)$ is a weight function and μ_i are constants. Now, using Equations 5.44 and 5.45, the constant a_{21} may easily be obtained as

$$a_{21} = -\frac{\iint_R \eta(x, y)(x^2 y)\phi_1(x, y)dx dy}{\iint_R \eta(x, y)[\phi_1(x, y)]^2 dx dy} \tag{5.46}$$

where the integration in the numerator and denominator is to be evaluated exactly if possible or by any numerical methods. After finding a_{21} from Equation 5.46 and then putting this in Equation 5.44, we finally get the second orthogonal polynomial $\phi_2(x, y)$. The next COP may now be written as

$$\phi_3(x, y) = xy^2 + a_{31}\phi_1(x, y) + a_{32}\phi_2(x, y) \tag{5.47}$$

Again, the constants a_{31} and a_{32} can be found in a similar way as above using the orthogonality property and those are obtained as

$$a_{31} = -\frac{\iint\limits_R \eta(x, y)(xy^2)\phi_1(x, y)dxdy}{\iint\limits_R \eta(x, y)[\phi_1(x, y)]^2 dxdy} \tag{5.48}$$

and

$$a_{32} = -\frac{\iint\limits_R \eta(x, y)(xy^2)\phi_2(x, y)dxdy}{\iint\limits_R \eta(x, y)[\phi_2(x, y)]^2 dxdy} \tag{5.49}$$

The above procedure of constructing the orthogonal polynomials may be continued until the required number of orthogonal polynomials is found. It is worth mentioning here that (till this point of discussion) we do not have any recurrence scheme for two or higher dimension as in one dimension, so we have to orthogonalize with all the previously generated orthogonal polynomials in the set, although we will introduce a recurrence scheme of higher dimensions for generating BCOPs in the next chapter. Bhat's procedure of generating the 2DCOPs may now be written in general form in the following manner.

For this, we will first designate the monomials in Equation 5.41 as

$$f_0 = 1; f_1 = x; f_2 = y; f_3 = x^2; f_4 = xy; f_5 = y^2; f_6 = x^3; f_7 = x^2y; f_8 = xy^2; \ldots \tag{5.50}$$

Let us also suppose that there are m boundary conditions to be satisfied, then the first polynomial will be

$$\phi_1(x, y) = \sum_{i=0}^{m} c_i f_i \tag{5.51}$$

where c_i, $i = 0, 1, \ldots, m$ are the constants to be determined by the specified boundary conditions. The two-dimensional COPs of the whole set may now be written as

$$\phi_i(x, y) = f_{i+m-1} + \sum_{k=1}^{i-1} a_{ik}\phi_k, \quad i = 2, 3, \ldots \tag{5.52}$$

where the constants a_{ik} of Equation 5.52 may easily be obtained by using the orthogonal property as

$$a_{ik} = -\frac{\iint_R \eta(x,y)(f_{i+m-1})\phi_k(x,y)dxdy}{\iint_R \eta(x,y)[\phi_k(x,y)]^2 dxdy} \tag{5.53}$$

After finding the double integrals in the numerator and denominator, one can get the constants a_{ik}. On substitution of these constants in Equation 5.52, the whole set of the required COPs is generated. For uniform and homogeneous plate, the weight function $\eta(x,y)$ in the above integrals is taken as unity. But, for variable thickness, non-homogeneity and other complicating cases of the plates, the weight functions have to be written accordingly, as will be shown in the later chapters.

5.5.1 Free Flexural Vibration of Triangular Plate Using Two-Dimensional Characteristic Orthogonal Polynomials

Again, the example of triangular plate will be discussed here from Bhat (1987). We define the deflection function of a triangular plate vibrating freely in terms of COPs as

$$W(x,y) = \sum_{i=1}^{n} c_i \phi_i(x,y) \tag{5.54}$$

where $x = \zeta/a$ and $y = \tau/a$, a and b are the sides of the triangle as shown in Figure 5.2 and ζ and τ are the Cartesian coordinates. Then, the maximum kinetic and strain energies may be written as

$$T_{max} = \frac{1}{2}\rho h a^2 \omega^2 \iint_R W^2(x,y)dxdy \tag{5.55}$$

$$U_{max} = \frac{1}{2}Da^2 \iint_R \left[W_{xx}^2 + W_{yy}^2 + 2\nu W_{xx}W_{yy} + 2(1-\nu)W_{xy}^2 \right] dxdy \tag{5.56}$$

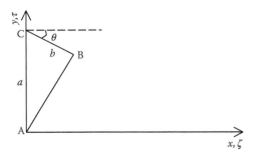

FIGURE 5.2
Triangular plate geometry.

Putting the deflection function from Equation 5.54 into the above maximum energies (Equations 5.55 and 5.56) and minimizing the Rayleigh quotient with respect to the coefficients c_i, we have the eigenvalue problem as (Bhat (1987)),

$$\sum_i [E_{ij} - \lambda F_{ij}^{(0,0,0,0)}]c_i = 0 \tag{5.57}$$

where $E_{ij} = F_{ij}^{(2,0,2,0)} + F_{ij}^{(0,2,0,2)} + \nu[F_{ij}^{(0,2,2,0)} + F_{ij}^{(2,0,0,2)}] + 2(1 - \nu)F_{ij}^{(1,1,1,1)}$ (5.58)

$$F_{ij}^{(m,n,r,s)} = \iint_R \left[\frac{d^m \phi_i(x,y)}{dx^m}\right]\left[\frac{d^n \phi_i(x,y)}{dy^n}\right]\left[\frac{d^r \phi_j(x,y)}{dx^r}\right]\left[\frac{d^s \phi_j(x,y)}{dy^s}\right]dxdy \tag{5.59}$$

where

$$\lambda = \frac{\rho h a^4 \omega^2}{D}$$
$$i,j = 1, 2, 3, \ldots$$
$$m, n, r, s = 0, 1, 2$$

Solution of the above eigenvalue problem will extract the vibration characteristics, viz., the natural frequencies and mode shapes of the said plate problem.

Different vibration problems, viz., plates with polygonal shape and with other complicating effects, have been solved by using the above-discussed 2DCOPs. Again, those are surveyed in Chakraverty et al. (1999). In Chapter 6, another form of the method, viz., BCOPs, will be introduced for efficient solution and simple computer implementation, which will be shown to render the vibration problem to a simpler form. The rest of the chapters in this book will use those in the solution for vibration behavior of complex plate problems.

Bibliography

Askey, R.A. 1975. *Theory and Application of Special Functions*, New York, Academic Press.

Bhat, R.B. 1985a. Natural frequencies of rectangular plates using characteristic orthogonal polynomials in Rayleigh-Ritz method. *Journal of Sound and Vibration*, 102(4): 493–499.

Bhat, R.B. 1985b. Plate deflections using orthogonal polynomials. *Journal of Engineering Mechanics, Trans. ASCE*, 111(11): 1301–1309.

Bhat, R.B. 1986a. Application of Rayleigh-Ritz method on the finite element model of a structure as a means of coordinate reduction. *Journal of Sound and Vibration*, 108(2): 355–356.

Bhat, R.B. 1986b. Natural frequencies of triangular plates using characteristic orthogonal polynomials in Rayleigh-Ritz method. *Proceedings of the Symposium on Recent Trends in Aeroelasticity, Structures and Structural Dynamics*, Gainesville, FL.

Bhat, R.B. 1987. Flexural vibration of polygonal plates using characteristic orthogonal polynomials in two variables. *Journal of Sound and Vibration*, 114(165): 71.

Bhat, R.B. and Chakraverty, S. 2003. *Numerical Analysis in Engineering*, Narosa, New Delhi.

Brown, R.E. and Stone, M.A. 1997. On the use of polynomial series with the Rayleigh-Ritz method. *Composite Structures*, 39(3–4): 191–196.

Chakraverty, S. 1992. Numerical solution of vibration of plates, PhD Thesis. University of Roorkee (Now IIT, Roorkee), Roorkee, India.

Chakraverty, S., Bhat, R.B., and Stiharu, I. 1999. Recent research on vibration of structures using boundary characteristic orthogonal polynomials in the Rayleigh-Ritz method. *The Shock and Vibration Digest*, 31(3): 187–194.

Chihara, T.S. 1978. *An Introduction to Orthogonal Polynomials*, Gordon and Breach, London.

Dickinson, S.M. and Blasio, A. Di. 1986. On the use of orthogonal polynomials in the Rayleigh-Ritz method for the study of the flexural vibration and buckling of isotropic and orthotropic rectangular plates. *Journal of Sound and Vibration*, 108(1): 51–62.

Freud, G. 1971. *Orthogonal Polynomials*, Pergamon Press, Oxford.

Kim, C.S. 1988. *The Vibration of Beams and Plates Studied Using Orthogonal Polynomials*, PhD Thesis. The University of Western Ontario, Canada.

Kim, C.S. and Dickinson, S.M. 1987. The flexural vibration of rectangular plates with point supports. *Journal of Sound and Vibration*, 117(2): 249–261.

Liew, K.M. 1990. *The Development of 2-D Orthogonal Polynomials for Vibration of Plates*, PhD Thesis. National University of Singapore, Singapore.

Meirovitch, L. 1967. *Analytical Methods in Vibrations*, Macmillan Co., New York.

Meirovitch, L. 1980. *Computational Methods in Structural Dynamics*, Sijthoff and Noordhoff, the Netherlands.

Singh, B. and Chakraverty, S. 1991. Transverse vibration of completely free elliptic and circular plates using orthogonal polynomials in Rayleigh-Ritz method. *International Journal of Mechanical Sciences*, 33(9): 741–751.

Singh, B. and Chakraverty, S. 1994. Boundary characteristic orthogonal polynomials in numerical approximation. *Communications in Numerical Methods in Engineering*, 10: 1027–1043.

Szego, G. 1967. *Orthogonal Polynomials*, 3rd ed., American Mathematical Society, New York.

Wendroff, B. 1961. On orthogonal polynomials. *American Mathematical Society*, 12: 554–555.

6

Boundary Characteristic Orthogonal Polynomials (BCOPs) in Vibration of Plates

The importance of orthogonal polynomials is well known in numerical approximation problems. Methods of least squares, Gauss quadrature, interpolation, and orthogonal collocation are a few areas in which the advantage of orthogonal polynomials is well documented. Recently, the orthogonal polynomials are being used in the Rayleigh–Ritz method to study vibration problems, particularly the vibration of plates. Furthermore, Chakraverty et al. (1999) provided a survey of literature that used these polynomials in vibration of structures. Chapter 5 addressed the vibration problems using characteristic orthogonal polynomials (COPs) in one and two dimensions. It is already known that for generating such polynomials in one variable, we can begin with a linearly independent set, such as $\{1, x, x^2, x^3, \ldots\}$ and apply the Gram–Schmidt process to generate them over the desired interval with an appropriate weight function. Fortunately, the whole procedure in this case is greatly simplified owing to the existence of three-term recurrence relation. This saves a lot of computation. As such, the problems in one dimension are well handled accordingly, as discussed in Chapter 5. When certain boundary conditions are to be satisfied, one can start with the set $f(x)$ $\{1, x, x^2, x^3, \ldots\}$ where $f(x)$ satisfies the boundary conditions of the problem.

Quite a large variety of one-dimensional problems have been solved using orthogonal polynomials. Subsequently, the application was extended to two-dimensional problems, either using one- or two-dimensional orthogonal polynomials, as briefed in Chapter 5. Chakraverty (1992) and Singh and Chakraverty (1994) generated and used two-dimensional orthogonal polynomials to study vibration problems of plates for a variety of geometries. These polynomials were named as boundary characteristic orthogonal polynomials (BCOPs) by Chakraverty (1992) and Singh and Chakraverty (1994). In this chapter, the method and procedure for generating these polynomials, specific to the vibration of plates, will be discussed. Furthermore, a type of recurrence scheme that was generated for the two-dimensional BCOPs by Bhat et al. (1998) will be addressed along with the illustration of some discussions on the refinement of the procedure.

The use of BCOPs in the Rayleigh–Ritz method for the study of vibration problems involves three steps. The first step is the generation of orthogonal polynomials over the domain occupied by the structural member and satisfying the appropriate boundary conditions. After the construction of the first member function that satisfies the boundary conditions, the higher members are constructed using the well-known Gram–Schmidt procedure. The second step is the use of these polynomials in the Rayleigh–Ritz method that renders the problem into a standard eigenvalue problem, rather than as a generalized one. This is the main advantage and beauty of this method, which converts the problem into a simple, straightforward, and computationally efficient form. The third and the last step involves the solution of this eigenvalue problem to get the vibration characteristics.

6.1 Boundary Characteristic Orthogonal Polynomials in n Dimensions

Let us consider that $\bar{x} = (x_1, x_2, x_3, \ldots, x_n)$ is a member in an n-dimensional domain D_n. We denote the boundary of D_n as ∂D_n. Then, as in Chapter 5, we can describe the inner product of the two functions $F_1(\bar{x})$ and $F_2(\bar{x})$ defined over the n-dimensional domain D_n as

$$\langle F_1(\bar{x}), F_2(\bar{x}) \rangle = \int_{D_n} \psi(\bar{x}) F_1(\bar{x}) F_2(\bar{x}) dD_n \tag{6.1}$$

where $\psi(\bar{x})$ is a suitably chosen weight function defined over the domain considered. Subsequently, the norm of a function for the above inner product can be written as

$$\| F_1 \| = \langle F_1, F_1 \rangle^{\frac{1}{2}} = \left[\int_{D_n} \psi(\bar{x}) \{ F_1(\bar{x}) \}^2 dD_n \right]^{\frac{1}{2}} \tag{6.2}$$

If we consider that a function $g(\bar{x})$ satisfies the desired boundary conditions on ∂D_n, then for generating the orthogonal sequence, we can start with the following set:

$$\{ g(\bar{x}) f_i(\bar{x}) \}, \quad i = 1, 2, 3, \ldots \tag{6.3}$$

where the functions $f_i(\bar{x})$ are linearly independent over the domain D_n. It is important to note that each member of the set in Equation 6.3 does satisfy the same boundary conditions. Thus, the final orthogonal polynomials generated are the BCOPs. Now, to generate the orthogonal polynomial sequence denoted as $\{\phi_i\}$, the Gram–Schmidt process is utilized:

$$\phi_1 = gf_1 \tag{6.4}$$

$$\phi_i = gf_i - \sum_{j=1}^{i-1} \alpha_{ij}\phi_j, \quad i = 2, 3, 4, \ldots \tag{6.5}$$

where

$$\alpha_{ij} = \frac{\langle gf_i, \phi_j \rangle}{\langle \phi_j, \phi_j \rangle} \tag{6.6}$$

The orthonormal polynomial sets may be obtained by dividing each ϕ_i by its norm, which can be written as

$$\hat{\phi}_i = \frac{\phi_i}{\|\phi_i\|} = \frac{\phi_i}{\langle \phi_i, \phi_i \rangle^{1/2}} \tag{6.7}$$

It may be pointed out that the problem of generating orthogonal polynomials is inherently an unstable process. Hence, small errors introduced at any stage of computation may lead to loss of significant digits and the results may become entirely absurd after a few steps. So, it is advisable to carry out all the calculations in double-precision arithmetic. There exists a variety of problems in different engineering applications, such as in fluid mechanics, elasticity, diffusion, electromagnetic theory, and electrostatic theory, where the problem can be formulated in the form of differential equation or in terms of variational formulation. If the BCOPs are used as the basic functions in the Ritz or other variational methods, the final equations are simplified greatly because the cross-terms vanish for the orthogonality. Thus, the solution of the problem becomes straightforward and easy. Moreover, the setback of ill-conditioning, which is invariably present in such problems, is also reduced to a great extent. Hence, in that case a few approximations are sufficient to provide reasonably accurate solutions. This book only includes the problems of vibration of plates, where the use of BCOPs has proved to be extremely powerful and useful. Accordingly, the generation of BCOPs in two variables will be discussed in general.

6.2 Boundary Characteristic Orthogonal Polynomials in Two Dimensions

To generate the BCOPs for any geometry, we start with a linearly independent set of the form

$$F_i(x, y) = g(x, y)f_i(x, y), \quad i = 1, 2, 3, \ldots \tag{6.8}$$

where

  $g(x, y)$ satisfies the essential boundary conditions of the problem
  $f_i(x, y)$ are suitably chosen linearly independent functions

These will be described in the subsequent discussions in this chapter. Let us define the inner product of two functions $p(x, y)$ and $q(x, y)$ in two dimensions, similar to n dimensions, over the domain occupied by the plate as follows:

$$\langle p, q \rangle = \iint_R \psi(x, y) p(x, y) q(x, y) dx dy \tag{6.9}$$

where $\psi(x,y)$ is a suitably chosen weight function depending on the problem. Now, the norm of a function $p(x,y)$ in two-dimensional case may be written as

$$\| p(x, y) \| = \langle p(x, y), p(x, y) \rangle^{1/2} = \left[\iint_R \psi(x, y) \{ p(x, y) \}^2 dx dy \right]^{1/2} \tag{6.10}$$

The orthogonal functions $\phi_i(x, y)$ can then be generated using the Gram–Schmidt orthogonalization process in the two-dimensional case as

$$\phi_1(x, y) = F_1(x, y) \tag{6.11}$$

$$\phi_i(x, y) = F_i(x, y) - \sum_{j=1}^{i-1} \alpha_{ij} \phi_j(x, y), \quad i = 2, 3, 4, \ldots \tag{6.12}$$

where the constants α_{ij} are determined by

$$\alpha_{ij} = \frac{\langle F_i(x, y), \phi_j(x, y) \rangle}{\langle \phi_j(x, y), \phi_j(x, y) \rangle}, \tag{6.13}$$

$$j = 1, 2, \ldots, (i - 1); \quad i = 2, 3, \ldots$$

Accordingly, the orthonormal functions can be obtained as

$$\hat{\phi}_i(x, y) = \frac{\phi_i(x, y)}{\| \phi_i(x, y) \|}, \quad i = 1, 2, 3, \ldots \tag{6.14}$$

Now, we will go a little further with respect to two dimensions, taking $i = 1, 2,$ and 3 and show the generation of the BCOPs, each of which satisfies the essential boundary conditions. Thus, from Equation 6.8, one can derive

$$F_1(x, y) = g(x, y) f_1(x, y) \tag{6.15}$$

$$F_2(x, y) = g(x, y) f_2(x, y) \tag{6.16}$$

and

$$F_3(x,y) = g(x,y) f_3(x,y) \tag{6.17}$$

where the function $g(x,y)$ satisfies the essential boundary condition. The orthogonal functions $\phi_1(x,y)$, $\phi_2(x,y)$, and $\phi_3(x,y)$ can then be generated using the Gram–Schmidt orthogonalization process as

$$\phi_1(x,y) = F_1(x,y) \tag{6.18}$$

$$\phi_2(x,y) = F_2(x,y) - \alpha_{21}\phi_1(x,y) \tag{6.19}$$

$$\phi_3(x,y) = F_3(x,y) - \alpha_{31}\phi_1(x,y) - \alpha_{32}\phi_2(x,y) \tag{6.20}$$

The above orthogonal polynomials can be written in terms of the functions $g(x,y)$ and $f_i(x,y)$ by utilizing Equations 6.15 through 6.17 as

$$\phi_1(x,y) = g(x,y) f_1(x,y) \tag{6.21}$$

$$\begin{aligned} \phi_2(x,y) &= g(x,y)f_2(x,y) - \alpha_{21}g(x,y)f_1(x,y) \\ &= g(x,y)[f_2(x,y) - \alpha_{21}f_1(x,y)] \end{aligned} \tag{6.22}$$

$$\begin{aligned} \phi_3(x,y) &= g(x,y)f_3(x,y) - \alpha_{31}g(x,y)f_1(x,y) - \alpha_{32}g(x,y)[f_2(x,y) - \alpha_{21}f_1(x,y)] \\ &= g(x,y)[f_3(x,y) - \alpha_{31}f_1(x,y) - \alpha_{32}(f_2(x,y) - \alpha_{21}f_1(x,y))] \end{aligned} \tag{6.23}$$

Thus, all the BCOPs generated will also satisfy the essential boundary conditions owing to the existence of the function $g(x,y)$ in each of the Equations 6.21 through 6.23. The constants α_{21}, α_{31}, and α_{32} are obtained from Equation 6.13. For example, α_{21} is given as

$$\begin{aligned} \alpha_{21} &= \frac{\langle F_2(x,y), \phi_1(x,y)\rangle}{\langle \phi_1(x,y), \phi_1(x,y)\rangle} \\ &= \frac{\langle g(x,y)f_2(x,y), g(x,y)f_1(x,y)\rangle}{\langle g(x,y)f_1(x,y), g(x,y)f_1(x,y)\rangle} \\ &= \frac{\iint\limits_{R} \psi(x,y)\{g(x,y)\}^2 f_1(x,y)f_2(x,y)dxdy}{\iint\limits_{R} \psi(x,y)\{g(x,y)\}^2 \{f_1(x,y)\}^2 dxdy} \end{aligned} \tag{6.24}$$

The functions $\phi_1(x,y)$, $\phi_2(x,y)$, and $\phi_3(x,y)$ can be expressed in terms of $f_1(x,y)$, $f_2(x,y)$, $f_3(x,y)$ from Equations 6.21 through 6.23 as

$$\phi_1(x,y) = g(x,y)\beta_{11}f_1(x,y) \tag{6.25}$$

$$\phi_2(x,y) = g(x,y)[\beta_{21}f_1(x,y) + \beta_{22}f_2(x,y)] \tag{6.26}$$

$$\phi_3(x,y) = g(x,y)[\beta_{31}f_1(x,y) + \beta_{32}f_2(x,y) + \beta_{33}f_3(x,y)] \tag{6.27}$$

where

$$\beta_{11} = 1, \beta_{21} = -\alpha_{21}, \beta_{22} = 1, \beta_{31} = \alpha_{21}\alpha_{32} - \alpha_{31}$$
$$\beta_{32} = -\alpha_{32}, \text{ and } \beta_{33} = 1 \tag{6.28}$$

Thus, the functions ϕ_i and $\hat{\phi}_i$ can be expressed, in general, in terms of $f_1, f_2, f_3, \ldots$ such that

$$\phi_i(x,y) = g(x,y) \sum_{j=1}^{i} \beta_{ij} f_j \tag{6.29}$$

and

$$\hat{\phi}_i = g(x,y) \sum_{j=1}^{i} \hat{\beta}_{ij} f_j \tag{6.30}$$

where the constants β_{ij} and $\hat{\beta}_{ij}$ can be written similarly as in Equation 6.28.

Now, we will proceed to outline the steps to generate the BCOPs for some important domains in R^2 (two dimensions), which are generally required for the study of vibration of plates. To use these BCOPs in the vibration of plates, one can consider only the following three cases to satisfy the geometrical boundary conditions

1. $\phi_i = 0$ on $\partial D, i = 1, 2, 3, \ldots$ \hfill (6.31)

2. $\phi_i = 0$ and $\dfrac{\partial \phi_i}{\partial n} = 0$ on $\partial D, i = 1, 2, 3, \ldots$ \hfill (6.32)

3. No conditions are imposed on ϕ_i \hfill (6.33)

where $\frac{\partial \phi_i}{\partial n}$ designates the normal derivatives of ϕ_i at the edges.

6.2.1 Elliptic and Circular Domains

Let us assume that the domain D occupied by an ellipse (Figure 6.1) is defined by

$$D = \left\{ (x,y) \text{ such that } \frac{x^2}{a^2} + \frac{y^2}{b^2} \leq 1 \right\}$$

where a and b are the semimajor and semiminor axes of the elliptic domain. When $a = b$, a particular case of circle will follow. Next, we will introduce a variable u so that

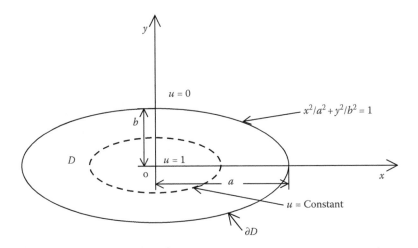

FIGURE 6.1
Geometry of the elliptic plate.

$$u = 1 - \left(\frac{x^2}{a^2} + \frac{y^2}{b^2}\right), \quad 0 \le u \le 1 \tag{6.34}$$

Thus, $u = 0$ would give the boundary of the elliptic domain, and $u = 1$ gives its center. Moreover, the curves $u = $ constant designate the concentric ellipses, as shown in Figure 6.1. For generating the BCOPs similar to that discussed earlier, we will start with the set $\{g(x,y)f_i(x,y)\}$, where

$$g(x, y) = u^s = \left[1 - \left(\frac{x^2}{a^2} + \frac{y^2}{b^2}\right)\right]^s \tag{6.35}$$

and f_i can be taken as the sequence

$$f_i = \{1, x, y, x^2, xy, y^2, \ldots\} \tag{6.36}$$

It is to be noted that we are having three geometrical boundary conditions, clamped, simply supported, and free, to be satisfied, as $s = 2$, 1, or 0, respectively, for vibration of elliptic and circular plates. We may note accordingly that

1. If $s = 2$, $g(x,y) = 0$ and $\frac{\partial g}{\partial n} = 0$ on ∂D i.e., clamped condition $\qquad$ (6.37)

2. If $s = 1$, $g(x,y) = 0$ on ∂D i.e., simply supported condition $\qquad$ (6.38)

3. If $s = 0$, $g(x,y) = 1$ on ∂D i.e., free condition $\qquad$ (6.39)

When the integrals in evaluating the inner products in the mentioned Gram–Schmidt process are determined either by closed form if possible or by better

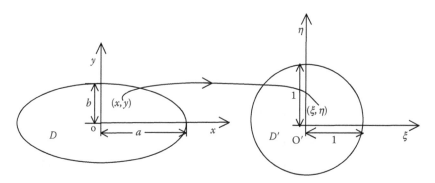

FIGURE 6.2
Mapping of the elliptic domain into a circular domain.

numerical methods, then the BCOPs may be generated and used in the Rayleigh–Ritz method for further analysis in the vibration behavior.

One can also generate the BCOPs by transforming the x–y coordinates to ξ–η coordinates such that

$$\xi = \frac{x}{a}, \quad \eta = \frac{y}{b} \tag{6.40}$$

so as to map the elliptic domain into a circular domain of radius unity, as shown in Figure 6.2. In that case, we can have the BCOPs in the circular domain (which is easier) and the rest of the calculations can be carried out in terms of x and y by using the relation (Equation 6.40). The variable u in this transformed case will be

$$u = 1 - (\xi^2 + \eta^2) = 1 - r^2 \tag{6.41}$$

where r is the radius of the unit circle in ξ–η plane. The BCOPs are then generated with the following set:

$$u^s\{1, \xi, \eta, \xi^2, \xi\eta, \eta^2, \ldots\} \tag{6.42}$$

$$(1 - r^2)^s\{1, \xi, \eta, \xi^2, \xi\eta, \eta^2, \ldots\} \tag{6.43}$$

The term in the first bracket again controls the boundary conditions at the edges, viz., clamped, simply supported, and free according as $s=2$, 1, or 0, respectively.

6.2.2 Triangular Domains

Let us assume that a general triangular domain as shown in Figure 6.3 is defined by the parameters a, b, and c. Here again, we will first transform the general triangular domain D in x–y plane to a standard right-angled triangle D' in ξ–η plane by the relation

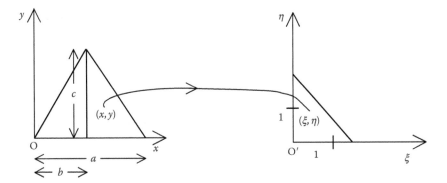

FIGURE 6.3
Mapping of the general triangular domain into a standard right-angled triangle.

$$x = a\xi + b\eta \tag{6.44}$$

$$y = c\eta \tag{6.45}$$

As in Section 6.2.1, it is now sufficient to generate the BCOPs in this standard right-angled triangle and then to convert them into the original coordinates using the above relations (Equations 6.44 and 6.45). Subsequently, one can use the BCOPs in further analysis of the vibration behavior.

For instance, we have three sides for a triangular region. Hence, we can have any of the three boundary conditions, viz., clamped, simply supported, and free, on each of the sides. One can define the three sides of the standard right-angled triangle from Figure 6.3 by

$$\xi = 0, \quad \eta = 0, \quad \text{and} \quad \xi + \eta = 1 \tag{6.46}$$

Accordingly, we can write

$$g(\xi,\eta) = \xi^r \eta^s [1 - (\xi + \eta)]^t \tag{6.47}$$

where r, s, and t can take any of the values 2, 1, or 0 becoming the clamped, simply supported, and free edges of the domain, respectively. For example, if $r = 2$, the approximating functions and their normal derivatives vanish for $\xi = 0$. If $r = 1$, the functions vanish for $\xi = 0$, and the side is free if $r = 0$. The parameters s and t will have similar interpretations for the boundary conditions as discussed for the parameter r.

Thus, we may start with the set

$$g(\xi,\eta)\{1,\xi,\eta,\xi^2,\xi\eta,\eta^2, \ldots\} \tag{6.48}$$

$$= \xi^r \eta^s (1 - \xi - \eta)^t \{1,\xi,\eta,\xi^2,\xi\eta,\eta^2, \ldots\} \tag{6.49}$$

to use them in the Gram–Schmidt orthogonalizing process as discussed earlier for generating the BCOPs in the triangular domain.

6.2.3 Parallelogram Domains

We define parallelogram domain using three parameters, viz., a, b, and α, as shown in Figure 6.4. Again, the general parallelogram domain D in x–y plane is transformed into a unit-square domain D' in ξ–η plane by the following relations:

$$x = a\xi + (b\cos\alpha)\eta \tag{6.50}$$

$$y = (b\sin\alpha)\eta \tag{6.51}$$

The BCOPs are first generated in ξ–η coordinates and then we can revert back to the original coordinates x–y, using the above relations (Equations 6.50 and 6.51). There are four sides in the square or parallelogram domain. Accordingly, each may have any of the clamped, simply supported, or free-boundary conditions. Here also, one can define four sides of the square as

$$\xi = 0, \quad \xi = 1, \quad \eta = 0, \quad \text{and} \quad \eta = 1 \tag{6.52}$$

Accordingly, we can again have

$$g(\xi,\eta) = \xi^{p}(1-\xi)^{q}\eta^{r}(1-\eta)^{s} \tag{6.53}$$

The boundary conditions of the sides are handled by plugging the values of 2, 1, or 0 to the parameters p, q, r, and s. For clarity, $p = 2$ will force the side $\xi = 0$ to be clamped; $p = 1$ designates the side $\xi = 0$ to be simply supported; and $p = 0$ signifies that the side $\xi = 0$ is free. Now, we may start with the following set to generate the BCOPs:

$$g(\xi,\eta)\{1,\xi,\eta,\xi^{2},\xi\eta,\eta^{2},\dots\} \tag{6.54}$$

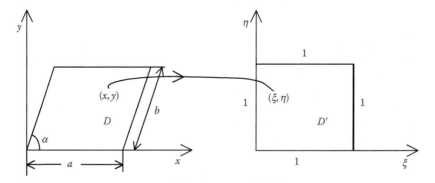

FIGURE 6.4
Mapping of the parallelogram domain into a unit-square domain.

$$\xi^p(1-\xi)^q\eta^r(1-\eta)^s\{1,\xi,\eta,\xi^2,\xi\eta,\eta^2,\ldots\} \tag{6.55}$$

It is worth mentioning here that the procedure for generating BCOPs in rectangular and square domains will be similar to the parallelogram domain as discussed earlier. In the rectangular case, one may take $\alpha = 90°$, while for the square case, we can take $a = b$ along with $\alpha = 90°$ and use the method as discussed above.

In the earlier discussions, the procedures of generating BCOPs in simple geometrical domains have been shown. The same methodology can be adopted for more complicated geometries and problems in two dimensions, such as annular circular, annular elliptic, parallelogram with hole or triangle with hole, and mixed boundary condition at the edges. The approach can also be extended for the generation of BCOPs in three dimensions. The BCOP generations also depend on the computation of the integrals in the inner product while using the Gram–Schmidt procedure, and also in the Rayleigh–Ritz method. For regular domains, closed-form integrals exist (which will be reported in the subsequent chapters). However, if this is not possible, then one may use some good numerical formulae. Similarly, in three-dimensional cases, in simple geometries such as ellipsoid, tetrahedron, and parallelepiped, one can find the integrals in closed form, but for complex geometries the numerical methods must be used. In this connection, one may refer to the paper by Singh and Chakraverty (1999), where some BCOPs are reported for regular geometrical domains. These BCOPs can be directly used for their specific problems.

We may now refer to Equations 5.183, 5.185, and 5.186 and owing to the orthogonality (actually orthonormal) one can easily observe the conversion of the generalized eigenvalue problem to a standard eigenvalue problem

$$\sum_{j=1}^{N}\left(a_{ij} - \lambda^2\delta_{ij}\right)c_j = 0, \quad i = 1, 2, \ldots, N \tag{6.56}$$

Thus, Equation 5.186 will become

$$b_{ij} = \iint_{R'} H\phi_i\phi_j \, \mathrm{d}X\mathrm{d}Y = \delta_{ij} \tag{6.57}$$

where δ_{ij} is the Kronecker delta and is defined as

$$\delta_{ij} = \begin{cases} 0, & \text{if } i \neq j \\ 1, & \text{if } i = j \end{cases} \tag{6.58}$$

This simplifies the problem to a great extent, and the convergence of the results is also improved significantly.

It is important to mention that (as pointed out earlier) for one-dimensional COPs, we have the three-term recurrence relation to generate them.

However, no such recurrence relation was employed in constructing the two-dimensional COPs/BCOPs. They were generated by orthogonalizing with all the previously generated orthogonal polynomials. Quite recently, Bhat et al. (1998) proposed a recurrence scheme for the generation of two-dimensional BCOPs involving three classes of polynomials, which will be discussed in the next section.

6.3   Recurrence Scheme for the BCOPs

In this scheme, the polynomials are first organized into the following classes forming a pyramid given by

$$
\phi_1^{(1)}
$$

$$
\phi_2^{(2)} \qquad \phi_3^{(2)}
$$

$$
\phi_4^{(3)} \qquad \phi_5^{(3)} \qquad \phi_6^{(3)}
$$

$$
\phi_7^{(4)} \qquad \phi_8^{(4)} \qquad \phi_9^{(4)} \qquad \phi_{10}^{(4)}
$$

$$
\phi_{\{\frac{n(n+1)}{2}-(n-1)\}}^{(n)} \qquad \phi_{\{\frac{n(n+1)}{2}-(n-2)\}}^{(n)} \qquad \cdots \qquad \phi_{\{\frac{n(n+1)}{2}-1\}}^{(n)} \qquad \phi_{\{\frac{n(n+1)}{2}\}}^{(n)}
$$

where superscript in $\phi_i^{(j)}$ denotes the class number j to which it belongs.

Now, in general, class j will have j orthogonal polynomials, which can be generated by the following recurrence scheme given by Bhat et al. (1998):

$$
\phi_i^{(j)} = \begin{cases} X\phi_{\{i-(j-1)\}}^{(j-1)} - \sum\limits_{k=l}^{i-1}\alpha_{ik}\phi_k^{(j)} - \sum\limits_{k=m}^{l-1}\alpha_{ik}\phi_k^{(j-1)} - \sum\limits_{k=p}^{m-1}\alpha_{ik}\phi_k^{(j-2)}; \ i=\{L-(j-1)\},\ldots,\{L-1\}, \\[4mm] Y\phi_{(i-j)}^{(j-1)} - \sum\limits_{k=l}^{i-1}\alpha_{ik}\phi_k^{(j)} - \sum\limits_{k=m}^{l-1}\alpha_{ik}\phi_k^{(j-1)} - \sum\limits_{k=p}^{m-1}\alpha_{ik}\phi_k^{(j-2)}; \ i=L, \end{cases}
$$

$$
j=2,3,\ldots,N \tag{6.59}
$$

where

$l = \left\{\frac{j(j+1)}{2} - (j-1)\right\}$, describes the first orthogonal polynomial of the jth class

$L = \frac{j(j+1)}{2}$, describes the last orthogonal polynomial of the jth class

$m = \left\{\frac{(j-1)j}{2} - (j-2)\right\}$, describes the first orthogonal polynomial of the $(j-1)$th class

$p = \left\{\frac{(j-2)(j-1)}{2} - (j-3)\right\} \geq 0$, describes the first orthogonal polynomial of the $(j-2)$th class

and

$$
\alpha_{ik} =
\begin{cases}
\dfrac{\langle X\phi_{(i-(j-1))}^{(j-1)}, \phi_k^{(r)}\rangle}{\langle \phi_k^{(r)}, \phi_k^{(r)}\rangle}, & i = \{L - (j-1)\}, \dots, \{L - 1\} \\[2ex]
\dfrac{\langle Y\phi_{(i-j)}^{(j-1)}, \phi_k^{(r)}\rangle}{\langle \phi_k^{(r)}, \phi_k^{(r)}\rangle}, & i = L
\end{cases}
\tag{6.60}
$$

where $r = j$, $(j-1)$, and $(j-2)$, corresponding to α_{ik} appearing in the first, second, and third summation terms, respectively, in Equation 6.59. These three summation terms in Equation 6.59 clearly demonstrate that all the orthogonal polynomials in the jth class can be constructed using only those previously generated in that class so far, and those of the previous two classes, i.e., $(j-1)$th and $(j-2)$th classes.

Thus, this method saves the undue labor of orthogonalizing with all the previous orthogonal polynomials. As reported by Bhat et al. (1998), this algorithm makes an efficient and straightforward generation, when compared with the previous methods and the execution time is also greatly reduced.

6.3.1 Recurrence Relations for Multidimensional Orthogonal Polynomials

For one-dimensional orthogonal polynomials $\{\phi_k(x)\}$, we have the three-term recurrence relation as follows:

$$
\phi_{k+1}(x) = (d_k x + e_k)\phi_k(x) + p_k\phi_{k-1}(x), \quad k = 0, 1, 2, \dots
\tag{6.61}
$$

where the coefficients d_k, e_k, p_k, $k = 0, 1, 2, \dots$ can be determined using the orthogonality property.

A similar three-term recurrence relation among the n-variables of the orthogonal polynomials was reported by Kowalski (1982a, 1982b). To develop such a recurrence scheme that can be numerically implemented for two variables, some preliminaries are discussed as follows:

Let $\prod_n^\infty$ be a vector space of all polynomials with real coefficients in n variables, and let $\prod_n^k$ be its subspace of polynomials whose total degree in n variables is not greater than k, then

$$
\dim \prod_n^k = \sum_{i=0}^k r_n^i = \binom{n+k}{k}
$$

where $r_n^k = \binom{n+k-1}{k}$ is the number of monomials in this basis whose degree is equal to k.

If a basis in $\prod_n^\infty$ is denoted by $\{\phi_i^k\}_{k=0,\, i=1}^{\infty,\, r_n^k}$, where each polynomial is of the degree indicated by its superscript, then we can define

$$\overline{\phi}_k(x) = \left[\phi_1^k(x), \phi_2^k(x), \dots, \phi_{r_n^k}^k(x)\right]^T \tag{6.62}$$

and

$$\overline{x\phi_k}(x) = \left[x_1\overline{\phi}_k(x)^T \,|\, x_2\overline{\phi}_k(x)^T \,|\, \dots \,|\, x_n\overline{\phi}_k(x)^T\right]^T \tag{6.63}$$

where

$$x = (x_1, x_2, \dots, x_n) \in R^n, \quad k = 0, 1, \dots$$

A recurrence formula with respect to $\overline{\phi}_{k-1}, \overline{\phi}_k$, and $\overline{\phi}_{k+1}$ can be written as

$$\begin{aligned} \overline{x\phi_k} &= A_k\overline{\phi}_{k+1} + B_k\overline{\phi}_k + C_k\overline{\phi}_{k-1} \\ \overline{\phi}_{k+1} &= D_k\overline{x\phi_k} + E_k\overline{\phi}_k + G_k\overline{\phi}_{k-1}, \quad k = 0, 1, \dots \end{aligned} \tag{6.64}$$

where $A_k, B_k, C_k, D_k, E_k,$ and G_k are matrices with

$$A_k\colon nr_n^k \times r_n^{k+1},\ B_k\colon nr_n^k \times r_n^k,\ C_k\colon nr_n^k \times r_n^{k-1}$$

$$D_k\colon r_n^{k+1} \times nr_n^k,\ E_k\colon r_n^{k+1} \times r_n^k,\ G_k\colon r_n^{k+1} \times r_n^{k-1}$$

and

$$\overline{\phi}_{-1} = 0, \quad C_0 = G_0 = 0$$

Further, if relations (Equation 6.64) hold true, then

$$\left. \begin{aligned} D_k A_k &= I \\ E_k &= -D_k B_k \\ G_k &= -D_k C_k \end{aligned} \right\} \tag{6.65}$$

6.3.2   Kowalski's Relations in Two Dimensions

It is interesting to note that in each class of the polynomials, the number of orthogonal polynomials is equal to the class number. For example, class number 2 contains two orthogonal polynomials that can be obtained as

$$\phi_2^{(2)} = F_2 = X\phi_1^{(1)} - \alpha_{21}\phi_1^{(1)} \tag{6.66}$$

and

$$\phi_3^{(2)} = F_3 = Y\phi_1^{(1)} - \alpha_{31}\phi_1^{(1)} - \alpha_{32}\phi_2^{(2)} \tag{6.67}$$

where

$$\alpha_{21} = \frac{\langle X\phi_1^{(1)}, \phi_1^{(1)} \rangle}{\langle \phi_1^{(1)}, \phi_1^{(1)} \rangle}$$

$$\alpha_{31} = \frac{\langle Y\phi_1^{(1)}, \phi_1^{(1)} \rangle}{\langle \phi_1^{(1)}, \phi_1^{(1)} \rangle}, \quad \alpha_{32} = \frac{\langle Y\phi_1^{(1)}, \phi_2^{(2)} \rangle}{\langle \phi_2^{(2)}, \phi_2^{(2)} \rangle},$$

and so on. It is to be noted here that the above relations (Equations 6.66 and 6.67) are the same as Kowalski's relations given in Equation 6.64, where

$$A_k = \begin{bmatrix} 1 & 0 \\ \alpha_{32} & 1 \end{bmatrix}, \quad B_k = \begin{bmatrix} \alpha_{21} \\ \alpha_{31} \end{bmatrix}$$

$$D_k = \begin{bmatrix} 1 & 0 \\ -\alpha_{32} & 1 \end{bmatrix}, \quad E_k = \begin{bmatrix} -\alpha_{21} \\ -\alpha_{31} + \alpha_{32}\alpha_{21} \end{bmatrix}$$

and

$$C_k = G_k = 0$$

Moreover, the above matrices also satisfy Equation 6.65. In general, class j will have j orthogonal polynomials that can be generated by the recurrence scheme given in Equation 6.59.

6.3.3 Matrix Form of Kowalski's Relations

Starting with the first polynomial, for example, $\phi_1^{(k)}$, we may write the polynomials in the next class by using Equation 6.64 as

$$\left\{ \begin{matrix} \phi_1^{(k+1)} \\ \phi_2^{(k+1)} \end{matrix} \right\} = \begin{bmatrix} 1 & 0 \\ D_{21} & 1 \end{bmatrix} \left\{ \begin{matrix} x\phi_1^{(k)} \\ y\phi_1^{(k)} \end{matrix} \right\} + \begin{bmatrix} E_{11} \\ E_{21} \end{bmatrix} \left\{ \phi_1^{(k)} \right\} \tag{6.68}$$

From the above matrix equation, the polynomials can be written as

$$\phi_1^{(k+1)} = x\phi_1^{(k)} + E_{11}\phi_1^{(k)} \tag{6.69}$$

$$\phi_2^{(k+1)} = D_{21}x\phi_1^{(k)} + y\phi_1^{(k)} + E_{21}\phi_1^{(k)} \tag{6.70}$$

Using Equation 6.69, we can obtain

$$\phi_2^{(k+1)} = y\phi_1^{(k)} + D_{21}\phi_1^{(k+1)} + (E_{21} - D_{21}E_{11})\phi_1^{(k)} \tag{6.71}$$

The constants D_{21}, E_{11}, E_{21} can be easily obtained by using orthogonality relations.

Similarly, we can write

$$\left\{\begin{array}{c} \phi_1^{(k+2)} \\ \phi_2^{(k+2)} \\ \phi_3^{(k+2)} \end{array}\right\} = \begin{bmatrix} 1 & 0 & 0 \\ D_{21} & 1 & 0 \\ D_{31} & D_{32} & 1 \end{bmatrix} \left\{\begin{array}{c} x\phi_1^{(k+1)} \\ x\phi_2^{(k+1)} \\ y\phi_2^{(k+1)} \end{array}\right\} + \begin{bmatrix} E_{11} & 0 \\ E_{21} & E_{22} \\ E_{31} & E_{32} \end{bmatrix} \left\{\begin{array}{c} \phi_1^{(k+1)} \\ \phi_2^{(k+1)} \end{array}\right\}$$

$$+ \begin{bmatrix} F_{11} \\ F_{21} \\ F_{31} \end{bmatrix} \left\{\phi_1^{(k)}\right\} \tag{6.72}$$

The three polynomials in this step may be written as

$$\phi_1^{(k+2)} = x\phi_1^{(k+1)} + E_{11}\phi_1^{(k+1)} + F_{11}\phi_1^{(k)} \tag{6.73}$$

$$\phi_2^{(k+2)} = D_{21}x\phi_1^{(k+1)} + x\phi_2^{(k+1)} + E_{21}\phi_1^{(k+1)} + E_{22}\phi_2^{(k+1)} + F_{21}\phi_1^{(k)} \tag{6.74}$$

Using Equation 6.73, we obtain

$$\phi_2^{(k+2)} = x\phi_2^{(k+1)} + D_{21}\phi_1^{(k+2)} + (E_{21} - D_{21}E_{11})\phi_1^{(k+1)} + E_{22}\phi_2^{(k+1)}$$
$$+ (F_{21} - D_{21}F_{11})\phi_1^{(k)} \tag{6.75}$$

Similarly, we can write the polynomial $\phi_3^{(k+2)}$ after some calculation as

$$\phi_3^{(k+2)} = y\phi_2^{(k+1)} + D_{32}\phi_2^{(k+2)} + (D_{31} - D_{32}D_{21})\phi_1^{(k+2)}$$
$$+ [E_{31} - D_{31}E_{11} - D_{32}(E_{21} - D_{21}E_{11})]\phi_1^{(k+1)} + (E_{32} - D_{32}E_{22})\phi_2^{(k+1)}$$
$$+ [F_{31} - D_{31}F_{11} - D_{32}(F_{21} - D_{21}F_{11})\phi_1^{(k)}] \tag{6.76}$$

Again, one can obtain the next step in the matrix form as

$$\left\{\begin{array}{c} \phi_1^{(k+3)} \\ \phi_2^{(k+3)} \\ \phi_3^{(k+3)} \\ \phi_4^{(k+3)} \end{array}\right\} = \begin{bmatrix} 1 & 0 & 0 & 0 \\ D_{21} & 1 & 0 & 0 \\ D_{31} & D_{32} & 1 & 0 \\ D_{41} & D_{42} & D_{43} & 1 \end{bmatrix} \left\{\begin{array}{c} x\phi_1^{(k+2)} \\ x\phi_2^{(k+2)} \\ x\phi_3^{(k+2)} \\ y\phi_3^{(k+2)} \end{array}\right\}$$

$$+ \begin{bmatrix} E_{11} & 0 & 0 \\ E_{21} & E_{22} & 0 \\ E_{31} & E_{32} & E_{33} \\ E_{41} & E_{42} & E_{43} \end{bmatrix} \left\{\begin{array}{c} \phi_1^{(k+2)} \\ \phi_2^{(k+2)} \\ \phi_3^{(k+2)} \end{array}\right\} + \begin{bmatrix} F_{11} & 0 \\ F_{21} & F_{22} \\ F_{31} & F_{32} \\ F_{41} & F_{42} \end{bmatrix} \left\{\begin{array}{c} \phi_1^{(k+1)} \\ \phi_2^{(k+1)} \end{array}\right\} \tag{6.77}$$

Thus, we can arrive at the following polynomials from the above matrix equations after little adjustments and calculations as:

$$\phi_1^{(k+3)} = x\phi_1^{(k+2)} + E_{11}\phi_1^{(k+2)} + F_{11}\phi_1^{(k+1)} \tag{6.78}$$

$$\phi_2^{(k+3)} = x\phi_2^{(k+2)} + D_{21}\phi_1^{(k+3)} + (E_{21} - D_{21}E_{11})\phi_1^{(k+2)} + E_{22}\phi_2^{(k+2)}$$
$$+ (F_{21} - D_{21}F_{11})\phi_1^{(k+1)} + F_{22}\phi_2^{(k+1)} \tag{6.79}$$

$$\phi_3^{(k+3)} = x\phi_3^{(k+2)} + D_{32}\phi_2^{(k+3)} + (D_{31} - D_{32}D_{21})\phi_1^{(k+3)}$$
$$+ [E_{31} - D_{31}E_{11} - D_{32}(E_{21} - D_{21}E_{11})]\phi_1^{(k+2)} + (E_{32} - D_{32}E_{22})\phi_2^{(k+2)} + E_{33}\phi_3^{(k+2)}$$
$$+ [F_{31} - D_{31}F_{11} - D_{32}(F_{21} - D_{21}F_{11})]\phi_1^{(k+1)} + [F_{32} - D_{32}F_{22}]\phi_2^{(k+1)} \tag{6.80}$$

$$\phi_4^{(k+3)} = y\phi_3^{(k+2)} + D_{43}\phi_3^{(k+3)} + (D_{42} - D_{43}D_{32})\phi_2^{(k+3)}$$
$$+ [D_{41} - D_{42}D_{21} - D_{43}(D_{31} - D_{32}D_{21})]\phi_1^{(k+3)} + [E_{41} - D_{41}E_{11}$$
$$- D_{42}(E_{21} - D_{21}E_{11}) - D_{43}\{E_{31} - D_{31}E_{11} - D_{32}(E_{21} - D_{21}E_{11})\}]\phi_1^{(k+2)}$$
$$+ [E_{42} - D_{42}E_{22} - D_{43}(E_{32} - D_{32}E_{22})]\phi_2^{(k+2)} + [E_{43} - D_{43}E_{33}]\phi_3^{(k+2)}$$
$$+ [F_{41} - D_{41}F_{11} - D_{42}(F_{21} - D_{21}F_{11}) - D_{43}\{F_{31} - D_{31}F_{11} - D_{32}(F_{21} - D_{21}F_{11})\}]\phi_1^{(k+1)}$$
$$+ [F_{42} - D_{42}F_{22} - D_{43}(F_{32} - D_{32}F_{22})]\phi_2^{(k+1)} \tag{6.81}$$

The above constants in the matrix equations can easily be written in terms of α_{ij}, as shown in Section 6.3.2.

6.3.4   BCOPs in Terms of the Original Functions

In this section, we will first consider that functions $f_1 = 1, f_2 = x, f_3 = y, f_4 = x^2$, $f_5 = xy, \ldots$ are given and the function satisfying boundary conditions is denoted by g. Let, the $\phi_1, \phi_2, \phi_3, \ldots$ are the BCOPs generated by the recurrence scheme as mentioned earlier. As such, the first six BCOPs can be written in terms of the original constants, $f_1, f_2, f_3, \ldots$. This would lead to easy computer implementation to generate the BCOPs. Accordingly, the first six BCOPs are

$$\phi_1 = gf_1 \tag{6.82}$$

$$\phi_2 = x\phi_1 - \alpha_{21}\phi_1 \tag{6.83}$$

$$\phi_3 = y\phi_1 - \alpha_{31}\phi_1 - \alpha_{32}\phi_2 \tag{6.84}$$

$$\phi_4 = x\phi_2 - \alpha_{41}\phi_1 - \alpha_{42}\phi_2 - \alpha_{43}\phi_3 \tag{6.85}$$

$$\phi_5 = x\phi_3 - \alpha_{51}\phi_1 - \alpha_{52}\phi_2 - \alpha_{53}\phi_3 - \alpha_{54}\phi_4 \tag{6.86}$$

$$\phi_6 = y\phi_3 - \alpha_{61}\phi_1 - \alpha_{62}\phi_2 - \alpha_{63}\phi_3 - \alpha_{64}\phi_4 - \alpha_{65}\phi_5 \tag{6.87}$$

where the constant α_{ij} can be obtained as discussed earlier. Thus, Equation 6.82 can be written as

$$\phi_1 = g\beta_{11}f_1 \tag{6.88}$$

where $[\beta_{11}] = [1.0]$. Substituting ϕ_1 from Equation 6.88 in Equation 6.83, ϕ_2 can be obtained as

$$\phi_2 = g(\beta_{21} f_1 - \beta_{22} f_2) \tag{6.89}$$

where $\beta_{21} = \alpha_{21}, \beta_{22} = 1$ or

$$\begin{bmatrix} \beta_{21} \\ \beta_{22} \end{bmatrix} = \begin{bmatrix} \alpha_{21} \\ 1 \end{bmatrix} \tag{6.90}$$

Similarly, we can easily obtain the BCOPs, ϕ_3 through ϕ_6, in terms of the functions $f_1 = 1, f_2 = x, f_3 = y, f_4 = x^2, f_5 = xy, \ldots$ as follows:

$$\phi_3 = g(\beta_{31} f_1 - \beta_{32} f_2 - \beta_{33} f_3) \tag{6.91}$$

where
$$\begin{bmatrix} \beta_{31} \\ \beta_{32} \\ \beta_{32} \end{bmatrix} = \begin{bmatrix} \alpha_{32}\beta_{21} + \alpha_{31}\beta_{11} \\ \alpha_{32}\beta_{22} \\ 1 \end{bmatrix} = \begin{bmatrix} \sum_{k=1}^{2} \alpha_{3k}\beta_{k1} \\ \sum_{k=2}^{2} \alpha_{3k}\beta_{k2} \\ 1 \end{bmatrix} \tag{6.92}$$

$$\phi_4 = g(\beta_{41} f_1 - \beta_{42} f_2 - \beta_{43} f_3 - \beta_{44} f_4) \tag{6.93}$$

where
$$\begin{bmatrix} \beta_{41} \\ \beta_{42} \\ \beta_{43} \\ \beta_{44} \end{bmatrix} = \begin{bmatrix} \alpha_{43}\beta_{31} + \alpha_{42}\beta_{21} + \alpha_{41}\beta_{11} \\ \alpha_{43}\beta_{32} + \alpha_{42}\beta_{22} + \beta_{21} \\ \alpha_{43}\beta_{33} \\ 1 \end{bmatrix} = \begin{bmatrix} \sum_{k=1}^{3} \alpha_{4k}\beta_{k1} \\ \sum_{k=2}^{3} \alpha_{4k}\beta_{k2} + \beta_{21} \\ \sum_{k=3}^{3} \alpha_{4k}\beta_{k3} \\ 1 \end{bmatrix} \tag{6.94}$$

$$\phi_5 = g(\beta_{51} f_1 - \beta_{52} f_2 - \beta_{53} f_3 - \beta_{54} f_4 - \beta_{55} f_5) \tag{6.95}$$

where
$$\begin{bmatrix} \beta_{51} \\ \beta_{52} \\ \beta_{53} \\ \beta_{54} \\ \beta_{55} \end{bmatrix} = \begin{bmatrix} \alpha_{54}\beta_{41} + \alpha_{53}\beta_{31} + \alpha_{52}\beta_{21} + \alpha_{51}\beta_{11} \\ \alpha_{54}\beta_{42} + \alpha_{53}\beta_{32} + \alpha_{52}\beta_{22} + \beta_{31} \\ \alpha_{54}\beta_{43} + \alpha_{53}\beta_{33} \\ \alpha_{54}\beta_{44} + \beta_{32} \\ 1 \end{bmatrix} = \begin{bmatrix} \sum_{k=1}^{4} \alpha_{5k}\beta_{k1} \\ \sum_{k=2}^{4} \alpha_{5k}\beta_{k2} + \beta_{31} \\ \sum_{k=3}^{4} \alpha_{5k}\beta_{k3} \\ \sum_{k=4}^{4} \alpha_{5k}\beta_{k4} + \beta_{32} \\ 1 \end{bmatrix} \tag{6.96}$$

and $\qquad \phi_6 = g(\beta_{61} f_1 - \beta_{62} f_2 - \beta_{63} f_3 - \beta_{64} f_4 - \beta_{65} f_5 - \beta_{66} f_6) \tag{6.97}$

where

$$
\begin{bmatrix} \beta_{61} \\ \beta_{62} \\ \beta_{63} \\ \beta_{64} \\ \beta_{65} \\ \beta_{66} \end{bmatrix} = \begin{bmatrix} \alpha_{65}\beta_{51} + \alpha_{64}\beta_{41} + \alpha_{63}\beta_{31} + \alpha_{62}\beta_{21} + \alpha_{61}\beta_{11} \\ \alpha_{65}\beta_{52} + \alpha_{64}\beta_{42} + \alpha_{63}\beta_{32} + \alpha_{62}\beta_{22} \\ \alpha_{65}\beta_{53} + \alpha_{64}\beta_{43} + \alpha_{63}\beta_{33} + \beta_{31} \\ \alpha_{65}\beta_{54} + \alpha_{64}\beta_{44} \\ \alpha_{65}\beta_{55} + \beta_{32} \\ 1 \end{bmatrix}
$$

$$
= \begin{bmatrix} \displaystyle\sum_{k=1}^{5} \alpha_{6k}\beta_{k1} \\ \displaystyle\sum_{k=2}^{5} \alpha_{6k}\beta_{k2} \\ \displaystyle\sum_{k=3}^{5} \alpha_{6k}\beta_{k3} + \beta_{31} \\ \displaystyle\sum_{k=4}^{5} \alpha_{6k}\beta_{k4} \\ \displaystyle\sum_{k=5}^{5} \alpha_{6k}\beta_{k5} + \beta_{32} \\ 1 \end{bmatrix} \tag{6.98}
$$

Thus, the required BCOPs may be obtained in terms of the original functions as discussed earlier, which are easy for computer implementation.

6.4 Generalization of the Recurrence Scheme for Two-Dimensional BCOPs

The BCOPs have been used by various researchers in the Rayleigh–Ritz method to compute natural frequencies and mode shapes of vibrating structures. As mentioned, the generation of the BCOPs for one-dimensional problems is easy owing to the existence of three-term recurrence relation. Recurrence scheme for two-dimensional BCOPs has also been proposed recently. However, this scheme sometimes does not converge for special geometrical shapes for evaluating all the modes at the same time. The method also requires a large number of terms for the convergence. This section addresses a procedure with a recurrence scheme for the two-dimensional BCOPs by undertaking the computation individually for the four types of modes of vibration.

The recurrence scheme, as employed in constructing the BCOPs (Bhat et al. (1998)), sometimes does not converge for special geometrical domains as the procedure requires, to evaluate all the modes at the same time. Moreover, it also requires large number of terms for the convergence. This may be clear from the studies of Singh and Chakraverty (1991, 1992a, 1992b), for example in the case of elliptical plates, where the procedure has been adopted by dividing it into four cases, viz., (1) symmetric–symmetric, (2) symmetric–antisymmetric, (3) antisymmetric–symmetric, and (4) antisymmetric–antisymmetric modes. This way, the rate of convergence was also increased and the numerical instability of the problem was overcome.

This section introduces a strategy to handle these four types of modes separately, using the proposed recurrence scheme for two-dimensional problems, thereby requiring less number of terms for the necessary convergence. The three-term recurrence relation to generate multidimensional orthogonal polynomials, presented by Kowalski (1982a, 1982b), did not consider these polynomials to satisfy any conditions at the boundary of the domain. However, in the proposed scheme, the two-dimensional orthogonal polynomials are constructed for each of the four different modes so as to satisfy the essential boundary conditions of the vibrating structure.

6.4.1 Numerical Procedure for Generalization of the Recurrence Scheme for Two-Dimensional BCOPs

For the implementation of the numerical scheme, the first polynomial is defined as

$$\phi_1^{(1)} = F_1 = g(x, y) f_1 \tag{6.99}$$

where $g(x, y)$ satisfies the essential boundary conditions.

To incorporate the recurrence scheme, the polynomials are written in the classes, forming a pyramid, as shown in Section 6.3.

The inner product of the two functions $\phi_i^{(j)}(X, Y)$ and $\phi_k^{(r)}(X, Y)$ is defined as

$$\langle \phi_i^{(j)} \phi_k^{(r)} \rangle = \iint\limits_R \phi_i^{(j)}(X, Y) \phi_k^{(r)}(X, Y) dX dY$$

The norm of $\Phi_i^{(j)}$ is thus given by

$$\| \phi_i^{(j)} \| = \langle \phi_i^{(j)}, \phi_i^{(j)} \rangle^{1/2}$$

It is already mentioned that in each class of polynomials, the number of orthogonal polynomials is equal to the class number.

The procedure to generate two-dimensional orthogonal polynomials for four types of models, viz., symmetric–symmetric, symmetric–antisymmetric, antisymmetric–symmetric, and antisymmetric–antisymmetric, is discussed as follows:

Symmetric–symmetric mode:
For this mode, f_1 is taken as unity in Equation 6.99 and the two orthogonal polynomials of class number 2, for instance, can be written as

$$\phi_2^{(2)} = F_2 = x^2\phi_1^{(1)} - \alpha_{21}\phi_1^{(1)} \tag{6.100}$$

$$\phi_3^{(2)} = F_3 = y^2\phi_1^{(1)} - \alpha_{31}\phi_1^{(1)} - \alpha_{32}\phi_2^{(2)} \tag{6.101}$$

where

$$\alpha_{21} = \frac{\langle X^2\phi_1^{(1)}, \phi_1^{(1)}\rangle}{\langle \phi_1^{(1)}, \phi_1^{(1)}\rangle}, \quad \alpha_{31} = \frac{\langle Y^2\phi_1^{(1)}, \phi_1^{(1)}\rangle}{\langle \phi_1^{(1)}, \phi_1^{(1)}\rangle}, \quad \alpha_{32} = \frac{\langle Y^2\phi_1^{(1)}, \phi_2^{(2)}\rangle}{\langle \phi_2^{(2)}, \phi_2^{(2)}\rangle} \tag{6.102}$$

Now, generally, class j will have j orthogonal polynomials that can be generated by following the recurrence scheme:

$$\phi_i^{(j)} = F_i = \begin{cases} X^2.\phi_{\{i-(j-1)\}}^{(j-1)} - \sum\limits_{k=\frac{i^2-5j+8}{2}}^{i-1} \alpha_{ik}F_k; & i = \{l-(j-1)\}, \dots, \{l-1\} \\ \\ Y^2\phi_{\{i-j\}}^{(j-1)} - \sum\limits_{k=\frac{i^2-5j+8}{2}}^{i-1} \alpha_{ik}F_k; & i = l \end{cases}$$

$$j = 2, 3, \dots, n \tag{6.103}$$

where

$$\alpha_{ik} = \begin{cases} \frac{<X^2\phi_{\{i-(j-1)\}}^{(j-1)}, F_k>}{<F_k, F_k>}, & i = \{l-(j-1)\}, \dots, \{l-1\} \\ \\ \frac{<Y^2\phi_{\{i-j\}}^{(j-1)}, F_k>}{<F_k, F_k>}, & i = l \end{cases}$$

$$k = \frac{(j^2-5j+8)}{2}, \dots, (i-1) \tag{6.104}$$

Thus, all the BCOPs in the nth class of symmetric–symmetric mode can be generated by using only the previous two classes, and the entire last generated orthogonal polynomials obtained in the nth class so far. orthogonal polynomials generated last can be derived from the nth class so far.

The procedures for generating the BCOPs for symmetric–antisymmetric, antisymmetric–symmetric, and antisymmetric–antisymmetric modes are similar to that of the symmetric–symmetric mode. The function f_1 in Equation 6.99 should be written according to the mode required and Equations 6.102 and 6.103 are valid for all these modes. Accordingly, the expressions for f_1 in the generation of the said orthogonal polynomials for various modes can be written as follows:

1. Symmetric–symmetric: $f_1 = 1$
2. Symmetric–antisymmetric: $f_1 = X$
3. Antisymmetric–symmetric: $f_1 = Y$
4. Antisymmetric–antisymmetric: $f_1 = XY$

When the above-mentioned f_1 is taken along with Equations 6.103 and 6.104, the BCOPs can be constructed as per the desired modes.

These are then normalized by the condition

$$\hat{\phi}_i^{(j)} = \frac{\phi_i^{(j)}}{\| \phi_i^{(j)} \|} \tag{6.105}$$

The above recurrence scheme is much convenient in generating the two-dimensional BCOPs, where separations of the modes are necessary to study the vibration of the structures. Inclusion of all the terms in the analysis, as studied previously, makes the procedure less efficient with slower rate of convergence. However, the procedure presented is quite beneficial for computer implementation, by separating the modes as discussed. The main characteristic of the method is that it makes the computation much efficient, straightforward, and also the time of execution is much faster.

Till now, the BCOPs are generated depending on the highest degree, say n. A monomial is of degree n, if the sum of its exponents is equal to n and there are consequently $(n+1)$ monomials of degree n, such that

$$x^n, x^{n-1}y, x^{n-2}y^2, \ldots, y^n \tag{6.106}$$

Hence, we started with the monomials according to the degree

$$f_1, f_2, f_3, f_4, f_5, f_6, \ldots \tag{6.107}$$

$$= 1, x, y, x^2, xy, y^2, \ldots \tag{6.108}$$

from which we have generated the BCOPs. In the following section generation of the BCOPs will be outlined for grade wise.

6.5   Generation of BCOPs as per Grades of the Monomials

Sometimes, one may generate the BCOPs grade-wise that can again be used in the Rayleigh–Ritz method for studying the vibration behavior of plates. A monomial $x^p y^q$ is of the nth grade, if $\max(p,q) = n$ and a polynomial is of nth grade, if the highest grade of its monomial is n. Degree-wise, there are $(2n+1)$ monomials of grade n, i.e.,

$$x^n, x^n y, \ldots, x^n y^n, \ldots, xy^n, y^n \tag{6.109}$$

If we order according to the grade, then the starting monomials (Equation 6.107) can be written as

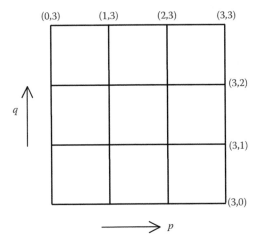

(0,3)   (1,3)   (2,3)   (3,3)

(3,2)

q

(3,1)

(3,0)

p

FIGURE 6.5
p–q Plane showing the monomial exponents for grade 3.

$$f_1, f_2, f_3, f_4, f_5, f_6, \cdots$$
$$= 1; \quad x, xy, y; \quad x^2, x^2y, x^2y^2, xy^2, y^2; \quad x^3, x^3y, x^3y^2, x^3y^3, x^2y^3, xy^3, y^3; \cdots \quad (6.110)$$

For example, it may be noted from the above expression that grade 3 will contain $(2*3+1)=7$ monomials and their exponents are located on the off-axis sides of a square with size 3 in the p–q plane, as shown in Figure 6.5. In general, we can say that polynomial of grade n will contain monomials whose exponents are located on or inside a square with size n in the p–q plane.

Accordingly, the BCOPs in the pyramid form can be written again and the BCOPs for the first four grades (grade 0, grade 1, grade 2, and grade 3) are shown as follows:

Grade 0 $\rightarrow$ $\phi_1^{(1)}$

Grade 1 $\rightarrow$ $\phi_2^{(2)}$ $\phi_3^{(2)}$ $\phi_4^{(2)}$

Grade 2 $\rightarrow$ $\phi_5^{(3)}$ $\phi_6^{(3)}$ $\phi_7^{(3)}$ $\phi_8^{(3)}$ $\phi_9^{(3)}$

Grade 3 $\rightarrow$ $\phi_{10}^{(4)}$ $\phi_{11}^{(4)}$ $\phi_{12}^{(4)}$ $\phi_{13}^{(4)}$ $\phi_{14}^{(4)}$ $\phi_{15}^{(4)}$ $\phi_{16}^{(4)}$

$\cdots$ $\cdots$ $\cdots$ $\cdots$ $\cdots$ $\cdots$

If $g(x, y)$ satisfies the boundary condition, then the recurrence scheme for generating the above BCOPs can be written as

$$\phi_1^{(1)} = F_1 = g(x, y)f_1 \quad (6.111)$$

$$\left.\begin{aligned} \phi_2^{(2)} &= x\phi_1^{(1)} - \alpha_{21}\phi_1^{(1)} \\ \phi_3^{(2)} &= xy\phi_1^{(1)} - \alpha_{31}\phi_1^{(1)} - \alpha_{32}\phi_2^{(2)} \\ \phi_4^{(2)} &= y\phi_1^{(1)} - \alpha_{41}\phi_1^{(1)} - \alpha_{42}\phi_2^{(2)} - \alpha_{43}\phi_3^{(2)} \end{aligned}\right\} \quad (6.112)$$

$$\left.\begin{aligned}
\phi_5^{(3)} &= x\phi_2^{(2)} - \alpha_{51}\phi_1^{(1)} - \alpha_{52}\phi_2^{(2)} - \alpha_{53}\phi_3^{(2)} - \alpha_{54}\phi_4^{(2)} \\
\phi_6^{(3)} &= xy\phi_2^{(2)} - \alpha_{61}\phi_1^{(1)} - \alpha_{62}\phi_2^{(2)} - \alpha_{63}\phi_3^{(2)} - \alpha_{64}\phi_4^{(2)} - \alpha_{65}\phi_5^{(3)} \\
\phi_7^{(3)} &= xy\phi_3^{(2)} - \alpha_{71}\phi_1^{(1)} - \alpha_{72}\phi_2^{(2)} - \alpha_{73}\phi_3^{(2)} - \alpha_{74}\phi_4^{(2)} - \alpha_{75}\phi_5^{(3)} - \alpha_{76}\phi_6^{(3)} \\
\phi_8^{(3)} &= xy\phi_4^{(2)} - \alpha_{81}\phi_1^{(1)} - \cdots\cdots\cdots\cdots\cdots - \alpha_{87}\phi_7^{(3)} \\
\phi_8^{(3)} &= y\phi_4^{(2)} - \alpha_{91}\phi_1^{(1)} - \cdots\cdots\cdots\cdots - \alpha_{98}\phi_8^{(3)}
\end{aligned}\right\}$$

$$(6.113)$$

$$\left.\begin{aligned}
\phi_{10}^{(4)} &= x\phi_5^{(3)} - \alpha_{10,2}\phi_2^{(2)} - \alpha_{10,3}\phi_3^{(2)} - \cdots - \alpha_{10,9}\phi_9^{(3)} \\
\phi_{11}^{(4)} &= xy\phi_5^{(3)} - \cdots - \alpha_{11,10}\phi_{10}^{(4)} \\
&\cdots \qquad\qquad \cdots \qquad\qquad \cdots \qquad \cdots \\
&\cdots \qquad\qquad \cdots \qquad\qquad \cdots \qquad \cdots \\
\phi_{16}^{(4)} &= y\phi_9^{(3)} - \alpha_{16,2}\phi_2^{(2)} - \cdots - \alpha_{16,15}\phi_{15}^{(4)}
\end{aligned}\right\}$$

$$(6.114)$$

where the constants α_{ij} in Equations 6.112 through 6.114 can be obtained using the property of orthogonality. In general, we can now write

$$\phi_i^{(j)} = \begin{cases}
x\phi_{(j-4j+5)}^{(j-1)} + \displaystyle\sum_{k=l}^{(i-1)}\alpha_{ik}\phi_k^{(j)} + \sum_{k=m}^{(l-1)}\alpha_{ik}\phi_k^{(j-1)} + \sum_{k=p}^{(m-1)}\alpha_{ik}\phi_k^{(j-2)}; & \text{for } i=(j^2-2j+2) \\[3ex]
xy\phi_{(i-2(j-1))}^{(j-1)} + \displaystyle\sum_{k=l}^{(i-1)}\alpha_{ik}\phi_k^{(j)} + \sum_{k=m}^{(l-1)}\alpha_{ik}\phi_k^{(j-1)} + \sum_{k=p}^{(m-1)}\alpha_{ik}\phi_k^{(j-2)}; & \text{for } i=(j^2-2j+3),\ldots,(j^2-1) \\[3ex]
y\phi_{(j-1)^2}^{(j-1)} + \displaystyle\sum_{k=l}^{(i-1)}\alpha_{ik}\phi_k^{(j)} + \sum_{k=m}^{(l-1)}\alpha_{ik}\phi_k^{(j-1)} + \sum_{k=p}^{(m-1)}\alpha_{ik}\phi_k^{(j-2)}; & \text{for } i=j^2
\end{cases}$$

$$j=2,3,\ldots,n \qquad\qquad (6.115)$$

where

$l = j^2 - 2j + 2$ describes the first BCOPs of the jth class
$m = (j-1)^2 - 2(j+1) + 2$ describes the first BCOPs of the $(j-1)$th class
$p = (j-2)^2 - 2(j-2) + 2$ describes the first BCOPs of the $(j-2)$th class

$$\alpha_{ik} = \begin{cases}
\dfrac{\langle x\phi_{(j-4j+5)}^{(j-1)}, \phi_k^{(r)}\rangle}{\langle \phi_k^{(r)}, \phi_k^{(r)}\rangle}; & i=(j-2j+2) \\[3ex]
\dfrac{\langle xy\phi_{(j-2(j+1))}^{(j-1)}, \phi_k^{(r)}\rangle}{\langle \phi_k^{(r)}, \phi_k^{(r)}\rangle}; & i=(j^2-2j+3),\ldots,(j^2-1) \\[3ex]
\dfrac{\langle y\phi_{(j-1)^2}^{(j-1)}, \phi_k^{(r)}\rangle}{\langle \phi_k^{(r)}, \phi_k^{(r)}\rangle}; & i=j^2
\end{cases}$$

$$(6.116)$$

where $r = j$, $(j-1)$, and $(j-2)$ for evaluating α_{ik} of Equation 6.115 for the first, second, and third summation terms, respectively.

References

Bhat, R.B., Chakraverty, S., and Stiharu, I. 1998. Recurrence scheme for the generation of two-dimensional boundary characteristic orthogonal polynomials to study vibration of plates. *Journal of Sound and Vibration*, 216(2): 321–327.

Chakraverty, S. 1992. Numerical solution of vibration of plates, PhD Thesis, University of Roorkee (Now IIT, Roorkee), Roorkee, India.

Chakraverty, S., Bhat, R.B., and Stiharu, I. 1999. Recent research on vibration of structures using boundary characteristic orthogonal polynomials in the Rayleigh-Ritz method. *The Shock and Vibration Digest*, 31(3): 187–194.

Chihara, T.S. 1978. *An Introduction to Orthogonal Polynomials*, Gordon and Breach, London.

Genin, Y. and Kamp, Y. 1976. Algebraic properties of two-variable orthogonal polynomials on the hypercircle. *Philips Research Report*, 31: 411–422.

Grossi, R.O. and Albarracin, C.M. 1998. A variant of the method of orthogonal polynomials. *Journal of Sound and Vibration*, 212(4): 749–752.

Kowalski, M.A. 1982a. The recursion formulas for orthogonal polynomials in n variables. *SIAM Journal of Mathematical Analysis*, 13: 309–315.

Kowalski, M.A. 1982b. Orthogonality and recursion formulas for polynomials in n variables. *SIAM Journal of Mathematical Analysis*, 13: 316–323.

Singh, B. and Chakraverty, S. 1991. Transverse vibration of completely free elliptic and circular plates using orthogonal polynomials in Rayleigh-Ritz method. *International Journal of Mechanical Sciences*, 33: 741–751.

Singh, B. and Chakraverty, S. 1992a. On the use of orthogonal polynomials in Rayleigh-Ritz method for the study of transverse vibration of elliptic plates. *Computers and Structures*, 43: 439–443.

Singh, B. and Chakraverty, S. 1992b. Transverse vibration of simply-supported elliptic and circular plates using orthogonal polynomials in two variables. *Journal of Sound and Vibration*, 152: 149–155.

Singh, B. and Chakraverty, S. 1994. Boundary characteristic orthogonal polynomials in numerical approximation. *Communications in Numerical Methods in Engineering*, 10: 1027–1043.

7

Transverse Vibration of Elliptic and Circular Plates

7.1 Introduction

As mentioned in earlier chapters, although a lot of information is available for rectangular and circular plates, comparatively little is known about the elliptic plates. Leissa (1969) provided excellent data on clamped and free elliptic plates in Chapter 3 of his monograph. Some important reference works that deal with clamped elliptic plates include Shibaoka (1956), McNitt (1962), Cotugno and Mengotti-Marzolla (1943), McLachlan (1947), and Mazumdar (1971). Singh and Tyagi (1985) and Singh and Goel (1985) studied the transverse vibrations of an elliptic plate of variable thickness with clamped boundary. The displacements were assumed analogous to axisymmetric vibrations of a circular plate. Leissa (1967) presented the fundamental frequencies for simply supported elliptic plates. With respect to the completely free elliptic plates, Waller (1938, 1950) provided few experimental results for the elliptical brass plates. Sato (1973) presented the experimental as well as the theoretical results for elliptic plates with free edge, and compared them with the results of Waller. Beres (1974) applied Ritz method by assuming the solution as a double-power series, and obtained approximate results for the first few frequencies when $a/b = 1.98$, where a and b are the semimajor and semiminor axes of the ellipse, respectively.

The basic aim of this chapter is to address the method of solution and the results for the elliptic and circular plates with uniform thickness. The boundary may be clamped, simply supported, or completely free. The method is based on the use of boundary characteristic orthogonal polynomials (BCOPs) satisfying the essential boundary conditions of the problem. In this method, the most general type of admissible functions have been used and the modes of vibration have been computed by segregating the modes as symmetric–symmetric, antisymmetric–symmetric, symmetric–antisymmetric, and antisymmetric–antisymmetric. Consequently, one can choose all the frequency parameters from these four cases in the increasing order.

The use of orthogonal polynomials in the study of vibration problems commenced only very recently. Some important reference works in this area of study include, Bhat (1985, 1987), Bhat et al. (1990), Laura et al.

(1989), Liew and Lam (1990), Liew et al. (1990), Dickinson and Blasio (1986), and Lam et al. (1990). The results from all these studies of boundaries with straight edges using characteristic orthogonal polynomials (COPs) were highly encouraging. Chakraverty (1992) and Singh and Chakraverty (1991, 1992a,b) developed the BCOPs to extend the analysis to the curved boundaries, such as ellipse and circle. Since the circular plates have been studied in detail, a comparison of their results with the known results serves as a test for the efficiency and powerfulness of the method. They (Singh and Chakraverty and co-authors) also made comparisons of special situations with the known results for the elliptic plate. One can refer the works of Singh and Chakraverty (1991, 1992a,b) for the detailed comparison and results of the elliptic plates with clamped, simply supported, and free boundary. One-dimensional BCOPs with modified polar coordinates have been used for circular and elliptic plates, by Rajalingham and Bhat (1991, 1993) and Rajalingham et al. (1991, 1993).

As has been mentioned in the previous chapters, this method involves three steps. The first step consists of the generation of orthogonal polynomials over the domain occupied by the plate in the *x–y* plane, satisfying at least the essential boundary conditions. For this, the well-known Gram–Schmidt procedure discussed earlier has been used. The second step is the use of the BCOPs in the Rayleigh–Ritz method that converts the problem into the standard eigenvalue problem. The third and the last step involves the solution of this eigenvalue problem for obtaining the vibration characteristics.

7.2 Generation of BCOPs for Elliptic and Circular Plates with Constant Thickness

For an elliptic plate with semimajor and semiminor axes *a* and *b*, as shown in Figure 7.1, we start by defining a variable *u* as

$$x^2 + \frac{y^2}{m^2} = 1 - u, \quad 0 \le u \le 1 \tag{7.1}$$

where *u* vanishes at the boundary and remains as unity at the center, and $m = b/a$. The curves $u = $ constant are concentric ellipses. As discussed in the previous chapters, we take the first starting function as

$$g(x, y) = \left(1 - x^2 - \frac{y^2}{m^2}\right)^s = u^s$$

Hence, we can write

$$F_i = u^s f_i(x, y), \quad i = 1, 2, 3, \ldots \tag{7.2}$$

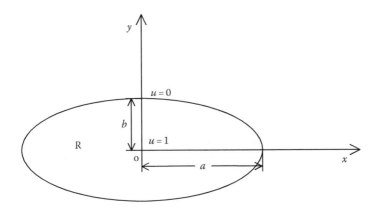

FIGURE 7.1
Geometry of elliptical plate.

Consequently, one may obtain the following cases depending on the boundary conditions:

Case 1. Clamped boundary:

For this case, we have $s=2$ and from Equation 7.2, we obtain

$$F_i = u^2 f_i(x, y) \tag{7.3}$$

The above function satisfies the essential boundary conditions for the clamped boundary.

Case 2. Simply supported boundary:

We will consider $s=1$ and in this case, Equation 7.2 will become

$$F_i = u f_i(x, y) \tag{7.4}$$

Case 3. Completely free boundary:

In this case, we consider $s=0$, and therefore,

$$F_i = f_i(x, y) \tag{7.5}$$

Again, it can be observed that the functions (Equations 7.4 and 7.5) satisfy the essential boundary conditions for the simply supported and completely free boundary, respectively. We define the inner product of the two functions $p(x, y)$ and $q(x, y)$ over the elliptic domain by taking the weight function $\psi(x, y)$ to be equal to unity. This gives

$$\langle p,q \rangle = \iint\limits_R p(x, y)q(x, y)dxdy \tag{7.6}$$

with the norm of the function as given in the earlier chapters. We subsequently generated the BCOPs $\phi_i(x, y)$ over R, the elliptic domain, and then normalized it as mentioned earlier.

If we consider $f_i(x, y)$ as the polynomials in x and y, then all the inner products involved in the above expressions can be found in the closed form. For this, the following result given by Singh et al. (1985) and Chakraverty (1992) was found to be very useful:

$$\iint\limits_R x^p y^q u^r dxdy = \frac{a^{p+1}b^{q+1}\overline{\Big(\frac{p+1}{2}\Big)}\,\overline{\Big(\frac{q+1}{2}\Big)}\,\overline{)r+1}}{\overline{\Big)\frac{p+q}{2}+r+2}} \tag{7.7}$$

where p and q are the non-negative even integers and $r+1>0$ and $\overline{)}(.)$ denotes the Gamma function. The integral vanishes if p or q is odd.

7.3 Rayleigh–Ritz Method for Elliptic and Circular Constant Thickness Plates

Proceeding as discussed in earlier chapters, the Rayleigh quotient for uniform thickness plate is given by

$$\omega^2 = \frac{D}{h\rho} \frac{\iint\limits_R (\nabla^2 W)^2 + 2(1-\nu)\Big\{W_{xy}^2 - W_{xx}W_{yy}\Big\}dxdy}{\iint\limits_R W^2 dxdy} \tag{7.8}$$

By inserting $W = \sum_{j=1}^{n} c_j \phi_j$ in Equation 7.8 and minimizing it as a function of c_j's, we get the standard eigenvalue problem

$$\sum_{j=1}^{n} (a_{ij} - \lambda^2 b_{ij})c_j = 0, \quad i = 1,\ldots,n \tag{7.9}$$

where
$$\lambda^2 = \frac{a^4 \omega^2 \rho h}{D} \tag{7.10}$$

$$a_{ij} = \iint\limits_R \Big[\phi_i^{xx}\phi_j^{xx} + \phi_i^{yy}\phi_j^{yy} + \nu\Big(\phi_i^{xx}\phi_j^{yy} + \phi_i^{yy}\phi_j^{xx}\Big) + 2(1-\nu)\phi_i^{xy}\phi_j^{xy}\Big]dxdy \tag{7.11}$$

$$b_{ij} = \iint\limits_R \phi_i \phi_j dxdy \qquad (7.12)$$

In what follows, we will provide some numerical results for all the possible boundary conditions, viz., clamped, simply supported, and completely free. But before that, we will write the terms in Equation 7.11 with respect to the orthonormal coefficients and $f_i(x, y)$ for the clamped boundary condition. It has already been shown in the previous chapters that the orthogonal polynomials ϕ_i may be written in terms of the function f_k as

$$\phi_i = u^2 \sum_{k=1}^{i} \alpha_{ik} f_k \qquad (7.13)$$

where

$$f_k = x^{m_k} y^{n_k} \qquad (7.14)$$

m_k, n_k are integers, and

$$u = 1 - x^2 - \frac{y^2}{m^2}$$

Accordingly, the first term of Equation 7.11 can easily be written as

$$\iint\limits_R \phi_i^{xx} \phi_j^{xx} dxdy = PX1 + PX2 + PX3 + PX4 + PX5 + PX6 + PX7 + PX8 + PX9$$

$$(7.15)$$

where

$$PX1 = \sum_{k=1}^{i} \alpha_{ik} m_k (m_k - 1) \sum_{\ell=1}^{j} \alpha_{j\ell} m_\ell (m_\ell - 1) \iint\limits_R x^{m_k + m_\ell - 4} y^{n_k + n_\ell} u^4 dxdy \qquad (7.16)$$

$$PX2 = -\sum_{k=1}^{i} \alpha_{ik} m_k (m_k - 1) \sum_{\ell=1}^{j} \alpha_{j\ell} (8m_\ell + 4) \iint\limits_R x^{m_k + m_\ell - 2} y^{n_k + n_\ell} u^3 dxdy \qquad (7.17)$$

$$PX3 = \sum_{k=1}^{i} \alpha_{ik} m_k (m_k - 1) \sum_{\ell=1}^{j} 8\alpha_{j\ell} \iint\limits_R x^{m_k + m_\ell} y^{n_k + n_\ell} u^2 dxdy \qquad (7.18)$$

$$PX4 = -\sum_{k=1}^{i} \alpha_{ik} (8m_k + 4) \sum_{\ell=1}^{j} \alpha_{j\ell} m_\ell (m_\ell - 1) \iint\limits_R x^{m_k + m_\ell - 2} y^{n_k + n_\ell} u^3 dxdy \qquad (7.19)$$

$$PX5 = \sum_{k=1}^{i} \alpha_{ik} (8m_k + 4) \sum_{\ell=1}^{j} \alpha_{j\ell} (8m_\ell + 4) \iint\limits_R x^{m_k + m_\ell} y^{n_k + n_\ell} u^2 dxdy \qquad (7.20)$$

$$PX6 = -\sum_{k=1}^{i} \alpha_{ik}(8m_k + 4) \sum_{\ell=1}^{j} 8\alpha_{j\ell} \iint_R x^{m_k+m_\ell-2} y^{n_k+n_\ell} u \, dx dy \qquad (7.21)$$

$$PX7 = \sum_{k=1}^{i} 8\alpha_{ik} \sum_{\ell=1}^{j} \alpha_{j\ell} m_\ell(m_\ell - 1) \iint_R x^{m_k+m_\ell} y^{n_k+n_\ell} u^2 \, dx dy \qquad (7.22)$$

$$PX8 = -\sum_{k=1}^{i} 8\alpha_{ik} \sum_{\ell=1}^{j} \alpha_{j\ell}(8m_\ell + 4) \iint_R x^{m_k+m_\ell+4} y^{n_k+n_\ell} u \, dx dy \qquad (7.23)$$

$$PX9 = \sum_{k=1}^{i} 8\alpha_{ik} \sum_{\ell=1}^{j} 8\alpha_{j\ell} \iint_R x^{m_k+m_\ell+4} y^{n_k+n_\ell} \, dx dy \qquad (7.24)$$

The second term of Equation 7.11 can then be written as

$$\iint_R \phi_i^{yy} \phi_j^{yy} \, dx dy = PY1 + PY2 + PY3 + PY4 + PY5 + PY6 + PY7 + PY8 + PY9$$

$$(7.25)$$

where

$$PY1 = \sum_{k=1}^{i} \alpha_{ik} n_k(n_k - 1) \sum_{\ell=1}^{j} \alpha_{j\ell} n_\ell(n_\ell - 1) \iint_R x^{m_k+m_\ell} y^{n_k+n_\ell-4} u^4 \, dx dy \qquad (7.26)$$

$$PY2 = -\sum_{k=1}^{i} \alpha_{ik} n_k(n_k - 1) \sum_{\ell=1}^{j} \alpha_{j\ell}(1 + 2n_\ell) \frac{4}{m^2} \iint_R x^{m_k+m_\ell} y^{n_k+n_\ell-2} u^3 \, dx dy \quad (7.27)$$

$$PY3 = \sum_{k=1}^{i} \alpha_{ik} n_k(n_k - 1) \sum_{\ell=1}^{j} \alpha_{j\ell} \frac{8}{m^4} \iint_R x^{m_k+m_\ell} y^{n_k+n_\ell} u^2 \, dx dy \qquad (7.28)$$

$$PY4 = -\sum_{k=1}^{i} \alpha_{ik}(1 + 2n_k) \frac{4}{m^2} \sum_{\ell=1}^{j} \alpha_{j\ell} n_\ell(n_\ell - 1) \iint_R x^{m_k+m_\ell} y^{n_k+n_\ell-2} u^3 \, dx dy \quad (7.29)$$

$$PY5 = \sum_{k=1}^{i} \alpha_{ik}(1 + 2n_k) \frac{4}{m^2} \sum_{\ell=1}^{j} \alpha_{j\ell}(1 + 2n_\ell) \frac{4}{m^2} \iint_R x^{m_k+m_\ell} y^{n_k+n_\ell} u^2 \, dx dy \quad (7.30)$$

$$PY6 = -\sum_{k=1}^{i} \alpha_{ik}(1 + 2n_k) \frac{4}{m^2} \sum_{\ell=1}^{j} \alpha_{j\ell} \frac{8}{m^4} \iint_R x^{m_k+m_\ell} y^{n_k+n_\ell+2} u \, dx dy \qquad (7.31)$$

$$PY7 = \sum_{k=1}^{i} \alpha_{ik} \frac{8}{m^4} \sum_{\ell=1}^{j} \alpha_{j\ell} n_\ell (n_\ell - 1) \iint_R x^{m_k+m_\ell} y^{n_k+n_\ell} u^2 dxdy \qquad (7.32)$$

$$PY8 = -\sum_{k=1}^{i} \alpha_{ik} \frac{8}{m^4} \sum_{\ell=1}^{j} \alpha_{j\ell} (n_\ell - 1) \frac{4}{m^2} \iint_R x^{m_k+m_\ell} y^{n_k+n_\ell+2} u dxdy \qquad (7.33)$$

$$PY9 = \sum_{k=1}^{i} \alpha_{ik} \frac{8}{m^4} \sum_{\ell=1}^{j} \alpha_{j\ell} \frac{8}{m^4} \iint_R x^{m_k+m_\ell} y^{n_k+n_\ell+4} dxdy \qquad (7.34)$$

The third term of Equation 7.11 can be obtained as

$$\iint_R \phi_i^{xx} \phi_j^{yy} dxdy = PXI1 + PXI2 + PXI3 + PXI4 + PXI5 + PXI6 + PXI7 + PXI8 + PXI9$$

$$(7.35)$$

where

$$PXI1 = \sum_{k=1}^{i} \alpha_{ik} m_k (m_k - 1) \sum_{\ell=1}^{j} \alpha_{j\ell} n_\ell (n_\ell - 1) \iint_R x^{m_k+m_\ell-2} y^{n_k+n_\ell-2} u^4 dxdy \quad (7.36)$$

$$PXI2 = -\sum_{k=1}^{i} \alpha_{ik} m_k (m_k - 1) \sum_{\ell=1}^{j} \alpha_{j\ell} (1 + 2n_\ell) \frac{4}{m^2} \iint_R x^{m_k+m_\ell-2} y^{n_k+n_\ell} u^3 dxdy \quad (7.37)$$

$$PXI3 = \sum_{k=1}^{i} \alpha_{ik} m_k (m_k - 1) \sum_{\ell=1}^{j} \alpha_{j\ell} \frac{8}{m^4} \iint_R x^{m_k+m_\ell-2} y^{n_k+n_\ell+2} u^2 dxdy \qquad (7.38)$$

$$PXI4 = -\sum_{k=1}^{i} \alpha_{ik} (8m_k + 4) \sum_{\ell=1}^{j} \alpha_{j\ell} n_\ell (n_\ell - 1) \iint_R x^{m_k+m_\ell} y^{n_k+n_\ell-2} u^3 dxdy \quad (7.39)$$

$$PXI5 = \sum_{k=1}^{i} \alpha_{ik} (8m_k + 4) \sum_{\ell=1}^{j} \alpha_{j\ell} (1 + 2n_\ell) \frac{4}{m^2} \iint_R x^{m_k+m_\ell} y^{n_k+n_\ell} u^2 dxdy \qquad (7.40)$$

$$PXI6 = -\sum_{k=1}^{i} \alpha_{ik} (8m_k + 4) \sum_{\ell=1}^{j} \alpha_{j\ell} \frac{8}{m^4} \iint_R x^{m_k+m_\ell} y^{n_k+n_\ell+2} u dxdy \qquad (7.41)$$

$$PXI7 = \sum_{k=1}^{i} 8\alpha_{ik} \sum_{\ell=1}^{j} \alpha_{j\ell} n_\ell (n_\ell - 1) \iint_R x^{m_k+m_\ell+2} y^{n_k+n_\ell-2} u^2 dxdy \qquad (7.42)$$

$$PXI8 = -\sum_{k=1}^{i} 8\alpha_{ik} \sum_{\ell=1}^{j} \alpha_{j\ell}(1 + 2n_\ell)\frac{4}{m^2} \iint_R x^{m_k+m_\ell+2} y^{n_k+n_\ell} u \, dxdy \qquad (7.43)$$

$$PXI9 = \sum_{k=1}^{i} 8\alpha_{ik} \sum_{\ell=1}^{j} \alpha_{j\ell}\frac{8}{m^4} \iint_R x^{m_k+m_\ell+2} y^{n_k+n_\ell+2} \, dxdy \qquad (7.44)$$

The fourth term of Equation 7.11 yields

$$\iint_R \phi_i^{yy} \phi_j^{xx} \, dxdy = PYI1 + PYI2 + PYI3 + PYI4 + PYI5 + PYI6 + PYI7 + PYI8 + PYI9$$

$$\qquad (7.45)$$

where

$$PYI1 = \sum_{k=1}^{i} \alpha_{ik} n_k(n_k - 1) \sum_{\ell=1}^{j} \alpha_{j\ell} m_\ell(m_\ell - 1) \iint_R x^{m_k+m_\ell-2} y^{n_k+n_\ell-2} u^4 \, dxdy \quad (7.46)$$

$$PYI2 = -\sum_{k=1}^{i} \alpha_{ik} n_k(n_k - 1) \sum_{\ell=1}^{j} \alpha_{j\ell}(8m_\ell + 4) \iint_R x^{m_k+m_\ell} y^{n_k+n_\ell-2} u^3 \, dxdy \quad (7.47)$$

$$PYI3 = \sum_{k=1}^{i} \alpha_{ik} n_k(n_k - 1) \sum_{\ell=1}^{j} 8\alpha_{j\ell} \iint_R x^{m_k+m_\ell+2} y^{n_k+n_\ell-2} u^2 \, dxdy \qquad (7.48)$$

$$PYI4 = -\sum_{k=1}^{i} \alpha_{ik}(1 + 2n_k)\frac{4}{m^2} \sum_{\ell=1}^{j} \alpha_{j\ell} m_\ell(m_\ell - 1) \iint_R x^{m_k+m_\ell-2} y^{n_k+n_\ell} u^3 \, dxdy \quad (7.49)$$

$$PYI5 = \sum_{k=1}^{i} \alpha_{ik}(1 + 2n_k)\frac{4}{m^2} \sum_{\ell=1}^{j} \alpha_{j\ell}(8m_\ell + 4) \iint_R x^{m_k+m_\ell} y^{n_k+n_\ell} u^2 \, dxdy \qquad (7.50)$$

$$PYI6 = -\sum_{k=1}^{i} \alpha_{ik}(1 + 2n_k)\frac{4}{m^2} \sum_{\ell=1}^{j} 8\alpha_{j\ell} \iint_R x^{m_k+m_\ell+2} y^{n_k+n_\ell} u \, dxdy \qquad (7.51)$$

$$PYI7 = \sum_{k=1}^{i} \alpha_{ik}\frac{8}{m^4} \sum_{\ell=1}^{j} \alpha_{j\ell} m_\ell(m_\ell - 1) \iint_R x^{m_k+m_\ell-2} y^{n_k+n_\ell+2} u^2 \, dxdy \qquad (7.52)$$

$$PYI8 = -\sum_{k=1}^{i} \alpha_{ik}\frac{8}{m^4} \sum_{\ell=1}^{j} \alpha_{j\ell}(8m_\ell + 4) \iint_R x^{m_k+m_\ell} y^{n_k+n_\ell+2} u \, dxdy \qquad (7.53)$$

$$PYI9 = \sum_{k=1}^{i} \alpha_{ik} \frac{8}{m^4} \sum_{\ell=1}^{j} 8\alpha_{j\ell} \iint_{R} x^{m_k+m_\ell+2} y^{n_k+n_\ell+2} dxdy \qquad (7.54)$$

The fifth term of Equation 7.11 can be obtained as

$$\iint_{R} \phi_i^{xy} \phi_j^{xy} dxdy = PXY1 + PXY2 + PXY3 + PXY4 + PXY5 + PXY6$$

$$+ PXY7 + PXY8 + PXY9 + PXY10 + PXY11$$
$$+ PXY12 + PXY13 + PXY14 + PXY15 + PXY16 \qquad (7.55)$$

where

$$PXY1 = \sum_{k=1}^{i} \alpha_{ik} m_k n_k \sum_{\ell=1}^{j} \alpha_{j\ell} m_\ell n_\ell \iint_{R} x^{m_k+m_\ell-2} y^{n_k+n_\ell-2} u^4 dxdy \qquad (7.56)$$

$$PXY2 = -\sum_{k=1}^{i} \alpha_{ik} m_k n_k \sum_{\ell=1}^{j} \alpha_{j\ell} \frac{4m_\ell}{m^2} \iint_{R} x^{m_k+m_\ell-2} y^{n_k+n_\ell} u^3 dxdy \qquad (7.57)$$

$$PXY3 = -\sum_{k=1}^{i} \alpha_{ik} m_k n_k \sum_{\ell=1}^{j} \alpha_{j\ell} 4n_\ell \iint_{R} x^{m_k+m_\ell} y^{n_k+n_\ell-2} u^3 dxdy \qquad (7.58)$$

$$PXY4 = \sum_{k=1}^{i} \alpha_{ik} m_k n_k \sum_{\ell=1}^{j} \alpha_{j\ell} \frac{8}{m^2} \iint_{R} x^{m_k+m_\ell} y^{n_k+n_\ell} u^2 dxdy \qquad (7.59)$$

$$PXY5 = -\sum_{k=1}^{i} \alpha_{ik} \frac{4m_k}{m^2} \sum_{\ell=1}^{j} \alpha_{j\ell} m_\ell n_\ell \iint_{R} x^{m_k+m_\ell-2} y^{n_k+n_\ell} u^3 dxdy \qquad (7.60)$$

$$PXY6 = \sum_{k=1}^{i} \alpha_{ik} \frac{4m_k}{m^2} n_k \sum_{\ell=1}^{j} \alpha_{j\ell} \frac{4m_\ell}{m^2} n_\ell \iint_{R} x^{m_k+m_\ell-2} y^{n_k+n_\ell+2} u^2 dxdy \qquad (7.61)$$

$$PXY7 = \sum_{k=1}^{i} \alpha_{ik} \frac{4m_k}{m^2} \sum_{\ell=1}^{j} \alpha_{j\ell} 4n_\ell \iint_{R} x^{m_k+m_\ell} y^{n_k+n_\ell} u^2 dxdy \qquad (7.62)$$

$$PXY8 = -\sum_{k=1}^{i} \alpha_{ik} \frac{4m_k}{m^2} \sum_{\ell=1}^{j} \alpha_{j\ell} \frac{8}{m^2} \iint_{R} x^{m_k+m_\ell} y^{n_k+n_\ell+2} u dxdy \qquad (7.63)$$

$$PXY9 = -\sum_{k=1}^{i} \alpha_{ik} 4n_k \sum_{\ell=1}^{j} \alpha_{j\ell} m_\ell n_\ell \iint_{R} x^{m_k+m_\ell} y^{n_k+n_\ell-2} u^3 dxdy \qquad (7.64)$$

$$PXY10 = \sum_{k=1}^{i} \alpha_{ik} 4n_k \sum_{\ell=1}^{j} \alpha_{j\ell} \frac{4m_\ell}{m^2} \iint_R x^{m_k+m_\ell} y^{n_k+n_\ell} u^2 dx dy \qquad (7.65)$$

$$PXY11 = \sum_{k=1}^{i} \alpha_{ik} 4n_k \sum_{\ell=1}^{j} \alpha_{j\ell} 4n_\ell \iint_R x^{m_k+m_\ell+2} y^{n_k+n_\ell-2} u^2 dx dy \qquad (7.66)$$

$$PXY12 = -\sum_{k=1}^{i} \alpha_{ik} 4n_k \sum_{\ell=1}^{j} \alpha_{j\ell} \frac{8}{m^2} \iint_R x^{m_k+m_\ell+2} y^{n_k+n_\ell} u \, dx dy \qquad (7.67)$$

$$PXY13 = \sum_{k=1}^{i} \alpha_{ik} \frac{8}{m^2} \sum_{\ell=1}^{j} \alpha_{j\ell} m_\ell n_\ell \iint_R x^{m_k+m_\ell} y^{n_k+n_\ell} u^2 dx dy \qquad (7.68)$$

$$PXY14 = -\sum_{k=1}^{i} \alpha_{ik} \frac{8}{m^2} \sum_{\ell=1}^{j} \alpha_{j\ell} \frac{4m_\ell}{m^2} \iint_R x^{m_k+m_\ell} y^{n_k+n_\ell+2} u \, dx dy \qquad (7.69)$$

$$PXY15 = -\sum_{k=1}^{i} \alpha_{ik} \frac{8}{m^2} \sum_{\ell=1}^{j} \alpha_{j\ell} 4n_\ell \iint_R x^{m_k+m_\ell+2} y^{n_k+n_\ell} u \, dx dy \qquad (7.70)$$

$$PXY16 = \sum_{k=1}^{i} \alpha_{ik} \frac{8}{m^2} \sum_{\ell=1}^{j} \alpha_{j\ell} \frac{8}{m^2} \iint_R x^{m_k+m_\ell+2} y^{n_k+n_\ell+2} dx dy \qquad (7.71)$$

All the terms of Equations 7.16 through 7.71 can easily be obtained by using the Integral formula given in Equation 7.7. Thus, by evaluating all the integrals and inserting them in Equation 7.11, we can obtain the standard eigenvalue problem from Equation 7.9, because the matrix b_{ij} in Equation 7.12 will be an identity matrix owing to the orthonormality condition.

7.4 Some Numerical Results of Elliptic and Circular Plates

7.4.1 Clamped Boundary

Various researchers have used different methods to determine the results for clamped elliptic plates. Here, the BCOPs have been used by choosing the functions $f_k(x, y)$ suitably, and then generating the corresponding orthogonal functions, thus carrying out an exhaustive study of the various modes of vibrations. By taking these functions as even in x and y, one can discuss all those modes that are symmetric about both the major and minor axes. When the functions are even in x and odd in y or vice versa, we can have symmetric–antisymmetric or antisymmetric–symmetric modes. Similarly, if

TABLE 7.1

First Three Frequency Parameters for Symmetric–
Symmetric Modes (Clamped)

	Frequency Parameters		
$m = b/a$	First	Second	Third
0.1	579.36	703.49	1026.1
0.2	149.66	198.55	295.71
0.3	69.147	104.80	166.16
0.4	40.646	71.414	123.61
0.5	27.377	55.985	105.17
0.6	20.195	47.820	93.980
0.7	15.928	43.050	70.776
0.8	13.229	39.972	55.784
0.9	11.442	37.628	45.800
1.0	10.216	34.878	39.773

we take the function as odd in both x and y, then we may have the case of antisymmetric–antisymmetric modes of vibration. One can use various types of combinations of these functions and obtain the corresponding results for various values of $m = b/a$, i.e., the aspect ratio of the elliptic plate.

Tables 7.1 and 7.2 give the representative results for symmetric–symmetric and antisymmetric–symmetric modes (first three), respectively, for the clamped plate. The aspect ratio of the domain is taken from 0.1 to 1.0. From these two tables, we can observe that the frequencies decrease as the aspect-ratio parameter is increased from 0.1 to 1.0. These tendencies can also

TABLE 7.2

First Three Frequency Parameters for Antisymmetric–
Symmetric Modes (Clamped)

	Frequency Parameters		
$m = b/a$	First	Second	Third
0.1	623.23	783.26	1180.1
0.2	171.10	229.81	350.50
0.3	84.979	128.70	205.26
0.4	53.982	93.202	158.49
0.5	39.497	77.037	138.36
0.6	31.736	68.506	115.91
0.7	27.204	63.367	90.972
0.8	24.383	59.650	75.165
0.9	22.532	55.875	65.562
1.0	21.260	51.033	60.844

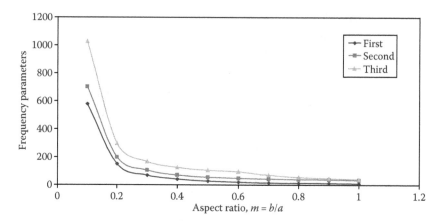

FIGURE 7.2
First three frequency parameters for symmetric–symmetric modes (clamped).

be observed from Figures 7.2 and 7.3. The last rows of these two tables give the results for clamped circular plate ($m = 1.0$). We have made sure to continue increasing the order of approximation till the results start agreeing to the desired accuracy. The only test on the accuracy of the results is the agreement with known results for $b/a = 1.0$ (circular plate). Just to get a feel of the convergence, Table 7.3 provides the results for the first three modes for symmetric–antisymmetric modes with $m = 0.2$, 0.6, and 0.8 from various approximations. The convergence results have been depicted for $m = 0.5$ in Figure 7.4. It may be worth mentioning that the first few modes of the overall may be chosen from the four different mode values. Moreover, for generating the four separate modes as discussed earlier, if one desires to use all the even

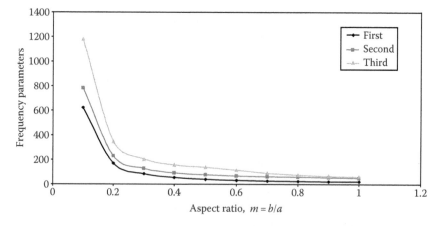

FIGURE 7.3
First three frequency parameters for antisymmetric–symmetric modes (clamped).

TABLE 7.3

Convergence Study for Symmetric–Antisymmetric Modes (Clamped)

	m = 0.2			m = 0.6			m = 0.8		
n	First	Second	Third	First	Second	Third	First	Second	Third
2	417.1	636.96		50.993	96.963		30.911	69.634	
3	409.79	635.49	1493.0	50.131	96.064	172.24	30.346	68.168	100.77
4	405.76	533.18	924.84	50.095	88.205	169.18	30.340	64.163	100.15
5	404.87	527.90	917.36	50.077	87.439	164.94	30.333	63.454	97.987
6	404.75	527.88	917.36	50.061	87.412	153.23	30.322	63.337	89.693
7	403.78	498.24	696.10	50.060	86.811	143.05	30.322	63.154	89.676
8	403.63	495.65	694.41	50.060	86.663	142.93	30.322	63.089	89.604
9	403.63	495.64	694.30	50.060	86.663	142.67	30.322	63.045	89.879
10	403.63	495.64	694.30	50.060	86.663	142.67	30.322	63.045	89.878

and odd functions at the same time and determine the result, then the obtained result may not be so accurate and acceptable. Accordingly, Table 7.4 incorporates a comparison between the first three frequency parameters when chosen from all the modes computed separately (denoted by SEP) as well as taken together at a time (denoted by AT). Both the cases used the 10th approximation results. The last row of this table gives the frequency parameters computed from the exact solution using Bessel's functions. It can be observed that the results match exactly when the procedure of separating and computing was carried out in terms of the different categories of modes. Hence, one should not use all even and odd functions at the same time for these problems.

Tables 7.5 through 7.7 incorporate the deflection on the domain for clamped circular modes for the first three modes, respectively. Here, the

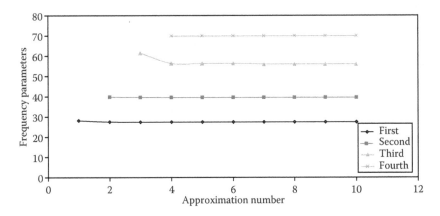

FIGURE 7.4
Convergence of results, $b/a = 0.5$ (clamped).

TABLE 7.4

Comparison of First Three Modes (Clamped) When Chosen from All the Modes Computed Separately (SEP) and All the Modes Taken together at a Time (AT) (After 10th Approximately)

	Frequency Parameters					
	First		Second		Third	
$m = b/a$	SEP	AT	SEP	AT	SEP	AT
0.1	579.36	594.12	623.2	671.18	703.49	971.45
0.2	149.66	151.46	171.10	177.34	198.55	255.69
0.3	69.147	69.461	84.979	86.145	104.80	124.93
0.4	40.646	40.713	53.982	54.268	71.414	80.765
0.5	27.377	27.395	39.497	39.594	55.985	61.455
0.6	20.195	20.201	31.736	31.780	47.820	50.131
0.7	15.928	15.931	27.204	27.230	38.087	38.125
0.8	13.229	13.231	24.383	24.402	30.322	30.346
0.9	11.442	11.444	22.532	22.547	25.021	25.039
1.0	10.216	10.217	21.260	21.275	34.878	36.275
1.0^a	10.216		21.260		34.878	

^{a} Computed from the exact solution.

domain in terms of x and y coordinates taking the values from -1 to 1 at the step of 0.2 are shown. The first three modes have been chosen from the sets of results of the symmetric–symmetric, antisymmetric–symmetric, symmetric–antisymmetric, and antisymmetric–antisymmetric modes. The deflection results of Table 7.5 show that there exist no nodal lines in the case of the

TABLE 7.5

Deflection on the Domain for Determining Nodal Lines of Circular Plate for First Mode of Vibration (Clamped): Shows No Nodal Lines, $m = 1.0$

$x \rightarrow$ $y \downarrow$	-1.0	-0.8	-0.6	-0.4	-0.2	0.0	0.2	0.4	0.6	0.8	1.0
-1.0						0.0					
-0.8			0.0	0.043	0.113	0.144	0.113	0.043	0.0		
-0.6		0.0	0.085	0.264	0.426	0.489	0.426	0.264	0.085	0.0	
-0.4		0.043	0.264	0.558	0.799	0.890	0.799	0.558	0.264	0.043	
-0.2		0.113	0.426	0.798	1.09	1.20	1.09	0.798	0.426	0.113	
0.0	0.0	0.144	0.489	0.890	1.20	1.32	1.20	0.890	0.489	0.144	0.0
0.2		0.113	0.426	0.798	1.09	1.20	1.09	0.798	0.426	0.113	
0.4		0.043	0.264	0.558	0.799	0.890	0.799	0.558	0.264	0.043	
0.6		0.0	0.085	0.264	0.426	0.489	0.426	0.264	0.085	0.0	
0.8			0.0	0.043	0.113	0.144	0.113	0.043	0.0		
1.0						0.0					

TABLE 7.6

Deflection on the Domain for Determining Nodal Lines of Circular Plate for Second Mode of Vibration (Clamped): Shows One Vertical Nodal Line, $m = 1.0$

$x\rightarrow$ $y\downarrow$	−1.0	−0.8	−0.6	−0.4	−0.2	0.0	0.2	0.4	0.6	0.8	1.0
−1.0						0.0					
−0.8			0.0	−0.053	−0.072	0.0	0.072	0.053	0.0		
−0.6		0.0	−0.162	−0.351	−0.293	0.0	0.293	0.351	0.162	0.0	
−0.4		−0.106	−0.527	−0.786	−0.582	0.0	0.582	0.786	0.527	0.106	
−0.2		−0.287	−0.879	−1.16	−0.824	0.0	0.824	1.16	0.879	0.287	
0.0	0.0	−0.371	−1.02	−1.31	−0.912	0.0	0.912	1.31	1.02	0.371	0.0
0.2		−0.287	−0.879	−1.16	−0.824	0.0	0.824	1.16	0.879	0.287	
0.4		−0.106	−0.527	−0.786	−0.582	0.0	0.582	0.786	0.527	0.106	
0.6		0.0	−0.162	−0.351	−0.293	0.0	0.293	0.351	0.162	0.0	
0.8			0.0	−0.053	−0.072	0.0	0.072	0.053	0.0		
1.0						0.0					

first mode for a clamped circular plate. One can observe from Table 7.6 that there is one vertical nodal lines, where zero values are obtained at $x = 0.0$ for various values of y. Similarly, there are two nodal lines for the third mode of vibration for a clamped circular plate. This is evident from the zero terms in Table 7.7, where the deflections are zero. For the clamped elliptic plate ($m = 0.8$), the results for deflection on the domain are depicted in Tables 7.8 through 7.10 for the first three modes of vibration. From Tables 7.8 through 7.10, we can observe that there are no nodal lines for the first mode, one vertical nodal line for the second mode, and one horizontal nodal line for the third mode, respectively.

TABLE 7.7

Deflection on the Domain for Determining Nodal Lines of Circular Plate for Third Mode of Vibration (Clamped): Shows Two Crossed Nodal Lines, $m = 1.0$

$x\rightarrow$ $y\downarrow$	−1.0	−0.8	−0.6	−0.4	−0.2	0.0	0.2	0.4	0.6	0.8	1.0
−1.0						0.0					
−0.8			0.0	0.109	0.384	0.535	0.384	0.109	0.0		
−0.6		0.0	0.0	0.328	0.905	1.20	0.905	0.328	0.0	0.0	
−0.4		−0.109	−0.328	0.0	0.715	1.09	0.715	0.0	−0.328	−00.109	
−0.2		−0.384	−0.905	−0.715	0.0	0.392	0.0	−00.715	−0.905	−00.384	
0.0	0.0	−0.535	−1.20	−1.09	−0.392	0.0	−0.392	−1.09	−1.20	−00.535	0.0
0.2		−0.384	−0.905	−0.715	0.0	0.392	0.0	−00.715	−0.905	−00.384	
0.4		−0.109	−0.328	0.0	0.715	1.09	0.715	0.0	−0.328	−00.109	
0.6		0.0	0.0	0.328	0.905	1.20	0.905	0.328	0.0	0.0	
0.8			0.0	0.109	0.384	0.535	0.384	0.109	0.0		
1.0						0.0					

TABLE 7.8

Deflection on the Domain for Determining Nodal Lines of Elliptic Plate for First Mode of Vibration (Clamped): Shows No Nodal Lines, $m = 0.8$

$x \rightarrow$ $y \downarrow$	−1.0	−0.8	−0.6	−0.4	−0.2	0.0	0.2	0.4	0.6	0.8	1.0
−1.0											
−0.8						0.0					
−0.6			0.007	0.096	0.206	0.254	0.206	0.096	0.007		
−0.4		0.013	0.185	0.459	0.699	0.793	0.699	0.459	0.185	0.013	
−0.2		0.099	0.420	0.827	1.16	1.29	1.16	0.827	0.420	0.997	
0.0	0.0	0.148	0.522	0.978	1.34	1.48	1.34	0.978	0.522	0.148	0.0
0.2		0.099	0.420	0.827	1.16	1.29	1.16	0.827	0.420	0.997	
0.4		0.013	0.185	0.459	0.699	0.793	0.699	0.459	0.185	0.013	
0.6			0.007	0.096	0.206	0.254	0.206	0.096	0.007		
0.8						0.0					
1.0											

7.4.2 Simply Supported Boundary

As mentioned previously, if we consider $s = 1$, then Equation 7.2 will become

$$F_i = u f_i(x, y)$$

For the purpose of computation, $f_i(x,y)$ was chosen of the form $x^{m_i} y^{n_i}$, where m_i and n_i are integers. For modes symmetric about both the axes, we can choose m_i and n_i as even integers. Similarly, by making other suitable

TABLE 7.9

Deflection on the Domain for Determining Nodal Lines of Elliptic Plate for Second Mode of Vibration (Clamped): Shows One Vertical Nodal Line, $m = 0.8$

$x \rightarrow$ $y \downarrow$	−1.0	−0.8	−0.6	−0.4	−0.2	0.0	0.2	0.4	0.6	0.8	1.0
−1.0											
−0.8						0.0					
−0.6			−0.014	−0.132	−0.146	0.0	0.46	0.132	0.014		
−0.4		−0.034	−0.378	−0.658	−0.517	0.0	0.517	0.658	0.378	0.034	
−0.2		−0.259	−0.879	−1.22	−0.881	0.0	0.881	1.22	0.879	0.259	
0.0	0.0	−0.386	−1.10	−1.45	−1.03	0.0	1.03	1.45	1.10	0.386	0.0
0.2		−0.259	−0.879	−1.22	−0.881	0.0	0.881	1.22	0.879	0.259	
0.4		−0.034	−0.378	−0.658	−0.517	0.0	0.517	0.658	0.378	0.034	
0.6			−0.014	−0.132	−0.146	0.0	0.46	0.132	0.014		
0.8						0.0					
1.0											

TABLE 7.10

Deflection on the Domain for Determining Nodal Lines of Elliptic Plate for Third Mode of Vibration (Clamped): Shows One Horizontal Nodal Line, $m = 0.8$

$x \rightarrow$ $y \downarrow$	−1.0	−0.8	−0.6	−0.4	−0.2	0.0	0.2	0.4	0.6	0.8	1.0
−1.0											
−0.8						0.0					
−0.6			−0.014	−0.219	−0.500	−0.628	−0.500	−0.219	−0.014		
−0.4		−0.017	−0.278	−0.762	−1.23	−1.43	−1.23	−0.762	−0.278	−0.017	
−0.2		−0.069	−0.333	−0.726	−1.08	−1.22	−1.08	−0.726	−0.333	−0.069	
0.0	0.0	0.0	0.0	0.0	0.0	0.0	0.0	0.0	0.0	0.0	0.0
0.2		0.069	0.333	0.726	1.08	1.22	1.08	0.726	0.333	0.069	
0.4		0.017	0.278	0.762	1.23	1.43	1.23	0.762	0.278	0.017	
0.6			0.014	0.219	0.500	0.628	0.500	0.219	0.014		
0.8						0.0					
1.0											

combinations, one may obtain other modes. Tables 7.11 and 7.12 give the first three frequency parameters for symmetric–symmetric, antisymmetric–symmetric modes for simply supported boundary with $\nu = 0.3$. The last row of these tables shows the results from the exact solution of the circular plates obtained using Bessel's functions, which are found to be in exact agreement with the solution obtained using BCOPs. These tables also depict the fact that as the aspect ratios of the elliptic plate are increased from 0.1 to 1.0 (circular),

TABLE 7.11

First Three Frequency Parameters for Symmetric–Symmetric Modes (Simply Supported) $\nu = 0.3$

	Frequency Parameters		
$m = b/a$	First	Second	Third
0.1	262.98	358.67	631.36
0.2	69.684	112.92	207.75
0.3	32.813	64.746	128.07
0.4	19.514	46.823	100.67
0.5	13.213	38.354	88.314
0.6	9.7629	33.777	70.271
0.7	7.7007	31.017	52.927
0.8	6.3935	29.139	41.677
0.9	5.5282	27.600	34.208
1.0	4.9351	25.613	29.720
1.0[a]	4.9351	25.613	29.720

[a] Computed from the exact solution.

TABLE 7.12

First Three Frequency Parameters for Antisymmetric–
Symmetric Modes (Simply Supported) $\nu = 0.3$

	Frequency Parameters		
$m = b/a$	First	Second	Third
0.1	299.24	419.95	766.22
0.2	88.792	141.23	264.65
0.3	46.830	86.882	170.57
0.4	31.146	66.937	138.65
0.5	23.641	57.625	122.27
0.6	19.566	52.527	90.816
0.7	17.157	49.269	71.655
0.8	15.634	46.687	59.503
0.9	14.615	43.783	52.181
1.0	13.898	39.981	48.582

the frequency parameters decrease. The effects of frequency parameters on the aspect ratios of the ellipse are shown in Figures 7.5 and 7.6.

To ensure the accuracy of the results, we carried out calculations for several values of n, till the first three frequencies in the different categories of modes converged to at least five significant digits in all the cases. To get the sense of convergence, the results are given for $m = 0.2$, 0.4, and 0.6 in Table 7.13. It is to be pointed out that for small m values, one should go up to 20 approximations, as sometimes they may not converge up to $n = 10$. Accordingly, the convergence results are shown in Figure 7.7 for $m = 0.5$.

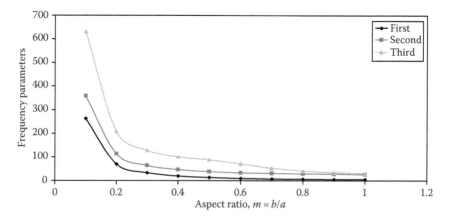

FIGURE 7.5

First three frequency parameters for symmetric–symmetric modes (simply supported).

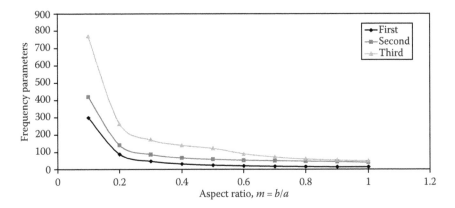

FIGURE 7.6
First three frequency parameters for antisymmetric–symmetric modes (simply supported).

Deflections on the domain of elliptic plate ($m = 0.4$, $\nu = 0.3$) with simply supported boundary for the first and second symmetric–symmetric modes are incorporated in Tables 7.14 and 7.15. Consequently, it can be observed that the first symmetric–symmetric modes will have no nodal lines, whereas the second symmetric–symmetric modes will have two curved-nodal lines. For the second case, Table 7.15 shows the changing values from positive to negative at $x = [-0.4, -0.2]$ and from negative to positive at $x = [0.2, 0, 4]$. Between these values of x, one may have the nodal lines. These tables may directly be used to draw the nodal lines and if one desires, may be used to plot the mode shapes.

Deflections on the domain of elliptic plate ($m = 0.4$, $\nu = 0.3$) with simply supported boundary for the first and second antisymmetric–symmetric modes are listed in Tables 7.16 and 7.17. Thus, it can be observed that the

TABLE 7.13

Convergence Study for Symmetric–Symmetric Modes (Simply Supported) $\nu = 0.3$

	$m = 0.2$			$m = 0.4$			$m = 0.6$		
n	First	Second	Third	First	Second	Third	First	Second	Third
2	77.885	191.09		21.256	69.159		10.519	49.408	
3	71.205	189.54	933.16	19.627	68.299	236.97	9.7838	48.342	108.42
4	70.262	130.72	376.25	19.541	50.783	180.13	9.7691	36.006	107.32
5	69.811	123.83	366.22	19.520	48.499	171.50	9.7661	34.562	100.73
6	69.793	123.82	366.21	19.515	48.488	156.36	9.7630	34.519	72.123
7	69.710	114.78	212.56	19.515	47.052	102.70	9.7630	33.912	72.118
8	69.684	112.93	207.80	19.514	46.834	100.83	9.7629	33.779	70.352
9	69.684	112.92	207.75	19.514	46.823	100.67	9.7629	33.777	70.272
10	69.684	112.92	207.75	19.514	46.823	100.67	9.7629	33.777	70.271

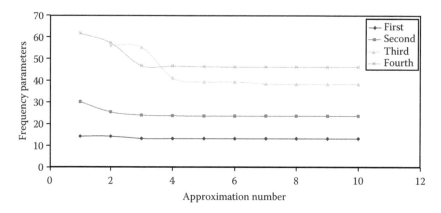

FIGURE 7.7
Convergence of results, $b/a = 0.5$ (simply supported).

first antisymmetric–symmetric modes will have one vertical nodal line, whereas the second antisymmetric–symmetric modes will have three nodal lines. For the second case, Table 7.17 shows one vertical nodal line, along with the changing values from negative to positive at $x = [-0.6, -0.4]$ and from negative to positive at $x = [0.4, 0.6]$, and between these values of x, one may find the other two nodal lines.

7.4.3 Completely Free Boundary

We have given the results of the first three frequency parameters of all the cases of symmetric–symmetric, symmetric–antisymmetric, antisymmetric–symmetric, and antisymmetric–antisymmetric modes in Tables 7.18

TABLE 7.14

Deflection on the Domain for Determining Nodal Lines of Elliptic Plate for First Symmetric–Symmetric Modes of Vibration (Simply Supported): Shows No Nodal Lines, $m = 0.4$, $\nu = 0.3$

$x \rightarrow$ $y \downarrow$	−1.0	−0.8	−0.6	−0.4	−0.2	0.0	0.2	0.4	0.6	0.8	1.0
−0.4						0.0					
−0.3			0.079	0.358	0.592	0.682	0.592	0.358	0.079		
−0.2		0.088	0.432	0.827	1.14	1.27	1.14	0.827	0.432	0.088	
−0.1		0.252	0.673	1.14	1.52	1.66	1.52	1.14	0.673	0.252	
0.0	0.0	0.311	0.759	1.26	1.65	1.80	1.65	1.26	0.759	0.311	0.0
0.1		0.252	0.673	1.14	1.52	1.66	1.52	1.14	0.673	0.252	
0.2		0.088	0.432	0.827	1.14	1.27	1.14	0.827	0.432	0.088	
0.3			0.079	0.358	0.592	0.682	0.592	0.358	0.079		
0.4						0.0					

TABLE 7.15

Deflection on the Domain for Determining Nodal Lines of Elliptic Plate for Second Symmetric–Symmetric Modes of Vibration (Simply Supported): Shows Two Curved Nodal Lines in $x = [-0.4, -0.2]$ and $[0.2, 0.4]$, $m = 0.4$, $\nu = 0.3$

$x\rightarrow$ $y \downarrow$	−1.0	−0.8	−0.6	−0.4	−0.2	0.0	0.2	0.4	0.6	0.8	1.0
−0.4						0.0					
−0.3			0.170	0.239	−0.276	−0.598	−0.276	0.239	0.170		
−0.2		0.320	1.01	0.619	−0.503	−1.10	−0.503	0.619	1.01	0.320	
−0.1		0.958	1.66	0.915	−0.638	−1.42	−0.638	0.915	1.66	0.958	
0.0	0.0	1.20	1.90	1.03	−0.683	−1.53	−0.683	1.03	1.90	1.20	0.0
0.1		0.958	1.66	0.915	−0.638	−1.42	−0.638	0.915	1.66	0.958	
0.2		0.320	1.01	0.619	−0.503	−1.10	−0.503	0.619	1.01	0.320	
0.3			0.170	0.239	−0.276	−0.598	−0.276	0.239	0.170		
0.4						0.0					

through 7.21. Here, ν was taken as 0.33 and the aspect ratios of the ellipse were increased from 0.1 to 1.0 (circular) at the step of 0.1. The last row in these four tables gives the exact solution from Bessel function with respect to completely free boundary. Excellent results in this special case prove the efficacy of the BCOPs method. Moreover, as in other boundary conditions, the frequency parameters were found to decrease as we increase the values of the aspect ratio m. These effects are shown in Figures 7.8 through 7.11. The convergence of results of the completely free elliptic plate ($m = 0.5$) as the

TABLE 7.16

Deflection on the Domain for Determining Nodal Lines of Elliptic Plate for First Antisymmetric–Symmetric Modes of Vibration (Simply Supported): Shows One Vertical Nodal Line, $m = 0.4$, $\nu = 0.3$

$x\rightarrow$ $y \downarrow$	−1.0	−0.8	−0.6	−0.4	−0.2	0.0	0.2	0.4	0.6	0.8	1.0
−0.4						0.0					
−0.3			−0.146	−0.479	−0.412	0.0	0.412	0.479	0.146		
−0.2		−0.199	−0.824	−1.13	−0.818	0.0	0.818	1.13	0.824	0.199	
−0.1		−0.580	−1.31	−1.59	−1.10	0.0	1.10	1.59	1.31	0.580	
0.0	0.0	−0.720	−1.48	−1.76	−1.20	0.0	1.20	1.76	1.48	0.720	0.0
0.1		−0.580	−1.31	−1.59	−1.10	0.0	1.10	1.59	1.31	0.580	
0.2		−0.199	−0.824	−1.13	−0.818	0.0	0.818	1.13	0.824	0.199	
0.3			−0.146	−0.479	−0.412	0.0	0.412	0.479	0.146		
0.4						0.0					

TABLE 7.17

Deflection on the Domain for Determining Nodal Lines of Elliptic Plate for Second Antisymmetric–Symmetric Modes of Vibration (Simply Supported): Shows Three Nodal Lines in $x = [-0.6, -0.4]$, $[0.4, 0.6]$ and One Vertical, $m = 0.4$, $\nu = 0.3$

$x \rightarrow$ $y \downarrow$	−1.0	−0.8	−0.6	−0.4	−0.2	0.0	0.2	0.4	0.6	0.8	1.0
−0.4						0.0					
−0.3			−0.123	0.206	0.551	0.0	−0.551	−0.206	0.123		
−0.2		−0.426	−0.821	0.400	1.06	0.0	−1.06	−0.400	0.821	0.426	
−0.1		−1.33	−1.43	0.480	1.38	0.0	−1.38	−0.480	1.43	1.33	
0.0	0.0	−1.69	−1.67	0.498	1.50	0.0	−1.50	−0.498	1.67	1.69	0.0
0.1		−1.33	−1.43	0.480	1.38	0.0	−1.38	−0.480	1.43	1.33	
0.2		−0.426	−0.821	0.400	1.06	0.0	−1.06	−0.400	0.821	0.426	
0.3			−0.123	0.206	0.551	0.0	−0.551	−0.206	0.123		
0.4						0.0					

number of approximation increased from 1 to 15 for the first four frequency parameters are as depicted in Figure 7.12.

From the above tables, one may choose, for example, the first four frequency parameters as 6.4923 (from symmetric–symmetric mode), 8.6797 (from antisymmetric–antisymmetric mode), 16.302 (from antisymmetric–symmetric mode), and 18.764 (from symmetric–antisymmetric mode) for an elliptic plate with aspect ratio, $m = 0.6$ ($\nu = 0.33$). Accordingly, the deflections

TABLE 7.18

First Three Frequency Parameters for Symmetric–Symmetric Modes (Completely Free) $\nu = 0.33$

	Frequency Parameters		
$m = b/a$	First	Second	Third
0.1	6.7077	32.414	78.058
0.2	6.7082	32.5307	78.3724
0.3	6.6959	32.408	71.688
0.4	6.6615	31.988	41.899
0.5	6.5969	27.763	31.223
0.6	6.4923	19.940	30.015
0.7	6.3321	15.198	28.273
0.8	6.0911	12.183	26.092
0.9	5.7377	10.256	23.763
1.0	5.2620	9.0689	21.5275
1.0[a]	5.2620	9.0689	21.5272

[a] Computed from the exact solution.

TABLE 7.19

First Three Frequency Parameters for Symmetric–
Antisymmetric Modes (Completely Free) $\nu = 0.33$

$m = b/a$	Frequency Parameters		
	First	Second	Third
0.1	92.294	180.74	299.25
0.2	47.625	100.08	173.31
0.3	33.096	74.772	134.90
0.4	25.945	62.329	107.54
0.5	21.661	54.598	70.535
0.6	18.764	48.970	50.190
0.7	16.629	37.814	44.367
0.8	14.940	29.770	40.281
0.9	13.517	24.310	36.522
1.0	12.243	20.513	33.062
1.0[a]	12.243	20.513	33.062

[a] Computed from the exact solution.

on the elliptic domain for the above first four frequency modes are given in Tables 7.22 through 7.25. Table 7.22 shows two curved-nodal lines along the changing values from positive to negative at $x = [-0.6, -0.4]$ and from negative to positive at $x = [0.4, 0.6]$, for the first mode of vibration of this particular elliptic plate, viz., with $m = 0.6$. There are two nodal lines, one vertical and

TABLE 7.20

First Three Frequency Parameters for Antisymmetric–
Symmetric Modes (Completely Free) $\nu = 0.33$

$m = b/a$	Frequency Parameters		
	First	Second	Third
0.1	17.188	52.404	109.32
0.2	17.223	52.639	109.64
0.3	17.184	52.332	94.993
0.4	17.036	51.429	60.028
0.5	16.754	42.854	49.834
0.6	16.302	33.170	47.355
0.7	15.620	27.343	44.012
0.8	14.665	23.814	40.243
0.9	13.487	21.734	36.522
1.0	12.243	20.513	33.064
1.0[a]	12.243	20.513	33.061

[a] Computed from the exact solution.

TABLE 7.21

First Three Frequency Parameters for Antisymmetric–
Antisymmetric Modes (Completely Free) $\nu = 0.33$

	Frequency Parameters		
$m = b/a$	First	Second	Third
0.1	50.097	135.43	229.06
0.2	25.207	72.300	131.47
0.3	16.948	52.162	101.24
0.4	12.826	42.300	86.226
0.5	10.346	36.319	76.591
0.6	8.6797	32.150	69.162
0.7	7.4765	28.923	55.009
0.8	6.5629	26.204	45.723
0.9	5.8437	23.7695	39.499
1.0	5.2620	21.527	35.242
1.0[a]	5.2620	21.527	35.242

[a] Computed from the exact solution.

one horizontal, where the deflections are zero, as shown in Table 7.23. Furthermore, Table 7.24 depicts the three nodal lines, showing one of them vertical at $x = 0.0$ and the other two between $x = [-0.8, -0.6]$ and $x = [0.6, 0.8]$. Similarly, Table 7.25 shows three nodal lines, depicting one of them as horizontal at $x = 0.0$ where the deflections are zero. The other two nodal

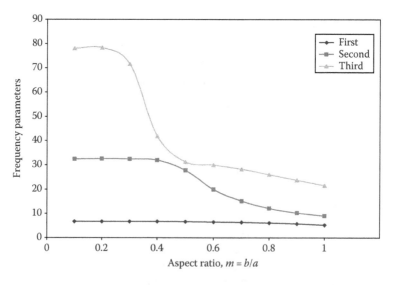

FIGURE 7.8
First three frequency parameters for symmetric–symmetric modes (completely free).

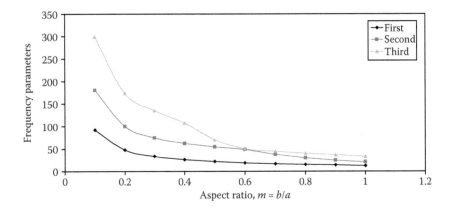

FIGURE 7.9
First three frequency parameters for symmetric–antisymmetric modes (completely free).

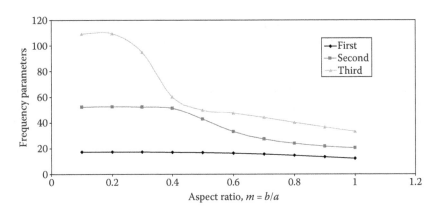

FIGURE 7.10
First three frequency parameters for antisymmetric–symmetric modes (completely free).

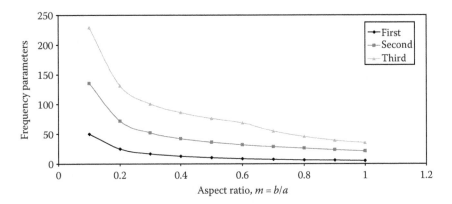

FIGURE 7.11
First three frequency parameters for antisymmetric–antisymmetric modes (completely free).

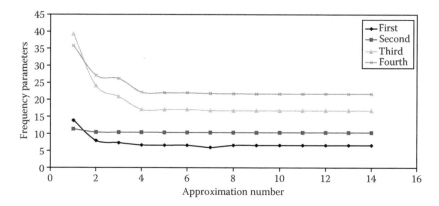

FIGURE 7.12

Convergence of results, $m = 0.5$ (completely free).

lines can be found at $x = [-0.4, -0.2]$ and $x = [0.2, 0.4]$. In this way, one may obtain the nodal lines and mode shapes for any of the elliptic plates for any of the three boundary conditions.

Lastly, the exact solutions of the frequency parameters for a circular plate with completely free boundary condition taking $\nu = 0.33$ were determined and are listed in Table 7.26. The exact solutions were found using the Bessel function solution as described by Leissa (1969) as well as in the earlier chapters of this book.

TABLE 7.22

Deflection on the Domain for Determining Nodal Lines of Elliptic Plate for First Mode of Vibration (Chosen from Symmetric–Symmetric Mode) (Completely Free)

$x \rightarrow$ $y \downarrow$	−1.0	−0.8	−0.6	−0.4	−0.2	0.0	0.2	0.4	0.6	0.8	1.0
−0.6						−1.03					
−0.5				−0.394	−0.777	−0.911	−0.777	−0.394			
−0.4			0.286	−0.297	−0.682	−0.817	−0.682	−0.297	0.286		
−0.3		1.07	0.364	−0.220	−0.607	−0.743	−0.607	−0.220	0.364	1.07	
−0.2		1.13	0.421	−0.164	−0.553	−0.689	−0.553	−0.164	0.421	1.13	
−0.1		1.16	0.456	−0.130	−0.520	−0.657	−0.520	−0.130	0.456	1.16	
0.0	0.0	1.17	0.468	−0.118	−0.509	−0.646	−0.509	−0.118	0.468	1.17	0.0
0.1		1.16	0.456	−0.130	−0.520	−0.657	−0.520	−0.130	0.456	1.16	
0.2		1.13	0.421	−0.164	−0.553	−0.689	−0.553	−0.164	0.421	1.13	
0.3		1.07	0.364	−0.220	−0.607	−0.743	−0.607	−0.220	0.364	1.07	
0.4			0.286	−0.297	−0.682	−0.817	−0.0682	−0.297	0.286		
0.5				−0.394	−0.777	−0.911	−0.777	−0.394			
0.6						−1.03					

Note: Shows two curved nodal lines in $x = [-0.6, -0.4]$ and $[0.6, 0.4]$, $m = 0.6$, $\nu = 0.33$.

TABLE 7.23

Deflection on the Domain for Determining Nodal Lines of Elliptic Plate for Second Mode of Vibration (Chosen from Antisymmetric–Antisymmetric Mode) (Completely Free)

$x \rightarrow$ $y \downarrow$	−1.0	−0.8	−0.6	−0.4	−0.2	0.0	0.2	0.4	0.6	0.8	1.0
−0.6						0.0					
−0.5				1.25	0.656	0.0	−0.656	−1.25			
−0.4			1.42	1.02	0.538	0.0	−0.538	−1.02	−1.42		
−0.3		1.29	1.08	0.784	0.413	0.0	−0.413	−0.784	−1.08	−1.29	
−0.2		0.871	0.731	0.532	0.280	0.0	−0.280	−0.532	−0.731	−0.871	
−0.1		0.439	0.369	0.269	0.142	0.0	−0.142	−0.269	−0.369	−0.439	
0.0	0.0	0.0	0.0	0.0	0.0	0.0	0.0	0.0	0.0	0.0	0.0
0.1		−0.439	−0.369	−0.269	−0.142	0.0	0.142	0.269	0.369	0.439	
0.2		−0.871	−0.731	−0.532	−0.280	0.0	0.280	0.532	0.731	0.871	
0.3		−1.29	−1.08	−0.784	−0.413	0.0	0.413	0.784	1.08	1.29	
0.4			−1.42	−1.02	−0.538	0.0	0.538	1.02	1.42		
0.5				−1.25	−0.656	0.0	0.656	1.25			
0.6						0.0					

Note: Shows two nodal lines (one vertical and one horizontal), $m = 0.6$, $\nu = 0.33$.

TABLE 7.24

Deflection on the Domain for Determining Nodal Lines of Elliptic Plate for Third Mode of Vibration (Chosen from Antisymmetric–Symmetric Mode) (Completely Free)

$x \rightarrow$ $y \downarrow$	−1.0	−0.8	−0.6	−0.4	−0.2	0.0	0.2	0.4	0.6	0.8	1.0
−0.6						0.0					
−0.5				1.15	0.777	0.0	−0.777	−1.15			
−0.4			0.636	0.972	0.687	0.0	−0.687	−0.972	−0.636		
−0.3		−0.544	0.415	0.823	0.614	0.0	−0.614	−0.823	−0.415	0.544	
−0.2		−0.748	0.248	0.711	0.558	0.0	−0.558	−0.711	−0.248	0.748	
−0.1		−0.877	0.144	0.640	0.523	0.0	−0.523	−0.640	−0.144	0.877	
0.0	0.0	−0.920	0.108	0.616	0.511	0.0	−0.511	−0.616	−0.108	0.920	0.0
0.1		−0.877	0.144	0.640	0.523	0.0	−0.523	−0.640	−0.144	0.877	
0.2		−0.748	0.248	0.711	0.558	0.0	−0.558	−0.711	−0.248	0.748	
0.3		−0.544	0.415	0.823	0.614	0.0	−0.614	−0.823	−0.415	0.544	
0.4			0.636	0.972	0.687	0.0	−0.687	−0.972	−0.636		
0.5				1.15	0.777	0.0	−0.777	−1.15			
0.6						0.0					

Note: Shows three nodal lines (one vertical and another two in $x = [−0.8, −0.6]$ and $[0.6, 0.8]$), $m = 0.6$, $\nu = 0.33$.

TABLE 7.25

Deflection on the Domain for Determining Nodal Lines of Elliptic Plate for Fourth Mode of Vibration (Chosen from Symmetric–Antisymmetric Mode) (Completely Free)

$x \rightarrow$ $y \downarrow$	−1.0	−0.8	−0.6	−0.4	−0.2	0.0	0.2	0.4	0.6	0.8	1.0
−0.6						1.58					
−0.5				−0.035	0.905	1.26	0.905	−0.035			
−0.4			−1.12	−0.111	0.680	0.979	0.680	−0.111	−1.12		
−0.3		−1.63	−0.922	−0.136	0.484	0.719	0.484	−0.136	−0.922	−1.63	
−0.2		−1.13	−0.659	−0.119	0.309	0.472	0.309	−0.119	−0.659	−1.13	
−0.1		−0.582	−0.343	−0.068	0.151	0.234	0.151	−0.068	−0.343	−0.582	
0.0	0.0	0.0	0.0	0.0	0.0	0.0	0.0	0.0	0.0	0.0	0.0
0.1		0.582	0.343	0.068	−0.151	−0.234	−0.151	0.068	0.343	0.582	
0.2		1.13	0.659	0.119	−0.309	−0.472	−0.309	0.119	0.659	1.13	
0.3		1.63	0.922	0.136	−0.484	−0.719	−0.484	0.136	0.922	1.63	
0.4			1.12	0.111	−0.680	−0.979	−0.680	0.111	1.12		
0.5				0.035	−0.905	−1.26	−0.905	0.035			
0.6						−1.58					

Note: Shows three nodal lines (one horizontal and another two in $x = [-0.4, -0.2]$ and $[0.2, 0.4]$), $m = 0.6$, $\nu = 0.33$.

7.5 Conclusion

The Rayleigh–Ritz method using BCOPs provides a highly accurate and computationally efficient scheme for finding the vibration characteristics of

TABLE 7.26

Exact Solution of Frequency Parameters for Circular Plate (Completely Free) $\nu = 0.33$

s				n			
	$n=0$	$n=1$	$n=2$	$n=3$	$n=4$	$n=5$	$n=6$
0	—	—	5.2620	12.243	21.527	33.061	46.808
1	9.0689	20.513	35.242	52.921	73.378	96.507	122.23
2	38.507	59.859	84.376	119.90	142.32	175.56	211.54
3	87.812	119.00	153.32	190.67	230.96	274.12	320.11
4	156.88	197.92	242.06	289.23	339.36	392.41	448.31
5	245.69	296.59	350.57	407.56	467.54	530.45	596.24
6	354.25	415.01	478.82	545.66	615.48	688.24	763.91
7	482.55	553.17	626.83	703.51	783.18	865.79	951.33
8	630.59	711.07	794.59	881.12	970.63	1063.1	1158.5
9	798.37	888.71	982.09	1078.4	1177.8	1280.1	1385.4
10	985.89	1086.1	1189.3	1295.5	1404.7	1516.9	1631.9

transverse vibration of elliptic and circular plates. This has already been demonstrated by several authors and the studies carried out by this author have been cited in the references. This chapter discusses this method for elliptic and circular plates. Various comparisons have already been mentioned in the author's previous publications, viz., in Singh and Chakraverty (1991, 1992a,b) and are not repeated here. The investigator worked with sufficiently large number of terms till the first three frequencies of all the categories of modes converged to at least five significant digits. Moreover, the results from the successive approximations acted as an indication of the rate of convergence.

References

Beres, D.P. 1974. Vibration analysis of a completely free elliptic plate. *Journal of Sound and Vibration*, 34(3): 441–443.

Bhat, R.B. 1985. Natural frequencies of rectangular plates using characteristic orthogonal polynomials in Rayleigh-Ritz method. *Journal of Sound and Vibration*, 102(4): 493–499.

Bhat, R.B. 1987. Flexural vibration of polygonal plates using characteristic orthogonal polynomials in two variables. *Journal of Sound and Vibration*, 114(1): 65–71.

Bhat, R.B., Laura, P.A.A., Gutierrez, R.G., Cortinez, V.H., and Sanzi, H.C. 1990. Numerical experiments on the determination of natural frequencies of transverse vibrations of rectangular plates of non-uniform thickness. *Journal of Sound and Vibration*, 138(2): 205–219.

Chakraverty, S. 1992. *Numerical Solution of Vibration of Plates*, PhD Thesis, University of Roorkee, Roorkee, India.

Cotugno, N. and Mengotti-Marzolla, C. 1943. Approssimazione per eccesso della piu bassa frequenza di una piastre ellittrica omogenea incastrata. Atti dell' Accedemia nazionale dei Lincei Memorie 5, Ser. 8, Sem. 2.

Dickinson, S.M. and Blasio, A. Di. 1986. On the use of orthogonal polynomials in the Rayleigh-Ritz method for the study of the flexural vibration and buckling of isotropic and orthotropic rectangular plates. *Journal of Sound and Vibration*, 108(1): 51–62.

Lam, K.Y., Liew, K.M., and Chow, S.T., 1990. Free vibration analysis of isotropic and orthotropic triangular plates. *International Journal of Mechanical Sciences*, 32(5): 455–464.

Laura, P.A.A., Gutierrez, R.H., and Bhat, R.B. 1989. Transverse vibrations of a trapezoidal cantilever plate of variable thickness. *AIAA Journal*, 27(7): 921–922.

Leissa, A.W. 1967. Vibration of a simply supported elliptic plate. *Journal of Sound and Vibration*, 6(1): 145–148.

Leissa, A.W. 1969. *Vibration of Plates*, NASA SP160, U.S. Government Printing Office, Washington, DC.

Liew, K.M. and Lam, K.Y. 1990. Application of 2-dimensional orthogonal plate function to flexural vibration of skew plates. *Journal of Sound and Vibration*, 139(2): 241–252.

Liew, K.M., Lam, K.Y., and Chow, S.T. 1990. Free vibration analysis of rectangular plates using orthogonal plate function. *Computers and Structures*, 34(1): 79–85.

Mazumdar, J. 1971. Transverse vibrations of elastic plates by the method of constant deflection lines. *Journal of Sound and Vibration*, 18, 147.

McLachlan, N.W. 1947. Vibration problems in elliptic coordinates. *Quarterly of Applied Mathematics*, 5(3): 289–297.

McNitt, R.P. 1962. Free vibrations of a clamped elliptic plate. *Journal of Aerospace Science*, 29(9): 1124–1125.

Rajalingham, C. and Bhat, R.B. 1991. Vibration of elliptic plates using characteristic orthogonal polynomials in the Rayleigh-Ritz method. *International Journal of Mechanical Sciences*, 33(9): 705–716.

Rajalingham, C. and Bhat, R.B. 1993. Axisymmetrical vibration of circular plates and its analog in elliptic plates using characteristic orthogonal polynomials. *Journal of Sound and Vibration*, 161(1): 109–118.

Rajalingham, C., Bhat, R.B., and Xistris, G.D. 1991. Natural frequencies and mode shapes of elliptic plates with boundary characteristic orthogonal polynomials as assumed shape functions, in *Proceedings of the 13th Biennial Conference on Mechanical Vibration and Noise*, Miami, FL.

Rajalingham, C., Bhat, R.B., and Xistris, G.D. 1993. Natural frequencies and mode shapes of elliptic plates with boundary characteristic orthogonal polynomials as assumed shape functions. *ASME Journal of Vibration Acoustics*, 115: 353–358.

Sato, K. 1973. Free flexural vibrations of an elliptical plate with free edge. *Journal of Acoustical Society of America*, 54: 547–550.

Shibaoka, Y. 1956. On the transverse vibration of an elliptic plate with clamped edge. *Journal of Physical Society of Japan*, 11(7): 797–803.

Singh, B. and Tyagi, D.K. 1985. Transverse vibration of an elliptic plate with variable thickness. *Journal of Sound and Vibration*, 99, 379–391.

Singh, B. and Goel, R. 1985. Transverse vibration of an elliptic plate with variable thickness. *Proceedings Workshop of Solid Mechanics*, Department of Mathematics, University of Roorkee, 13–16 March.

Singh, B. and Chakraverty, S. 1991. Transverse vibration of completely free elliptic and circular plates using orthogonal polynomials in Rayleigh-Ritz method. *International Journal of Mechanical Sciences*, 33(9): 741–751.

Singh, B. and Chakraverty, S. 1992a. Transverse vibration of simply-supported elliptic and circular plates using boundary characteristic orthogonal polynomials in two dimensions. *Journal of Sound and Vibration*, 152(1): 149–155.

Singh, B. and Chakraverty, S. 1992b. On the use of orthogonal polynomials in Rayleigh-Ritz method for the study of transverse vibration of elliptic plates. *International Journal of Computers and Structures*, 43(3): 439–443.

Waller, M.D. 1938. Vibrations of free circular plates. *Proceedings of the Physical Society (London)*, 50, 70–76.

Waller, M.D. 1950. Vibration of free elliptic plates. *Proceedings of the Physical Society (London)*, Ser. B, 63, 451–455.

8

Triangular Plates

8.1 Introduction

Various researchers around the globe have studied free vibration of triangular plates. The work of Leissa (1969), being an excellent monograph, has information related to triangular plates. Further information may also be obtained from the subsequent review papers by Leissa (1977, 1981, 1987). In this chapter, the descriptions about the references mentioned in those review papers will not be repeated. The methods in those investigations range from experimental to approximate by employing a few term approximations with Rayleigh–Ritz or collocation methods. In some cases, the finite difference methods were also employed. In most of the cases, the plate was considered to be either equilateral or isosceles or a right-angled triangle. Moreover, majority of the investigations deal with the cantilever triangular plates, i.e., having one edge clamped and the other two free.

Gorman (1983, 1986, 1987, 1989) gave an accurate method, viz., superposition method, for the free vibration analysis of right-angled triangular plates that are either simply supported or clamped–simply supported, or having one edge free. Bhat (1987) had used characteristic orthogonal polynomials with two variables to discuss the flexural vibration of polygonal plates, but the results were given for a triangular isosceles or a right-angled plate with one edge clamped and the other two free. Kim and Dickinson (1990) studied the free flexural vibration of right triangular isotropic and orthotropic plates by using simple polynomials as admissible functions. Lam et al. (1990) used two-dimensional orthogonal plate functions as admissible functions in the Rayleigh–Ritz method. Numerical results were given for an isotropic right-angled triangular plate.

In this chapter, free vibration of triangular plates is discussed in the most general form, viz., for triangular domain of arbitrary shape and size and with all possible types of boundary conditions on the three edges. In this context, it is worth mentioning that for a general triangle, there exist 27 different cases of boundary conditions, of which 10 will be similar (but not the same). Rayleigh–Ritz method with boundary characteristics orthogonal polynomials (BCOPs) as the basis functions is applied to generate a sequence of approximations (Singh and Chakraverty, 1992). The process is terminated when the required number of frequencies is converged to the desired accuracy. It may be noted that there will be different sets of polynomials over triangles of different shapes. Accordingly, a mapping function is used to map the given general triangle onto a standard triangle. Then, a set of orthogonal polynomials (BCOPs) is generated on this standard triangle satisfying the essential boundary conditions and was used in the Rayleigh–Ritz method. The procedure, however, will be clear in the following sections.

8.2 Mapping of General Triangle onto a Standard Triangle

Let us consider that the three numbers a, b, and c determine a triangle completely, as shown in Figure 8.1. Then, the following transformation is used to map the interior of a general triangle R onto an isosceles right-angled triangle T, as shown in Figure 8.2

$$\begin{cases} x = a\xi + b\eta \\ y = c\eta \end{cases} \tag{8.1}$$

or

$$\begin{cases} \eta = \frac{y}{c} \\ \xi = \frac{(x-by/c)}{a} \end{cases} \tag{8.2}$$

It may be noticed here that the transformation is similar to that used in the finite element method where one element is transformed into another element of similar shape. But, here the transformation used in Equations 8.1 and 8.2 is globally used for the whole domain of the plate. We now use the above transformation and then generate the orthogonal polynomials satisfying the essential boundary conditions, viz., BCOPs, by using the Gram–Schmidt process.

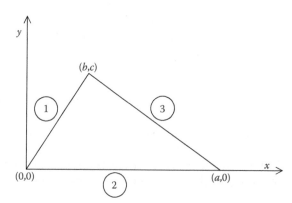

FIGURE 8.1
A general triangular domain.

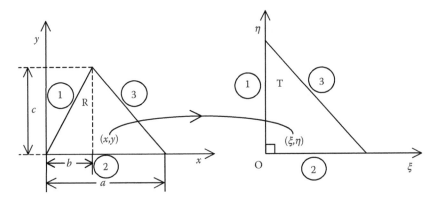

FIGURE 8.2
Mapping of a general triangle onto the standard triangle.

8.3 Generation of the BCOPs over the Triangular Domain

For this, we start with a linearly independent set

$$F_i = f(\xi,\eta) \, f_i(\xi,\eta), \quad i = 1, 2, \ldots \tag{8.3}$$

where

$$f(\xi,\eta) = \xi^p \eta^q (1 - \xi - \eta)^r \tag{8.4}$$

satisfies the essential boundary conditions of the edges of triangular plate and

$$f_i(\xi,\eta) = \{1, \xi, \eta, \xi^2, \xi\eta, \eta^2, \ldots\} \tag{8.5}$$

The superscripted numbers p, q, and r in Equation 8.4 determine the boundary conditions of the sides of the plate. These numbers can take the value 0, 1, or 2. For example, $p = 0$, 1, or 2 designate whether the side $\xi = 0$ happens to be free, simply supported, or clamped boundary. Similarly, q and r control the boundary conditions on the sides $\eta = 0$ and $\xi + \eta = 1$, respectively. The inner product of two functions u and v over the standard triangular domain T is defined as

$$\langle u,v \rangle = \iint_T u(\xi,\eta)v(\xi,\eta)d\xi d\eta \tag{8.6}$$

Norm of the function u may then be given by

$$\| u \| = \langle u,u \rangle^{1/2} = \left[\iint_T u^2(\xi,\eta)d\xi d\eta \right]^{1/2} \tag{8.7}$$

Proceeding as in the previous chapters, the Gram–Schmidt process can be described to generate the BCOPs as

$$
\left.
\begin{aligned}
&\phi_1 = F_1 \\
&\phi_i = F_i - \sum_{j=1}^{i-1} \alpha_{ij}\phi_j \\
&\text{where} \\
&\alpha_{ij} = \frac{\langle F_i, \phi_j \rangle}{\langle \phi_j, \phi_j \rangle}, \quad j = 1, 2, \ldots, (i-1)
\end{aligned}
\right\} \quad i = 2, 3, 4, \ldots \qquad (8.8)
$$

The functions ϕ_i may then be normalized by using

$$
\hat{\phi}_i = \frac{\phi_i}{\| \phi_i \|} \qquad (8.9)
$$

Next, these BCOPs are used in the Rayleigh–Ritz method to extract the free vibration characteristics.

8.4   Rayleigh–Ritz Method in Triangular Plates

We will use the usual procedure and use the N-term approximation

$$
W(x,y) = \sum_{j=1}^{N} c_j \phi_j \qquad (8.10)
$$

for the deflection in the Rayleigh quotient for the plate after equating the maximum kinetic and strain energy as

$$
\omega^2 = \frac{D}{h\rho} \frac{\underset{R}{\iint} \left[(\nabla^2 W)^2 + 2(1 - \nu)\left\{ W_{xy}^2 - W_{xx}W_{yy} \right\} \right] dxdy}{\underset{R}{\iint} W^2 dxdy} \qquad (8.11)
$$

Then, the above Rayleigh quotient is extremized as a function of c_j. The variables x and y are finally changed to ξ and η and this leads to the following eigenvalue problem:

$$
\sum_{j=1}^{N} (a_{ij} - \lambda^2 b_{ij})c_j = 0, \quad i = 1, 2, \ldots, N \qquad (8.12)
$$

where

$$
a_{ij} = \iint_T \left[A_1 \phi_i^{\xi\xi} \phi_j^{\xi\xi} + A_2 \left(\phi_i^{\xi\eta} \phi_j^{\xi\xi} + \phi_i^{\xi\xi} \phi_j^{\xi\eta} \right) + A_3 \left(\phi_i^{\eta\eta} \phi_j^{\xi\xi} + \phi_i^{\xi\xi} \phi_j^{\eta\eta} \right) \right.
$$

$$
\left. + A_4 \phi_i^{\xi\eta} \phi_j^{\xi\eta} + A_5 \left(\phi_i^{\eta\eta} \phi_j^{\xi\eta} + \phi_i^{\xi\eta} \phi_j^{\eta\eta} \right) + A_6 \phi_i^{\eta\eta} \phi_j^{\eta\eta} \right] d\xi d\eta \qquad (8.13)
$$

$$
b_{ij} = \iint_T \phi_i \phi_j \, d\xi d\eta \qquad (8.14)
$$

and

$$
\lambda^2 = \frac{a^4 \omega^2 \rho h}{D} \qquad (8.15)
$$

and $A_1, A_2, A_3, \ldots, A_6$ are given by

$$
\left.
\begin{aligned}
A_1 &= (1 + \theta^2)^2 \\
A_2 &= \frac{-2\theta(1 + \theta^2)}{\mu} \\
A_3 &= \frac{(\nu + \theta^2)}{\mu^2} \\
A_4 &= \frac{2(1 - \nu + 2\theta^2)}{\mu^2} \\
A_5 &= -\frac{2\theta}{\mu^3} \\
A_6 &= \frac{1}{\mu^4}
\end{aligned}
\right\} \qquad (8.16)
$$

$$
\mu = \frac{c}{a} \quad \text{and} \quad \theta = \frac{b}{c} \qquad (8.17)
$$

The integrals involved are available in closed form by using the following result (Singh and Chakraverty, 1992):

$$
\iint_T \xi^i \eta^j (1 - \xi - \eta)^k \, d\xi d\eta = \frac{i! \, j! \, k!}{(i + j + k + 2)!} \qquad (8.18)
$$

8.5 Some Numerical Results and Discussions for Triangular Plates

By using the shape functions in terms of the BCOPs, Equation 8.12 will become a standard eigenvalue problem. The solution of this equation

TABLE 8.1

Fundamental Frequencies for a Right-Triangular Plate and Sides with Clamped (C) or Free (F) Boundary Conditions

S. No.	Sides 1 2 3	Values of c/a				
		1.0	1.5	2.0	2.5	3.0
1	C F F	6.173	5.796	5.511	5.294	5.126
2	F C F	6.173	2.878	1.658	1.075	0.7528
3	F F C	12.64	8.895	7.363	6.553	6.054
4	C C F	29.09	19.78	15.43	12.94	11.34
5	C F C	41.12	33.93	30.86	29.19	28.14
6	F C C	41.12	24.13	17.52	14.12	12.07

Note: $\theta = 0$, $\mu = 1.0, 1.5, 2.0, 2.5, 3.0$, and $\nu = 0.3$.

will give the frequency parameter λ and the associated mode shapes. When we increase N, a sequence of approximation is got to obtain the vibration characteristics. The computations are carried out up to $N = 20$, and it was found that in almost all the cases, the first five frequencies converged to four significant digits. Computations were carried out for several values of parameters μ and θ with all the possible combinations of boundary conditions. The value of Poisson's ratio has been chosen to be 0.3. The boundary conditions refer to sides 1, 2, and 3. Moreover, C, S, and F denote boundary conditions of the edges as clamped, simply supported, and free. Three different geometry of triangular plates viz. Figures 8.3 to 8.5 are studied here.

Tables 8.1 through 8.5 give results for fundamental frequency parameters of a right-triangular plate (Figure 8.3) for different values of μ, i.e., c/a along with all the possible boundary conditions. All the boundary conditions are incorporated in terms of the above five tables. In all of these tables, the value of μ has been taken as 1.0, 1.5, 2.0, 2.5, and 3.0. Accordingly, Table 8.1 shows the fundamental frequencies when the sides are either clamped or free only. There are six possible combinations in this case as given in Table 8.1.

TABLE 8.2

Fundamental Frequencies for a Right-Triangular Plate and Sides with Clamped (C) or Simply Supported (S) Boundary Conditions

S. No.	Sides 1 2 3	Values of c/a				
		1.0	1.5	2.0	2.5	3.0
1	C S S	60.54	43.78	36.41	32.30	29.68
2	S C S	60.54	40.40	31.84	27.17	24.25
3	S S C	65.81	45.96	37.60	33.04	30.15
4	C C S	73.40	50.99	41.31	35.96	32.57
5	C S C	78.89	57.04	47.69	42.57	39.31
6	S C C	78.89	53.19	42.44	36.64	33.04

Note: $\theta = 0$, $\mu = 1.0, 1.5, 2.0, 2.5, 3.0$, and $\nu = 0.3$.

TABLE 8.3

Fundamental Frequencies for a Right-Triangular Plate and Sides with Free (F) or Simply Supported (S) Boundary Conditions

S. No.	Sides 1 2 3	Values of c/a				
		1.0	1.5	2.0	2.5	3.0
1	F S S	17.31	9.386	6.303	4.720	3.769
2	S F S	17.31	15.00	13.92	13.28	12.85
3	S S F	9.798	6.535	4.905	3.297	3.274
4	F F S	19.61	12.60	9.030	7.000	5.711
5	F S F	14.56	8.062	4.861	3.199	2.253
6	S F F	14.56	10.32	7.875	6.352	5.321

Note: $\theta=0$, $\mu=1.0, 1.5, 2.0, 2.5, 3.0$, and $\nu=0.3$.

TABLE 8.4

Fundamental Frequencies for a Right-Triangular Plate and Sides with Any of Clamped (C), Simply Supported (S), or Free (F) Boundary Conditions, with no Repetition of the Conditions

S. No.	Sides 1 2 3	Values of c/a				
		1.0	1.5	2.0	2.5	3.0
1	C S F	17.96	13.75	11.51	10.12	9.178
2	S C F	17.96	10.56	7.301	5.510	4.395
3	S F C	31.78	25.45	22.77	21.32	20.41
4	F S C	31.78	19.35	14.46	11.93	10.39
5	C F S	23.93	21.36	20.14	19.42	18.94
6	F C S	23.93	12.48	8.121	5.924	4.630

Note: $\theta=0$, $\mu=1.0, 1.5, 2.0, 2.5, 3.0$, and $\nu=0.3$.

TABLE 8.5

Fundamental Frequencies for a Right-Triangular Plate and Sides with All of Clamped (C), Simply Supported (S), or Free (F) Boundary Conditions

S. No.	Sides 1 2 3	Values of c/a				
		1.0	1.5	2.0	2.5	3.0
1	C C C	93.80	65.46	53.46	46.91	42.81
2	S S S	49.36	34.29	27.79	24.18	21.89
3	F F F	19.17	10.80	6.466	4.429	2.992

Note: $\theta=0$, $\mu=1.0, 1.5, 2.0, 2.5, 3.0$, and $\nu=0.3$.

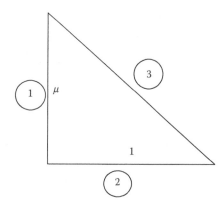

FIGURE 8.3
Right-angled triangle.

Table 8.2 incorporates the results for right triangular plates when its sides are either clamped or simply supported only. In this case also, there exist only six combinations as shown in Table 8.2. The cases with the sides either free or simply supported only are introduced in Table 8.3. The possible six combinations are shown in Table 8.3. When the sides of a right-triangular plate are clamped, simply supported, or free (i.e., with no repetition), there are six cases of combinations that are incorporated in Table 8.4. Table 8.5 contains the fundamental frequency parameters with all the three sides as clamped, simply supported, or free, respectively. Tables 8.1 through 8.5 show all the possible 27 combinations of boundary conditions of this triangular geometry.

As such, effect of $\mu = c/a$ on the fundamental frequency parameter of right triangular plate and sides with only clamped and free boundary conditions is shown in Figure 8.6. It may be seen in Figure 8.6 that at $\mu = 1$, the fundamental frequency is same for boundary conditions CFF and FCF and CFC and FCC, respectively. Moreover, this figure also reveals that frequency decreases as c/a is increased for all the boundary conditions considered in this figure. Figure 8.7 shows the effect of $\mu = c/a$ on the fundamental frequency parameter with sides being clamped or simply supported only. One may note in Figure 8.7 that the fundamental frequency at $\mu = 1$ for boundary conditions CSS and SCS, and CSC and SCC are exactly the same. The effect of $\mu = c/a$ on the fundamental frequency parameter for right-triangular plate and sides with free or simply supported only is depicted in Figure 8.8.

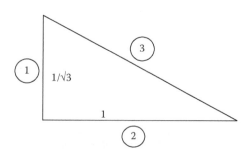

FIGURE 8.4
Right-angled triangle with angles 30°, 60°, and 90°.

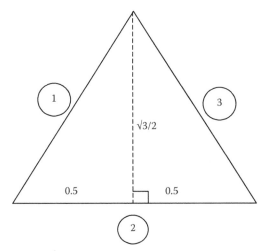

FIGURE 8.5
Equilateral triangle.

Here, the fundamental frequency for FSS and SFS, and FSF and SFF is same for $\mu = 1$. The other two cases of the boundary conditions as per Tables 8.4 and 8.5 are shown in Figures 8.9 and 8.10, respectively. In Figure 8.9, one may note that the fundamental frequency parameters at $\mu = 1$ is same in boundary conditions CSF and SCF, SFC and FSC, and CFS and FCS, respectively. All the plots in Figures 8.6 through 8.10 reveal that the fundamental frequency decreases as $\mu = c/a$ is increased.

Tables 8.6 through 8.10 give the first five modes for a right-angled triangle with angles 30°, 60°, and 90°, as shown in Figure 8.4. Sides 1, 2, and 3 have

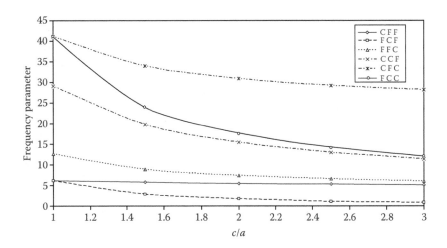

FIGURE 8.6
Effect of c/a on the fundamental frequencies of right-triangular plate and sides with clamped or free boundary conditions.

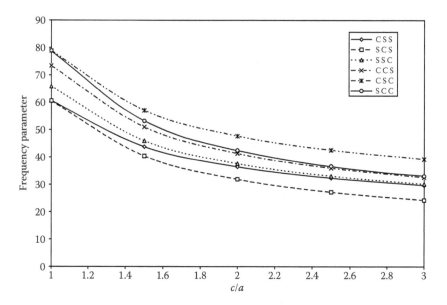

FIGURE 8.7
Effect of c/a on the fundamental frequencies of right-triangular plate for sides with clamped or simply supported boundary conditions.

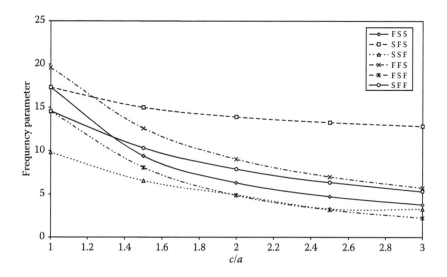

FIGURE 8.8
Effect of c/a on the fundamental frequencies of right-triangular plate and sides with free or simply supported boundary conditions.

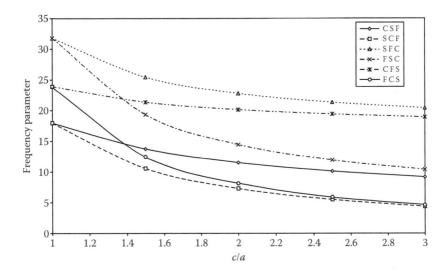

FIGURE 8.9
Effect of c/a on the fundamental frequencies of right-triangular plate with any of clamped, simply supported or free boundary conditions.

lengths $1/\sqrt{3}$, 1, and $2/\sqrt{3}$, respectively, in nondimensional form. The parameters θ and μ take the values 0 and $1/\sqrt{3}$, respectively. Combining the results of Tables 8.6 through 8.10 counts to a total of 27 different cases of boundary conditions. Table 8.6 gives the first five frequencies for the said right-triangular plate with sides being clamped or free boundary conditions.

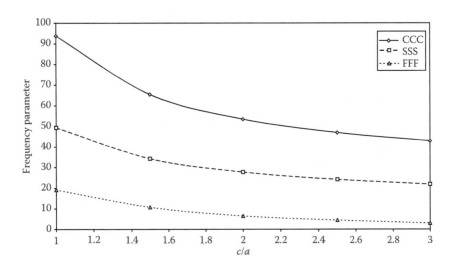

FIGURE 8.10
Effect of c/a on the fundamental frequencies of right-triangular plate for CCC, SSS, and FFF boundary conditions.

TABLE 8.6

First Five Frequencies for a Right-Triangular Plate with Angles 30°, 60°, and 90° and Sides with Clamped (C) or Free (F) Boundary Conditions

S. No.	Sides 1 2 3	λ_1	λ_2	λ_3	λ_4	λ_5
1	C F F	6.5608	27.962	44.443	68.716	108.97
2	F C F	16.960	49.737	87.604	108.46	180.78
3	F F C	24.139	62.262	106.66	133.86	206.53
4	C C F	52.270	106.77	168.81	192.94	286.47
5	C F C	61.320	122.97	192.37	207.96	319.84
6	F C C	96.710	182.35	247.97	304.01	400.49

Note: $\theta = 0$, $\mu = 1/\sqrt{3}$, and $\nu = 0.3$.

TABLE 8.7

First Five Frequencies for a Right-Triangular Plate with Angles 30°, 60°, and 90° and Sides with Clamped (C) or Simply Supported (S) Boundary Conditions

S. No.	Sides 1 2 3	λ_1	λ_2	λ_3	λ_4	λ_5
1	C S S	107.07	192.24	280.80	317.06	442.91
2	S C S	119.20	210.04	298.79	339.48	463.29
3	S S C	124.01	216.47	305.39	343.79	471.95
4	C C S	137.05	231.20	324.00	362.31	493.97
5	C S C	141.79	237.07	334.00	367.94	505.28
6	S C C	155.65	255.85	354.68	401.22	516.89

Note: $\theta = 0$, $\mu = 1/\sqrt{3}$, and $\nu = 0.3$.

TABLE 8.8

First Five Frequencies for a Right-Triangular Plate with Angles 30°, 60°, and 90° and Sides with Free (F) or Simply Supported (S) Boundary Conditions

S. No.	Sides 1 2 3	λ_1	λ_2	λ_3	λ_4	λ_5
1	F S S	43.229	112.07	159.06	209.60	293.46
2	S F S	22.988	65.721	128.26	141.26	223.44
3	S S F	16.984	55.040	107.13	134.18	191.42
4	F F S	31.843	68.904	88.635	150.70	167.90
5	F S F	27.051	64.182	70.245	129.58	165.09
6	S F F	18.897	34.965	55.685	89.030	117.91

Note: $\theta = 0$, $\mu = 1/\sqrt{3}$, and $\nu = 0.3$.

TABLE 8.9

First Five Frequencies for a Right-Triangular Plate with Angles 30°, 60°, and 90° and Sides with Any of Clamped (C), Simply Supported (S), or Free (F) Boundary Conditions

S. No.	Sides 1 2 3	λ_1	λ_2	λ_3	λ_4	λ_5
1	C S F	26.326	69.907	126.41	151.56	230.23
2	S C F	37.706	88.151	147.47	176.61	247.31
3	S F C	49.876	106.24	173.57	194.72	289.98
4	F S C	71.883	149.41	207.71	259.29	355.43
5	C F S	30.103	78.874	143.43	157.87	254.86
6	F C S	62.112	141.00	193.83	252.54	337.32

Note: $\theta = 0$, $\mu = 1/\sqrt{3}$, and $\nu = 0.3$.

Similarly, Tables 8.7 and 8.8 incorporate the first five frequencies for the above plate geometry when the sides are clamped–simply supported and free–simply supported, respectively. When the sides have any of clamped, simply supported, and free, the corresponding first five frequencies are introduced in Table 8.9. Next, Table 8.10 gives the first five frequencies for the mentioned right-triangular plate when all the sides are clamped, simply supported, or free.

Table 8.11 gives the results for an equilateral triangular plate as shown in Figure 8.5, with sides of unity each in nondimensional form. The parameters θ and μ take the values $1/\sqrt{3}$ and $\sqrt{3}/2$, respectively. By taking all possible boundary conditions, we have, in all, 10 different cases for which the first five frequencies have been reported. All combinations of the boundary condition are categorized as mentioned in Table 8.11. Finally, Table 8.12 incorporates the convergence of results for this equilateral triangular plate. This table shows how the frequencies converge as the value of N, i.e., the number of approximation is increased from 2 to 20.

The present method provides an efficient way for studying the triangular plate of arbitrary shape and any set of boundary conditions. The shape and boundary conditions are controlled by five parameters, viz., p, q, r, θ, and μ.

TABLE 8.10

First Five Frequencies for a Right-Triangular Plate with Angles 30°, 60°, and 90° and Sides with All of Clamped (C), Simply Supported (S), or Free (F) Boundary Conditions

S. No.	Sides 1 2 3	λ_1	λ_2	λ_3	λ_4	λ_5
1	C C C	176.58	280.95	380.92	416.53	558.48
2	S S S	92.140	173.94	255.33	288.19	421.06
3	F F F	25.202	53.333	69.258	97.985	123.98

Note: $\theta = 0$, $\mu = 1/\sqrt{3}$, and $\nu = 0.3$.

TABLE 8.11

First Five Frequencies for an Equilateral Triangular Plate with the Boundary Condition of the Sides as Mentioned

S. No.	Sides 1 2 3	λ_1	λ_2	λ_3	λ_4	λ_5
		Sides with Clamped (C) or Free (F) Boundary Conditions				
1	C C F	40.022	95.891	101.85	174.46	197.11
2	F C F	8.9219	35.155	38.503	91.624	96.725
		Sides with Clamped (C) or Simply Supported (S) Boundary Conditions				
3	C C S	81.604	165.12	165.52	271.30	286.95
4	S C S	66.189	142.96	143.73	243.89	262.04
		Sides with Free (F) or Simply Supported (S) Boundary Conditions				
5	F F S	22.666	26.717	71.033	74.867	91.959
6	S F S	16.092	57.709	68.593	123.67	151.32
		Sides with Clamped (C), Simply Supported (S), and Free (F) Boundary Conditions				
7	S C F	26.565	75.360	84.432	147.45	175.07
		Sides with all Clamped (C), Simply Supported (S), or Free (F) Boundary Conditions				
8	C C C	99.022	189.05	189.22	296.85	316.83
9	S S S	52.638	122.91	124.11	218.28	235.79
10	F F F	34.962	36.331	36.337	89.276	92.660

Note: $\theta = 1/\sqrt{3}$, $\mu = \sqrt{3}/2$, and $\nu = 0.3$.

TABLE 8.12

Convergence of Fundamental Frequency for an Equilateral-Triangular Plate with the Boundary Condition of the Sides as Mentioned

S. No.	Sides 1 2 3	2	5	10	15	19	20
		Sides with Clamped (C) or Free (F) Boundary Conditions					
1	C C F	46.57	40.78	40.05	40.03	40.02	40.02
2	F C F	10.02	9.255	8.945	8.924	8.921	8.921
		Sides with Clamped (C) or Simply Supported (S) Boundary Conditions					
3	C C S	101.8	83.32	81.71	81.60	81.60	81.60
4	S C S	89.71	68.16	66.33	66.19	66.18	66.18
		Sides with Free (F) or Simply Supported (S) Boundary Conditions					
5	F F S	28.63	24.99	22.92	22.67	22.66	22.66
6	S F S	17.98	16.77	16.11	16.09	16.09	16.09
		Sides with Clamped (C), Simply Supported (S) and Free (F) Boundary Conditions					
7	S C F	30.09	27.29	26.64	26.57	26.56	26.56

TABLE 8.12 (continued)

Convergence of Fundamental Frequency for an Equilateral-Triangular Plate with the Boundary Condition of the Sides as Mentioned

S. No.	Sides 1 2 3	2	5	10	15	19	20
	Sides with all Clamped (C), Simply Supported (S), or Free (F) Boundary Conditions						
8	C C C	103.3	101.3	99.05	99.03	99.02	99.02
9	S S S	66.93	59.71	52.84	52.63	52.63	52.63
10	F F F	—	40.98	38.76	35.27	34.96	34.96

Note: $\theta = 1/\sqrt{3}$, $\mu = \sqrt{3}/2$, and $\nu = 0.3$.

In this chapter, the BCOPs only need to be generated over the standard triangle. They may then be used over an arbitrary triangle by mapping it onto the standard triangle, which saves a lot of computations because the orthogonal polynomials only need to be generated once and can be used for triangles of arbitrary shape.

Bibliography

Bhat, R.B. 1987. Flexural vibration of polygonal plates using characteristic orthogonal polynomials in two variables. *Journal of Sound and Vibration*, 114: 65.

Christensen, R.M. 1963. Vibration of 45° right triangular cantilever plate by a grid-work method. *AIAA Journal*, 1: 1790.

Cowper, G.R., Kosko, E., Lindberg, G.M. and Olson, M.D. 1969. Static and dynamic application of a high-precision triangular plate bending element. *AIAA Journal*, 7: 1957.

Gorman, D.J. 1983. A highly accurate analytical solution for free vibration analysis of simply supported right triangular plates. *Journal of Sound and Vibration*, 89: 107.

Gorman, D.J. 1986. Free vibration analysis of right triangular plates with combinations of clamped-simply supported boundary conditions. *Journal of Sound and Vibration*, 106: 419.

Gorman, D.J. 1987. A modified superposition method for the free vibration analysis of right triangular plates. *Journal of Sound and Vibration*, 112: 173.

Gorman, D.J. 1989. Accurate free vibration analysis of right triangular plates with one free edge. *Journal of Sound and Vibration*, 131: 115.

Gustafson, P.N., Stokey, W.F., and Zorowski, C.F. 1953. An experimental study of natural vibrations of cantilever triangular plates. *Journal of Aeronautical Society*, 20: 331.

Kim, C.S. and Dickinson, S.M. 1990. The free flexural vibration of right triangular isotropic and orthotropic plates. *Journal of Sound and Vibration*, 141: 291.

Kuttler, J.R. and Sigillito, V.G. 1981. Upper and lower bounds for frequencies of trapezoidal and triangular plates. *Journal of Sound and Vibration*, 78: 585.

Lam, K.Y., Liew, K.M. and Chow, S.T. 1990. Free vibration analysis of isotropic and orthotropic triangular plates. *International Journal of Mechanical Sciences*, 32: 455.

Leissa, A.W. 1969. *Vibration of Plates*. NASA SP-160, Washington, DC.

Leissa, A.W. 1977. Recent research in plate vibrations: 1973–1976: Classical theory. *Shock and Vibration Digest* 9, No. 10, 13–24.

Leissa, A.W. 1981. Plate vibration research, 1976–1980: Classical theory. *Shock and Vibration Digest* 13, No. 9, 11–22.

Leissa, A.W. 1987. Recent studies in plate vibrations: 1981–85. Part I – Classical theory. *Shock and Vibration Digest* 19, No. 2, 11–18.

Mirza, S. and Bijlani, M. 1985. Vibration of triangular plates of variable thickness. *Computers and Structures*, 21: 1129.

Singh, B. and Chakraverty, S. 1992. Transverse vibration of triangular plates using characteristic orthogonal polynomials in two variables. *International Journal of Mechanical Sciences*, 34: 947.

9

Rectangular and Skew Plates

9.1 Introduction

Free vibration of rectangular, skew, and square plates has been studied extensively throughout the globe using different methods of solutions. Among them, Leissa (1969, 1977, 1981, 1987) and Gorman (1982) provided excellent sources of information regarding the studies carried out by various researchers in the field of vibration of plate geometries. In fact, Leissa (1969) included most of the results of rectangular and skew plates till that date. Moreover, Gorman (1982) published a book, which in particular, is entirely devoted to the vibrations of rectangular plates. Some other important reference works, including the studies of Leissa (1973), Barton (1951), Dickinson and Li (1982), Mizusawa (1986), Mizusawa et al. (1979), Nair and Durvasula (1973), Durvasula (1968, 1969), Durvasula and Nair (1974), Kuttler and Sigillito (1980), and many others are provided in the References section in this chapter. It is worth mentioning that the list of the works related to the said geometry is very long and hence, citing the names of all the researchers who worked in this area is quite difficult.

In this chapter, the main focus, as usual, is in the use of the method of boundary characteristic orthogonal polynomials (BCOPs) in the study of the vibration of skew plates. Notable references in the use of this method have already been mentioned in the previous chapters. In particular, the rectangular plates using orthogonal polynomials have been studied by Bhat (1985), Bhat et al. (1990), Dickinson and di Blasio (1986), and Lam and Hung (1990). Liew et al. (1990) and Liew and Lam (1990) used the two-dimensional orthogonal polynomials for skew plates. The two-dimensional BCOPs (2D BCOPs) were used by Singh and Chakraverty (1994) for skew plates with various boundary conditions at the edges.

However, Bhat (1987) introduced the two-dimensional orthogonal polynomials for the first time, but they lacked the property of satisfying the essential boundary conditions during their generation. Hence, we will not term those two-dimensional orthogonal polynomials as BCOPs. In another study, Bhat (1985) generated one-dimensional orthogonal polynomials in both x and y directions along the edges of the plate, and used their product to approximate the displacement. A total of 6×6 terms (i.e., 36 terms) were used in the expansion. Numerical results were given for SSSS, CCCC, SSFF, and CCFF cases. Dickinson and di Blassio (1986) also used one-dimensional orthogonal polynomials in both x and y directions to study the flexural vibrations of isotropic and orthotropic plates. Lam and Hung (1990) used the product of one-dimensional orthogonal polynomials to study the vibrations of a rectangular plate with centrally located square notch and centrally located internal crack. Liew et al. (1990) analyzed the free vibrations of rectangular plates using orthogonal plate functions in two variables. Out of the 21 distinct cases, the numerical results were given for 12 cases. In another paper, Liew and Lam (1990) followed the same procedure to study the vibrations of skew plates. Out of 22 cases, results were given for four cases only for different skew plates.

In the above studies, the rectangular and skew plates were studied using either one- or two-dimensional orthogonal polynomials, directly by generating the polynomials over their particular domains. However, in this case the rectangular or skew domains are first mapped into a standard unit-square domain (Singh and Chakraverty, 1994). The orthogonal polynomials satisfying the essential boundary conditions, viz., the BCOPs are then generated over this standard square domain. These are then used in the approximation by the Rayleigh–Ritz method. Thus, the procedure minimizes the computations for generating the BCOPs for each and every skew plate. In this method, the process of generation of the BCOPs is done only once, irrespective of the sides and skew angle of the domain. Furthermore, the boundary conditions of the skew plates are taken care of by the four parameters. It may also be noted that there exist 21 distinct cases for rectangular and square plates and 22 distinct cases for a skew plate with various skew angles. The results corresponding to all of these boundary conditions are computed for some selected domains to have a better understanding of the methodology.

9.2 Mapping of General Skew Domain into a Standard Unit Square

Let us consider a skew plate R, as shown in Figure 9.1a, which is defined by three parameters a, b, and α. It can be easily noted that this particular case of rectangular domain follows by substituting the angle $\alpha = 90°$. Now, a transformation is used to map the general skew domain into a unit square domain, say S and the same can be written using the following relation:

$$x = a\xi + b(\cos \alpha)\eta \tag{9.1}$$

$$y = b(\sin \alpha)\eta \tag{9.2}$$

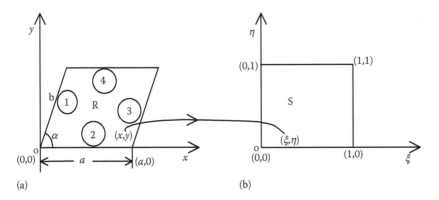

FIGURE 9.1
Mapping of the skew domain into a unit square domain.

where ξ and η are the new coordinates, as shown in Figure 9.1b The above transformation can now be used and subsequently, the BCOPs can be generated on the standard square domain using the Gram–Schmidt process.

9.3 Generation of the BCOPs in the Standard Square Domain

For generating the orthogonal polynomials over the standard unit-square domain S satisfying the essential boundary conditions, we begin with the linearly independent set as carried out in the previous chapters

$$F_i = f\, f_i, \quad i = 1, 2, \ldots \tag{9.3}$$

where the boundary conditions are controlled by the first term f on the right-hand side of Equation 9.3 and is given by

$$f = \xi^p (1 - \xi)^q \eta^r (1 - \eta)^s \tag{9.4}$$

The parameters p, q, r, and s can take any of the values of 0, 1, or 2 indicating the free, simply supported, or clamped boundary conditions, respectively. For example, the parameter p can take the values of 0, 1, or 2 accordingly, as the side $\xi = 0$ is free, simply supported, or clamped. As such, similar interpretations are given to the parameters q, r, and s corresponding to the sides $\xi = 1$, $\eta = 0$, and $\eta = 1$, respectively. The functions f_i in Equation 9.3 are suitably chosen and are taken of the form $\xi^{m_i} \eta^{n_i}$ where m_i and n_i are non-negative integers. As mentioned in the earlier chapters, the choice of m_i and n_i determines the mode shapes. Accordingly, the even values of m_i and n_i give the symmetric modes, and their odd values give the antisymmetric modes and so on.

Next, we define the inner product of the two functions f and g over the standard square domain as

$$\langle f,g \rangle = \iint_S f(\xi,\eta)g(\xi,\eta)\mathrm{d}\xi\mathrm{d}\eta \tag{9.5}$$

and the norm of f is written as

$$\| f \| = \langle f,f \rangle^{1/2} = \left[\iint_S f^2(\xi,\eta)\mathrm{d}\xi\mathrm{d}\eta \right]^{1/2} \tag{9.6}$$

The BCOPs are generated as described in the previous chapters. As mentioned earlier, all the polynomials will automatically satisfy the essential boundary conditions. Thus, the procedure for the generation of BCOPs can be summarized as

$$\left.\begin{aligned} &\phi_1 = F_1 \\ &\phi_i = F_i - \sum_{j=1}^{i-1} \alpha_{ij}\phi_j, \\ \text{where} \qquad &\alpha_{ij} = \frac{\langle F_i,\phi_j \rangle}{\langle \phi_j,\phi_j \rangle}, j = 1, 2, \ldots (i-1) \end{aligned}\right\}, \quad i = 2, 3, 4, \ldots \tag{9.7}$$

The functions ϕ_i may then be normalized by using

$$\hat{\phi}_i = \frac{\phi_i}{\| \phi_i \|} \tag{9.8}$$

The procedure is independent of the skew geometry of the plate domain. The parameters p, q, r, and s will control the various boundary conditions of the plate, and the computations will depend on these four parameters. It can be observed that the procedure for generating these BCOPs does not require repetition for other values of a, b, and α and is computed only once. These BCOPs are then used in the Rayleigh–Ritz method to extract the free-vibration characteristics.

9.4 Rayleigh–Ritz Method for Skew Plates

The usual procedure is followed as described earlier, and the N-term approximation is taken as

$$W(x,y) = \sum_{j=1}^{N} c_j \phi_j \tag{9.9}$$

for deflection in the Rayleigh quotient of the plate, (after equating the maximum kinetic and strain energy)

$$\omega^2 = \frac{D}{h\rho} \frac{\iint\limits_{R} \left[(\nabla^2 W)^2 + 2(1-\nu)\left\{ W_{xy}^2 - W_{xx}W_{yy} \right\} \right] dxdy}{\iint\limits_{R} W^2 dxdy} \tag{9.10}$$

Consequently, the above Rayleigh-quotient is extremized as a function of the c_j's. The variables x and y are finally changed to ξ and η, and this leads to the following eigenvalue problem

$$\sum_{j=1}^{N} \left(a_{ij} - \lambda^2 b_{ij} \right) c_j = 0, \quad i = 1, 2, \ldots, N \tag{9.11}$$

where

$$a_{ij} = \iint\limits_{T} \left[\phi_i^{\xi\xi}\phi_j^{\xi\xi} + B_1\left(\phi_i^{\xi\eta}\phi_j^{\xi\xi} + \phi_i^{\xi\xi}\phi_j^{\xi\eta} \right) + B_2\left(\phi_i^{\eta\eta}\phi_j^{\xi\xi} + \phi_i^{\xi\xi}\phi_j^{\eta\eta} \right) \right.$$

$$\left. + B_3\phi_i^{\xi\eta}\phi_j^{\xi\eta} + B_4\left(\phi_i^{\eta\eta}\phi_j^{\xi\eta} + \phi_i^{\xi\eta}\phi_j^{\eta\eta} \right) + B_5\phi_i^{\eta\eta}\phi_j^{\eta\eta} \right] d\xi d\eta \tag{9.12}$$

$$b_{ij} = \iint\limits_{T} \phi_i \phi_j d\xi d\eta \tag{9.13}$$

and

$$\lambda^2 = \frac{a^4 \omega^2 \rho h}{D} \tag{9.14}$$

The symbols $B_1, B_2, B_3, \ldots, B_5$ are given by

$$\left.\begin{array}{l} B_1 = -2\mu \cos\alpha \\ B_2 = \mu^2(\nu \sin^2\alpha + \cos^2\alpha) \\ B_3 = 2\mu^2(1 + \cos^2\alpha - \nu \sin^2\alpha) \\ B_4 = -2\mu^3 \cos\alpha \\ B_5 = \mu^4 \end{array}\right\} \tag{9.15}$$

and

$$\mu = \frac{a}{b} \tag{9.16}$$

The integrals involved are available in closed form by using the following result (Singh and Chakraverty, 1994)

$$\iint_T \xi^i (1 - \xi)^j \eta^k (1 - \eta)^\ell \, d\xi d\eta = \frac{i! \, j! \, k! \, \ell!}{(i+j+1)!(k+\ell+1)!} \qquad (9.17)$$

9.5 Some Numerical Results and Discussions for Rectangular and Skew Plates

The generated BCOPs have been used in the Rayleigh–Ritz method for determining the vibration frequencies. There are a number of parameters controlling the domain as well as the boundary conditions of the plate. Numerical results have been given here for $\alpha = 30°, 60°,$ and $90°$ covering all the possible boundary conditions, and Poisson's ratio has been taken as 0.3 for all the computations. This chapter comprises the results of square and rhombic plates $(b/a = 1.0)$ with different skew angles, and rectangular plate with $b/a = 0.5$.

It has already been mentioned that there exist 21 possible combinations of boundary conditions for square and rectangular plates, while the skew plates may have 22 distinct cases of boundary conditions. Here, these sets of possible boundary conditions are divided into five categories of subsets involving, (1) sides with only clamped and free boundary conditions; (2) sides with only clamped and simply supported boundary conditions; (3) sides with only simply supported and free boundary conditions; (4) sides with any of the clamped, simply supported, and free boundary conditions; and finally, (5) all sides with clamped, simply supported, or free boundary conditions.

Accordingly, Tables 9.1 through 9.5 give the first five frequencies for a square plate, covering all 21 distinct cases depending on the boundary conditions on sides 1, 2, 3, and 4 of Figure 9.1a. The letters C, S, and F denote the clamped, simply supported, and completely free boundary. As such, Tables 9.1 through 9.5 list the results for the five categories of

TABLE 9.1

First Five Frequencies for a Square Plate and Sides with Clamped or Free Boundary Conditions

S. No.	Sides 1 2 3 4	λ_1	λ_2	λ_3	λ_4	λ_5
1	C C C F	23.960	40.021	63.291	78.139	80.689
2	C F F F	3.4864	8.5443	21.325	27.576	31.199
3	C C F F	6.9365	23.948	26.603	47.763	63.386
4	C F C F	22.223	26.556	44.072	61.265	67.649

TABLE 9.2

First Five Frequencies for a Square Plate and Sides with Clamped or Simply Supported Boundary Conditions

S. No.	Sides 1 2 3 4	λ_1	λ_2	λ_3	λ_4	λ_5
1	C C C S	31.827	63.348	71.083	100.83	118.96
2	C S S S	23.646	51.813	58.650	86.252	101.80
3	C C S S	27.055	60.550	60.550	92.914	115.04
4	C S C S	28.950	54.873	69.327	94.703	103.71

TABLE 9.3

First Five Frequencies for a Square Plate and Sides with Simply Supported or Free Boundary Conditions

S. No.	Sides 1 2 3 4	λ_1	λ_2	λ_3	λ_4	λ_5
1	S S S F	11.684	27.757	41.220	59.360	62.461
2	S F F F	6.6444	14.931	25.631	26.298	49.534
3	S S F F	3.3670	17.333	19.318	38.408	51.925
4	S F S F	9.6317	16.135	37.180	39.134	47.280

TABLE 9.4

First Five Frequencies for a Square Plate and Sides with Any of the Clamped, Simply Supported, or Free Boundary Conditions

S. No.	Sides 1 2 3 4	λ_1	λ_2	λ_3	λ_4	λ_5
1	S C SF	12.687	33.067	41.714	63.260	73.870
2	C F S F	15.218	20.613	40.210	49.543	56.671
3	C F F S	5.3600	19.084	24.694	43.296	54.293
4	C F S C	17.555	36.036	51.862	71.218	75.789
5	C F S S	16.811	31.173	51.454	65.735	67.794
6	C S C F	23.411	35.595	62.960	67.355	77.540

TABLE 9.5

First Five Frequencies for a Square Plate with All Sides as Clamped, Simply Supported, or Free Boundary Conditions

S. No.	Sides 1 2 3 4	λ_1	λ_2	λ_3	λ_4	λ_5
1	C C C C	35.988	73.398	73.398	108.26	131.89
2	S S S S	19.739	49.348	49.348	79.400	100.17
3	F F F F	13.714	19.774	24.566	35.596	35.628

TABLE 9.6

First Five Frequencies for a Skew Plate ($\alpha = 60°$, $b/a = 1.0$) and Sides with Clamped or Free Boundary Conditions

S. No.	Sides 1 2 3 4	λ_1	λ_2	λ_3	λ_4	λ_5
1	C C C F	29.026	48.139	77.489	84.895	113.70
2	C F F F	3.9454	9.6209	26.011	26.396	42.218
3	C C F F	6.3106	24.842	29.154	49.957	73.459
4	C F C F	27.606	30.933	50.391	74.899	85.538

subsets of the possible boundary conditions. However, Table 9.1 contains four different distinct cases of boundary conditions with only clamped and free sides. Table 9.2 also exhibits four cases of sides with clamped and simply supported boundaries. Sides with only simply supported and free boundaries can again have four distinct cases, which are presented in Table 9.3. However, the sides of the square plate that do have any of the clamped, simply supported, or free boundaries have six cases of conditions (Table 9.4). The last case, i.e., the fifth category of the subset of the boundary conditions, is when all the sides are clamped, simply supported, or free, as shown in Table 9.5.

Subsequently, the examples for a skew plate with skew angle 60° and $b/a = 1.0$ are presented next. In this case, a total of 22 cases of boundary conditions are possible. Accordingly, Tables 9.6 through 9.10 present the first five frequencies for the five categories of boundary conditions. Here, it is to be noted that seven distinct cases are possible in the fourth category of the boundary condition, compared with the six distinct cases of square plates. The results of skew plates for $\alpha = 30°$ and $b/a = 1.0$ with all the categories of boundary conditions, including 22 cases are given in Tables 9.11 through 9.15.

Tables 9.16 through 9.20 give the first five frequencies of a rectangular plate with $b/a = 0.5$. Here, the possible cases of boundary conditions are 21 and as such the above five tables present the results of all the five categories of the boundary conditions. Similar results of the frequencies for the five categories of boundary conditions for a skew plate with skew angles 60° and 30° with $b/a = 1.0$ are shown in Tables 9.21 through 9.30, respectively.

TABLE 9.7

First Five Frequencies for a Skew Plate ($\alpha = 60°$, $b/a = 1.0$) and Sides with Clamped or Simply Supported Boundary Conditions

S. No.	Sides 1 2 3 4	λ_1	λ_2	λ_3	λ_4	λ_5
1	C C C S	40.762	73.132	98.310	110.85	157.51
2	C S S S	30.269	58.812	81.843	94.264	136.09
3	C C S S	34.306	66.412	88.145	101.76	147.12
4	C S C S	31.193	64.390	93.626	103.46	144.11

TABLE 9.8

First Five Frequencies for a Skew Plate ($\alpha = 60°$, $b/a = 1.0$)
and Sides with Simply Supported or Free Boundary Conditions

S. No.	Sides 1 2 3 4	λ_1	λ_2	λ_3	λ_4	λ_5
1	S S S F	14.370	30.969	53.635	61.025	89.704
2	S F F F	6.7253	19.169	22.076	36.740	45.175
3	S S F F	4.3739	18.078	24.814	38.953	61.981
4	S F S F	12.246	17.875	36.424	50.352	65.496

TABLE 9.9

First Five Frequencies for a Skew Plate ($\alpha = 60°$, $b/a = 1.0$) with
Any of Clamped, Simply Supported, or Free Boundary Conditions

S. No.	Sides 1 2 3 4	λ_1	λ_2	λ_3	λ_4	λ_5
1	C C S F	18.906	40.973	62.782	75.127	104.41
2	S C F C	28.689	42.925	73.032	80.701	104.67
3	S S C F	22.452	37.333	65.886	70.575	98.825
4	S F S C	15.314	36.091	56.002	66.155	97.035
5	F F S C	5.5528	20.903	27.907	44.111	68.871
6	C F S C	23.434	42.740	68.883	75.906	106.37
7	F S F C	18.072	24.543	43.139	62.105	73.690

TABLE 9.10

First Five Frequencies for a Skew Plate ($\alpha = 60°$, $b/a = 1.0$) with All
Sides as Clamped, Simply Supported, or Free Boundary Conditions

S. No.	Sides 1 2 3 4	λ_1	λ_2	λ_3	λ_4	λ_5
1	C C C C	46.166	81.613	105.56	119.98	167.16
2	S S S S	25.314	52.765	73.006	87.478	130.25
3	F F F F	11.690	22.926	27.254	36.512	46.214

TABLE 9.11

First Five Frequencies for a Skew Plate ($\alpha = 30°$, $b/a = 1.0$)
and Sides with Clamped or Free Boundary Conditions

S. No.	Sides 1 2 3 4	λ_1	λ_2	λ_3	λ_4	λ_5
1	C C C F	64.924	109.17	168.55	221.98	251.20
2	C F F F	6.1126	17.223	35.275	59.757	69.554
3	C C F F	7.8269	32.033	59.423	72.101	129.60
4	C F C F	68.472	70.707	107.86	146.78	227.16

TABLE 9.12

First Five Frequencies for a Skew Plate ($\alpha = 30°$, $b/a = 1.0$) and Sides with Clamped or Simply Supported Boundary Conditions

S. No.	Sides 1 2 3 4	λ_1	λ_2	λ_3	λ_4	λ_5
1	C C C S	112.48	167.66	255.91	303.31	400.32
2	C S S S	85.458	130.61	220.47	257.90	359.21
3	C C S S	92.906	151.39	221.70	278.46	407.42
4	C S C S	104.53	148.64	248.26	286.66	368.22

TABLE 9.13

First Five Frequencies for a Skew Plate ($\alpha = 30°$, $b/a = 1.0$) and Sides with Simply Supported or Free Boundary Conditions

S. No.	Sides 1 2 3 4	λ_1	λ_2	λ_3	λ_4	λ_5
1	S S S F	29.934	62.845	96.151	169.66	189.13
2	S F F F	8.1666	23.860	42.965	53.424	88.411
3	S S F F	6.6964	27.003	56.998	58.071	115.19
4	S F S F	30.130	31.102	56.743	84.838	161.87

TABLE 9.14

First Five Frequencies for a Skew Plate ($\alpha = 30°$, $b/a = 1.0$) with Any of Clamped, Simply Supported, or Free Boundary Conditions

S. No.	Sides 1 2 3 4	λ_1	λ_2	λ_3	λ_4	λ_5
1	C C S F	33.310	76.995	128.02	189.42	193.63
2	S C F C	66.486	103.90	152.42	217.67	244.96
3	S S C F	58.915	85.896	132.03	208.25	251.66
4	S F S C	29.736	69.995	109.66	170.48	187.76
5	F F S C	7.3343	30.330	60.006	65.538	124.06
6	C F S C	58.212	95.386	149.26	209.69	244.99
7	F S F C	36.264	59.554	77.697	114.88	191.10

TABLE 9.15

First Five Frequencies for a Skew Plate ($\alpha = 30°$, $b/a = 1.0$) with All Sides as Clamped, Simply Supported, or Free Boundary Conditions

S. No.	Sides 1 2 3 4	λ_1	λ_2	λ_3	λ_4	λ_5
1	C C C C	127.06	185.00	282.94	322.61	385.49
2	S S S S	73.135	112.64	209.84	233.52	323.51
3	F F F F	9.8235	23.140	40.033	52.669	78.110

TABLE 9.16

First Five Frequencies for a Rectangular Plate ($b/a = 0.5$)
and Sides with Clamped or Free Boundary Conditions

S. No.	Sides 1 2 3 4	λ_1	λ_2	λ_3	λ_4	λ_5
1	C C C F	31.150	70.235	103.42	129.43	144.21
2	C F F F	3.4578	14.829	21.486	48.444	60.380
3	C C F F	17.159	36.426	73.607	91.007	115.53
4	C F C F	22.114	36.080	60.896	82.836	110.65

TABLE 9.17

First Five Frequencies for a Rectangular Plate ($b/a = 0.5$)
and Sides with Clamped or Simply Supported Boundary Conditions

S. No.	Sides 1 2 3 4	λ_1	λ_2	λ_3	λ_4	λ_5
1	C C C S	73.397	108.22	165.19	210.60	242.77
2	C S S S	69.327	94.593	141.38	208.44	218.18
3	C C S S	71.078	100.80	152.22	209.46	233.75
4	C S C S	95.262	115.80	157.40	224.57	254.21

TABLE 9.18

First Five Frequencies for a Rectangular Plate ($b/a = 0.5$)
and Sides with Simply Supported or Free Boundary Conditions

S. No.	Sides 1 2 3 4	λ_1	λ_2	λ_3	λ_4	λ_5
1	S S S F	16.134	46.746	75.417	97.623	111.43
2	S F F F	13.049	14.858	43.154	49.283	86.857
3	S S F F	6.6438	25.377	59.295	65.442	89.515
4	S F S F	9.5124	27.579	38.531	65.016	89.501

TABLE 9.19

First Five Frequencies for a Rectangular Plate ($b/a = 0.5$) with Any
of Clamped, Simply Supported, or Free Boundary Conditions

S. No.	Sides 1 2 3 4	λ_1	λ_2	λ_3	λ_4	λ_5
1	S C S F	22.815	50.756	99.779	100.52	132.40
2	C F S F	15.091	31.374	49.112	73.241	103.37
3	C S F F	8.5179	30.990	64.264	71.163	93.587
4	C C S F	26.294	59.789	101.44	113.61	137.89
5	C S S F	20.605	56.334	77.474	111.10	117.69
6	C S C F	26.451	67.276	79.964	124.73	127.18

TABLE 9.20

First Five Frequencies for a Rectangular Plate ($b/a = 0.5$) with All
Sides as Clamped, Simply Supported, or Free Boundary Conditions

S. No.	Sides 1 2 3 4	λ_1	λ_2	λ_3	λ_4	λ_5
1	C C C C	98.317	127.31	179.27	254.39	256.02
2	S S S S	49.348	78.957	129.68	168.42	198.63
3	F F F F	21.683	26.923	60.312	61.572	89.219

TABLE 9.21

First Five Frequencies for a Skew Plate ($\alpha = 60°$, $b/a = 0.5$)
and Sides with Clamped or Free Boundary Conditions

S. No.	Sides 1 2 3 4	λ_1	λ_2	λ_3	λ_4	λ_5
1	C C C F	37.198	79.534	135.84	138.12	190.91
2	C F F F	3.7331	16.154	25.813	47.332	75.980
3	C C F F	17.234	37.658	77.393	105.22	143.04
4	C F C F	114.50	115.49	136.87	171.55	316.22

TABLE 9.22

First Five Frequencies for a Skew Plate ($\alpha = 60°$, $b/a = 0.5$)
and Sides with Clamped or Simply Supported Boundary Conditions

S. No.	Sides 1 2 3 4	λ_1	λ_2	λ_3	λ_4	λ_5
1	C C C S	95.268	133.39	194.40	273.05	289.69
2	C S S S	66.684	104.94	163.94	223.85	260.61
3	C C S S	92.416	124.61	179.37	272.11	289.10
4	C S C S	70.165	114.46	179.91	225.15	261.81

TABLE 9.23

First Five Frequencies for a Skew Plate ($\alpha = 60°$, $b/a = 0.5$)
and Sides with Simply Supported or Free Boundary Conditions

S. No.	Sides 1 2 3 4	λ_1	λ_2	λ_3	λ_4	λ_5
1	S S S F	52.737	71.954	109.97	167.59	216.75
2	S F F F	11.126	21.278	37.791	67.885	76.123
3	S S F F	8.5779	28.427	64.299	80.835	112.40
4	S F S F	50.882	57.037	82.222	121.99	210.44

TABLE 9.24

First Five Frequencies for a Skew Plate ($\alpha = 60°$, $b/a = 0.5$) with
Any of Clamped, Simply Supported, or Free Boundary Conditions

S. No.	Sides 1 2 3 4	λ_1	λ_2	λ_3	λ_4	λ_5
1	C C S F	28.782	64.157	117.63	132.70	179.51
2	S C F C	114.12	129.36	158.26	213.27	319.91
3	S S C F	26.927	65.253	104.44	115.50	163.52
4	S F S C	27.273	57.793	108.68	130.55	173.81
5	F F S C	17.056	35.076	70.934	106.17	129.17
6	C F S C	34.350	71.459	124.53	134.36	183.89
7	F S F C	74.093	86.214	106.76	150.21	253.48

TABLE 9.25

First Five Frequencies for a Skew Plate ($\alpha = 60°$, $b/a = 0.5$) with All
Sides as Clamped, Simply Supported, or Free Boundary Conditions

S. No.	Sides 1 2 3 4	λ_1	λ_2	λ_3	λ_4	λ_5
1	C C C C	128.90	159.72	215.29	291.45	341.33
2	S S S S	64.069	96.558	153.76	218.69	237.12
3	F F F F	18.874	34.716	48.554	79.361	107.46

TABLE 9.26

First Five Frequencies for a Skew Plate ($\alpha = 30°$, $b/a = 0.5$)
and Sides with Clamped or Free Boundary Conditions

S. No.	Sides 1 2 3 4	λ_1	λ_2	λ_3	λ_4	λ_5
1	C C C F	84.566	147.70	275.70	382.00	426.66
2	C F F F	5.1282	22.308	55.989	61.458	112.63
3	C C F F	24.459	65.946	117.05	197.36	247.39
4	C F C F	294.59	306.86	370.29	406.41	889.41

TABLE 9.27

First Five Frequencies for a Skew Plate ($\alpha = 30°$, $b/a = 0.5$)
and Sides with Clamped or Simply Supported Boundary Conditions

S. No.	Sides 1 2 3 4	λ_1	λ_2	λ_3	λ_4	λ_5
1	C C C S	268.38	334.84	473.74	700.38	839.16
2	C S S S	186.73	257.83	399.85	631.10	692.85
3	C C S S	262.94	322.02	422.83	751.71	839.39
4	C S C S	192.94	271.56	444.55	607.31	689.23

TABLE 9.28

First Five Frequencies for a Skew Plate ($\alpha = 30°$, $b/a = 0.5$)
and Sides with Simply Supported or Free Boundary Conditions

S. No.	Sides 1 2 3 4	λ_1	λ_2	λ_3	λ_4	λ_5
1	S S S F	125.86	177.76	228.66	362.36	637.60
2	S F F F	11.012	38.998	49.340	78.362	153.53
3	S S F F	15.346	41.689	95.023	192.28	218.73
4	S F S F	129.00	137.63	176.42	215.47	426.01

TABLE 9.29

First Five Frequencies for a Skew Plate ($\alpha = 30°$, $b/a = 0.5$) with
Any of Clamped, Simply Supported, or Free Boundary Conditions

S. No.	Sides 1 2 3 4	λ_1	λ_2	λ_3	λ_4	λ_5
1	C C S F	62.368	108.47	202.88	312.80	393.35
2	S C F C	288.88	367.79	409.80	510.87	904.29
3	S S C F	61.008	135.55	216.79	312.41	425.53
4	S F S C	62.492	107.26	185.80	335.29	394.28
5	F F S C	25.444	66.730	108.21	222.84	257.33
6	C F S C	82.128	150.97	251.01	383.33	447.38
7	F S F C	165.35	244.96	263.92	307.79	608.52

TABLE 9.30

First Five Frequencies for a Skew Plate ($\alpha = 30°$, $b/a = 0.5$) with
All Sides as Clamped, Simply Supported, or Free Boundary Conditions

S. No.	Sides 1 2 3 4	λ_1	λ_2	λ_3	λ_4	λ_5
1	C C C C	372.52	416.35	552.09	707.17	1010.4
2	S S S S	182.44	240.11	394.64	562.85	675.53
3	F F F F	16.479	41.070	64.904	98.417	136.49

TABLE 9.31

Convergence of First Five Frequencies for a Square Plate with
All Sides Clamped

N	λ_1	λ_2	λ_3	λ_4	λ_5
2	36.000	74.296	—	—	—
5	36.000	74.296	74.296	108.59	137.97
10	36.000	73.432	73.432	108.59	137.29
15	35.988	73.432	73.428	108.26	131.89
17	35.988	73.398	73.398	108.26	131.89
19	35.988	73.398	73.398	108.26	131.89
20	35.988	73.398	73.398	108.26	131.89

In these two skew angles, there again exist 22 distinct cases of boundary conditions as discussed earlier. The comparison of the results of special cases can be found in the work of Singh and Chakraverty (1994).

Convergences of the results for the first five frequencies of a square plate, when all the sides are clamped are shown in Table 9.31. Furthermore, in Table 9.32, the convergence of the first five frequencies for a skew plate with skew angle 60° and $b/a = 1.0$ having all sides clamped are also presented. These two tables give a pattern of how the results converge as we increase the number of approximations, i.e., the BCOPs, and it may be noted that the results do converge up to 20th approximation. Hence, all the computations have been carried out taking the number of approximations as 20.

Now, we will examine how the frequencies behave for some particular boundary conditions when the skew angles are increased. Figure 9.2 depicts the effect of skew angle on the first five frequencies of a CCCC plate with $b/a = 1.0$. We can observe from this figure that as the skew angle is increased, the frequencies decrease. Figures 9.3 and 9.4 show the effect of skew angles on the first five frequencies for SSSS and FFFF plates, respectively, for $b/a = 1.0$. One can note that as we increase the skew angle, the behavior of the frequencies remain the same as that of the CCCC, i.e., the frequencies increase as the skew angles are increased. The third case, i.e., the FFFF case has been depicted in Figure 9.4, which was observed to have a completely different behavior. Here, the first frequency increased as the skew angle increased. The second frequency first increased and then, again decreased as we increased the skew angle. However, the third frequency first decreased and then increased a little bit as the skew angle increased. The fourth frequency had a similar behavior as that of the third frequency. But, the fifth frequency decreased smoothly as the angle of skew increased.

The above behavior for the three cases, viz., CCCC, SSSS, and FFFF, with $b/a = 0.5$ are shown in Figures 9.5 through 9.7. The cases of CCCC and SSSS

TABLE 9.32

Convergence of First Five Frequencies for a Skew Plate ($\alpha = 60°$, $b/a = 1.0$) with All Sides Clamped

N	λ_1	λ_2	λ_3	λ_4	λ_5
2	50.596	103.692	—	—	—
5	46.742	89.084	116.48	144.76	199.15
10	46.524	81.759	107.13	135.80	187.99
15	46.166	81.759	107.11	119.98	167.78
17	46.166	81.613	107.58	119.98	167.17
19	46.166	81.613	107.56	119.98	167.16
20	46.166	81.613	107.56	119.98	167.16

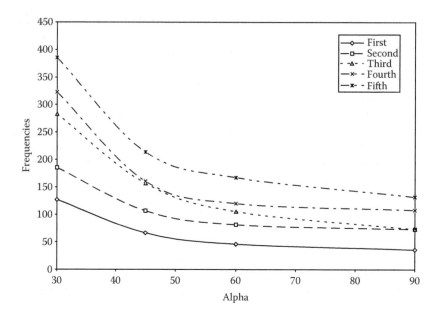

FIGURE 9.2
Effect of alpha on the first five frequencies of CCCC plates ($b/a = 1.0$).

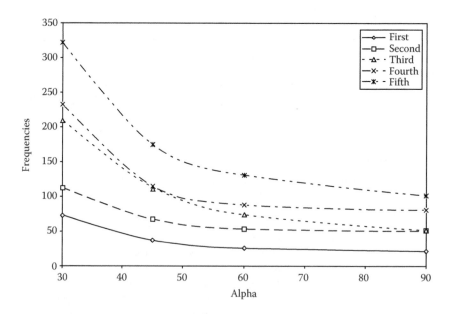

FIGURE 9.3
Effect of alpha on the first five frequencies of SSSS plates ($b/a = 1.0$).

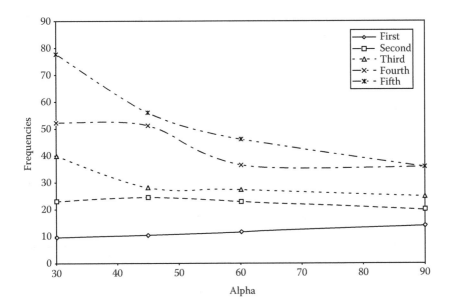

FIGURE 9.4
Effect of alpha on the first five frequencies of FFFF plates ($b/a = 1.0$).

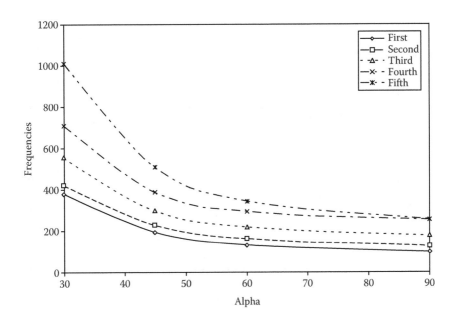

FIGURE 9.5
Effect of alpha on the first five frequencies of CCCC plates ($b/a = 0.5$).

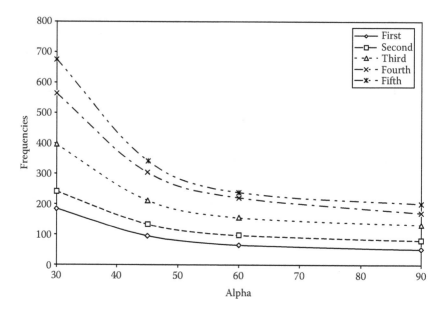

FIGURE 9.6
Effect of alpha on the first five frequencies of SSSS plates ($b/a = 0.5$).

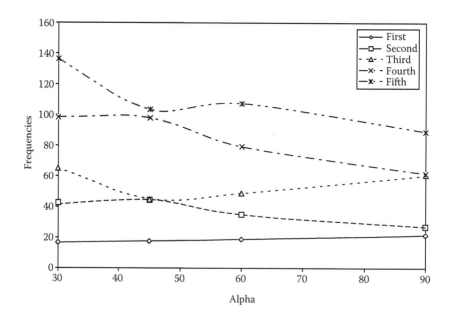

FIGURE 9.7
Effect of alpha on the first five frequencies of FFFF plates ($b/a = 0.5$).

plates in Figures 9.5 and 9.6 show the decreasing tendency of the first five frequencies as the skew angles are increased. Figure 9.7 illustrates the case of FFFF depicting that the first and third frequencies increase as we increase the skew angle. However, the second frequency shows the decreasing behavior when the skew angle is increased. The fourth and fifth frequencies in this case show different trends, as shown in Figure 9.7.

Lastly, the behaviors of different boundary conditions as we increase the skew angle are studied for fundamental frequency alone, as shown in Figures 9.8 through 9.11. The effect of skew angle on the fundamental frequencies of CCCC, SSSS, and FFFF plate are shown in Figure 9.8. As usual, this figure illustrates that the FFFF plate will have the lowest, and the CCCC plate will have highest values of the fundamental frequencies for all the skew angles considered. Figure 9.9 depicts the behavior of fundamental frequencies for the plates when the sides are only with clamped and free boundaries, as well as their effect on the skew angles. Here, we can observe that the CFFF has the lowest and the CFCF has the highest fundamental frequencies for all the considered skew angles. The effect of skew angles on the fundamental frequencies with clamped, simply supported, and simply supported, free sides are shown in Figures 9.10 and 9.11, respectively. The cited tables can be referred if readers would like to know any other details of the behavior depending on the boundary conditions of the skew plates.

The beauty of this method, as pointed out earlier, is the use of computationally efficient procedure of BCOPs, which may easily be implemented on a

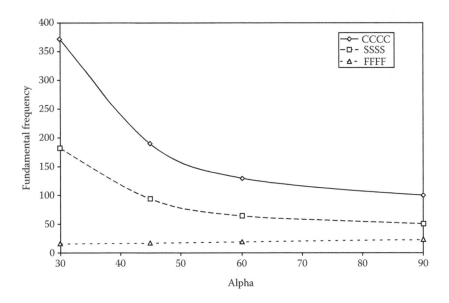

FIGURE 9.8
Effect of alpha on the fundamental frequencies with all sides as clamped, simply supported, or free boundary conditions ($b/a = 0.5$).

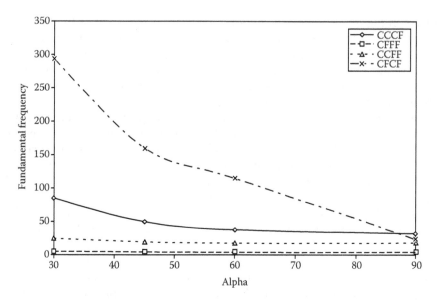

FIGURE 9.9
Effect of alpha on the fundamental frequencies and sides with clamped or free boundary conditions ($b/a = 0.5$).

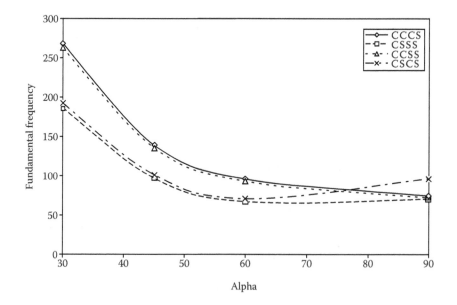

FIGURE 9.10
Effect of alpha on the fundamental frequencies and sides with clamped, or simply supported boundary conditions ($b/a = 0.5$).

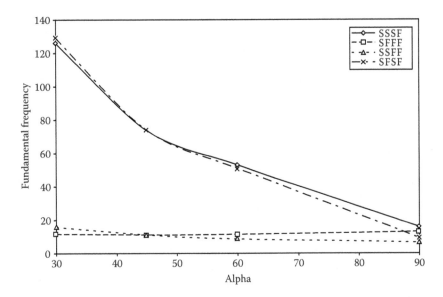

FIGURE 9.11
Effect of alpha on the fundamental frequencies and sides with simply supported or free boundary conditions ($b/a = 0.5$).

computer. Shape of the plate and the boundary conditions may be controlled by the six parameters, μ, α, p, q, r, and s. The mapping of the plate into the standard square can also save a lot of computation time. The orthogonal polynomials can be generated only once for the standard square, and may be used for plates with different values of b/a and α, i.e., different shapes of skew plates.

References

Barton, M.V. 1951. Vibration of rectangular and skew cantilever plates. *ASCE Journal of Applied Mechanics*, 18: 129–134.

Bhat, R.B. 1985. Natural frequencies of rectangular plates using characteristic orthogonal polynomials in Rayleigh–Ritz method. *Journal of Sound and Vibration*, 102: 493–499.

Bhat, R.B. 1987. Flexural vibration of polygonal plates using characteristic orthogonal polynomials in two variables. *Journal of Sound and Vibration*, 114: 65–71.

Bhat, R.B., Laura, P.A.A., Gutierrez, R.C., Cortinez, V.H., and Sanzi, H.C. 1990. Numerical experiments on the determination of natural frequencies of transverse vibrations of rectangular plates of non-uniform thickness. *Journal of Sound and Vibration*, 138: 205–219.

Dickinson, S.M. and Di Blasio, A. 1986. On the use of orthogonal polynomials in the Rayleigh–Ritz method for the study of the flexural vibration and buckling of isotropic and orthotropic rectangular plates. *Journal of Sound and Vibration*, 108: 51–62.

Dickinson, S.M. and Li, E.K.H. 1982. On the use of simply supported plate functions in the Rayleigh–Ritz method applied to the flexural vibration of rectangular plates. *Journal of Sound and Vibration*, 80: 292–297.

Durvasula, S. 1968. Natural frequencies and modes of skew membranes. *Journal of the Acoustical Society of America*, 44: 1636–1646.

Durvasula, S. 1969. Natural frequencies and modes of clamped skew plates. *American Institute of Aeronautics and Astronautics Journal*, 7: 1164–1167.

Durvasula, S. and Nair, P.S. 1974. Application of partition method to vibration problems of plates. *Journal of Sound and Vibration*, 37: 429–445.

Gorman, D.J. 1982. *Free Vibration Analysis of Rectangular Plates*, Elsevier, Amsterdam, the Netherlands.

Kuttler, J.R. and Sigillito, V.G. 1980. Upper and lower bounds for frequencies of clamped rhombic plates. *Journal of Sound and Vibration*, 68: 597–607.

Lam, K.Y. and Hung, K.C. 1990. Vibration study on plates with stiffened openings using orthogonal polynomials and partitioning method. *Computers and Structures*, 37: 295–301.

Leissa, A.W. 1969. *Vibration of Plates*, NASA SP-160, U.S. Government Printing Office, Washington DC.

Leissa, A.W. 1973. The free vibration of rectangular plates. *Journal of Sound and Vibration*, 31: 257–293.

Leissa, A.W. 1977. Recent research in plate vibrations, 1973–1976: Classical theory. *The Shock and Vibration Digest*, 9, No. 10, 13–24.

Leissa, A.W. 1981. Plate vibration research, 1976–1980: Classical theory. *The Shock and Vibration Digest*, 13, No. 9, 11–22.

Leissa, A.W. 1987. Recent studies in plate vibrations, No. 10,13–24: 1981–85. Part I—Classical theory. *The Shock and Vibration Digest*, 19, No. 2, 11–18.

Liew, K.M. and Lam, K.Y. 1990. Application of two-dimensional orthogonal plate functions to flexural vibration of skew plates. *Journal of Sound and Vibration*, 139: 241–252.

Liew, K.M., Lam, K.Y., and Chow, S.T. 1990. Free vibration analysis of rectangular plates using orthogonal plate function. *Computers and Structures*, 34: 79–85.

Mizusawa, T. 1986. Natural frequency of rectangular plates with free edges. *Journal of Sound and Vibration*, 105: 451–459.

Mizusawa, T., Kajita, T., and Naruoka, M. 1979. Vibration of skew plates by using B-spline functions. *Journal of Sound and Vibration*, 62: 301–308.

Nair, P.S. and Durvasula, S. 1973. Vibration of skew plates. *Journal of Sound and Vibration*, 26: 1–19.

Singh, B. and Chakraverty, S. 1994. Flexural vibration of skew plates using boundary characteristic orthogonal polynomials in two variables. *Journal of Sound and Vibration*, 173: 157–178.

10

Circular Annular and Elliptic Annular Plates

10.1 Introduction

In the recent decades, lightweight plate structures have been widely used in many engineering and practical applications. Vibration analyses of plates of different geometries and shapes have been carried out extensively by various researchers throughout the globe. The annular elliptic and circular plates are used quite often in aeronautical and ship structures, and in several other industrial applications. Therefore, the vibration analyses of these shapes are becoming more important.

Annular circular plates are special cases of annular elliptic plates and are quite simple to analyze using polar coordinates. Furthermore, the solution is found to be in the form of Bessel functions for all the nine cases of inner and outer boundary conditions. A survey of literature on the vibration of annular circular plates and the results of several cases are provided in a monograph by Leissa (1969). When compared with the amount of information available for circular plates, studies reported on the vibration of annular elliptic plates are limited. The main difficulty in studying elliptic plates is the choice of coordinates. Elliptic coordinates can be used with the exact mode shape in the form of Mathieu functions (Leissa 1969); however, they are quite cumbersome to handle. Rajalingham and Bhat (1991, 1992) and Rajalingham et al. (1993, 1994) studied the vibration of elliptic plates using modified polar coordinates by employing one-dimensional characteristic orthogonal polynomials (COPs)-shape functions, originally suggested by Bhat (1985) in the Rayleigh–Ritz method. Chakraverty (1992) and Singh and Chakraverty (1991, 1992a,b) analyzed the vibration of elliptic plates using two-dimensional boundary characteristic orthogonal polynomials (2D BCOPs) in the Rayleigh–Ritz method.

However, the annular elliptic plates also have not been studied in detail until recently, owing to the difficulty of studying elliptical plates. Chakraverty (1992) and Singh and Chakraverty (1993) reported the fundamental frequency for different aspect ratios of the annular elliptic plates for various conditions at the inner and outer boundaries. Recently, Chakraverty et al. (2001) also studied the free vibration of the annular elliptic and circular plates, and reported the first 12 frequency parameters.

This chapter provides the free vibration analysis of annular elliptic plates for all the nine boundary conditions at the inner and outer edges. Similar to the earlier studies, the results are reported for the four types of modes separately, viz., symmetric–symmetric, symmetric–antisymmetric, antisymmetric–symmetric, and antisymmetric–antisymmetric. The analysis was carried out using 2D BCOPs in the Rayleigh–Ritz method.

10.2 Domain Definition

The outer boundary of the elliptic plate, as shown in Figure 10.1, is defined as

$$R = \left\{ (x,y), \frac{x^2}{a^2} + \frac{y^2}{b^2} \le 1, x, y \in \mathbf{R} \right\} \tag{10.1}$$

where a and b are the semimajor and semiminor axes, respectively. Families of concentric ellipses are thus defined by introducing a variable C, as

$$x^2 + \frac{y^2}{m^2} = 1 - C, \quad 0 \le C \le C_0 < 1 \tag{10.2}$$

where $m = b/a$ and C_0 defines the inner boundary of the ellipse. The eccentricity of the inner boundary is defined by k where

$$k = \sqrt{1 - C_0} \tag{10.3}$$

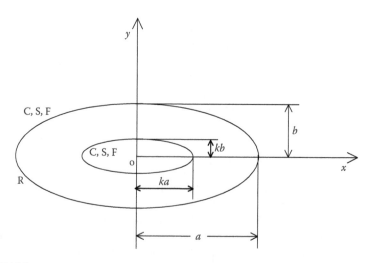

FIGURE 10.1
Domain of the annular elliptical plate.

However, $\dot{K}=0$ will give the full elliptic plate and as m takes the value of 1, the geometry will turn into a circular plate. Furthermore, the values of m for $0<m<1$, along with different values of k (which is the inner-boundary parameter) will turn the geometry to annular elliptic plates. When $m=1.0$ with different values of the inner-boundary parameter k, then the resulting geometry will be annular circular plates.

10.3   Governing Equations and Method of Solution

When the plate structure undergoes simple harmonic motion, the maximum strain energy, V_{max}, and the maximum kinetic energy, T_{max}, of the deformed annular elliptic plate may be given by

$$V_{max} = \frac{D}{2} \iint_R \left[W_{xx}^2 + 2\nu W_{xx} W_{yy} + W_{yy}^2 + 2(1-\nu)W_{xy}^2 \right] dy dx \qquad (10.4)$$

$$T_{max} = \frac{\rho h \omega^2}{2} \iint_R W^2 dy dx \qquad (10.5)$$

where
   $W(x, y)$ is the deflection of the plate, the subscripts on W denote
      differentiation with respect to the subscripted variable
   $D = Eh^3/(12(1-\nu^2))$ is the flexural rigidity
   E is Young's modulus
   ν is the Poisson's ratio
   h is the uniform plate thickness
   ρ is the density of the plate material
   ω is the radian frequency of vibration.

   Now, similar to the earlier chapters, the maximum strain and kinetic energies can be equated to obtain the Rayleigh quotient as

$$\omega^2 = \frac{D \iint_R \left[W_{xx}^2 + 2\nu W_{xx} W_{yy} + W_{yy}^2 + 2(1-\nu)W_{xy}^2 \right] dy dx}{\rho h \iint_R W^2 dy dx} \qquad (10.6)$$

By substituting the following N-term approximation for the deflection of annular plate in Equation 10.6

$$W(x, y) = \sum_{j=1}^{N} c_j \phi_j(x,y) \tag{10.7}$$

and applying the condition for stationarity of ω^2 with respect to the coefficients c_j, we can obtain

$$\sum_{j=1}^{N} \left(a_{ij} - \lambda^2 b_{ij} \right) c_j = 0, \quad i = 1, 2, \ldots, N \tag{10.8}$$

where

$$a_{ij} = \iint_{R'} \left[(\phi_i)_{XX}(\phi_j)_{XX} + (\phi_i)_{YY}(\phi_j)_{YY} + \nu\{(\phi_i)_{XX}(\phi_j)_{YY} + (\phi_i)_{YY}(\phi_j)_{XX}\} \right.$$

$$\left. + 2(1 - \nu)(\phi_i)_{XY}(\phi_j)_{XY} \right] dY dX \tag{10.9}$$

$$b_{ij} = \iint_{R'} \phi_i \phi_j \, dY dX \tag{10.10}$$

$$\lambda^2 = \frac{a^4 \rho h \omega^2}{D} \tag{10.11}$$

which is the frequency parameter of the said plate elements,

and

$$X = \frac{x}{a}, \quad Y = \frac{y}{a}$$

Further, $(\phi_i)_{XX}$, $(\phi_i)_{YY}$ etc., are second derivatives of ϕ_i with respect to X and Y. The new domain R' is defined as

$$R' = \left\{ (X,Y), \; X^2 + \frac{Y^2}{m^2} \leq 1, \quad X, Y \in R \right\}$$

The ϕ_i's are the orthogonal polynomials, and the procedure for generating them is described in the following section. Since the ϕ_i's are orthogonal, Equation 10.8 can be reduced to

$$\sum_{j=1}^{N} \left(a_{ij} - \lambda^2 \delta_{ij} \right) c_j = 0, \quad i = 1, 2, \ldots, N \tag{10.12}$$

where

$$\delta_{ij} = 0, \quad \text{if} \quad i \neq j$$
$$= 1, \quad \text{if} \quad i = j$$

Equation 10.12 is a standard eigenvalue problem and can be solved to obtain the vibration characteristics.

10.4 Generation of the BCOPs in Annular Domains

For the generation of 2D orthogonal polynomials, the following linearly independent set of functions is employed:

$$F_i(X,Y) = g(X,Y)\{f_i(X,Y)\}, \quad i = 1, 2, \ldots, N \tag{10.13}$$

where $g(X,Y)$ satisfies the essential boundary conditions, and $f_i(X,Y)$ are taken as the combinations of terms of the form $x^{\ell_i} y^{n_i}$, where ℓ_i and n_i are nonnegative positive integers. Depending on the even and odd values of ℓ_i and n_i, one can get the symmetric and antisymmetric modes of the vibrations. The function $g(X,Y)$ is defined for the annular plate as

$$g(X,Y) = C^s(C_0 - C)^t \tag{10.14}$$

where s takes the value of 0, 1, or 2, defining free, simply supported, or clamped conditions, respectively, at the outer boundary of the annular elliptic plate. Similarly, $t = 0, 1,$ or 2 will define the corresponding boundary conditions at the inner edge of the annular elliptic plate.

From $F_i(X,Y)$, an orthogonal set can be generated by the well-known Gram–Schmidt process. In this case, the inner product of the two functions $\phi_i(X,Y)$ and $F_i(X,Y)$ is defined as

$$<\phi_i, F_i> = \iint\limits_R \phi_i(X,Y)F_i(X,Y)dxdy \tag{10.15}$$

The norm of ϕ_i is, therefore, given by

$$\|\phi_i\| = \langle \phi_i, \phi_i \rangle^{1/2} \tag{10.16}$$

As done in the previous chapters (Singh and Chakraverty 1993, Chakraverty et al. 2001), the Gram–Schmidt orthogonalization process can be written as

$$\phi_1 = F_1$$

$$\left.\begin{array}{l} \phi_i = F_i - \displaystyle\sum_{j=1}^{i-1} \alpha_{ij}\phi_j \\[2ex] \alpha_{ij} = \dfrac{\langle F_i, \phi_j \rangle}{\langle \phi_j, \phi_j \rangle}, \quad j = 1, 2, \ldots, (i-1) \end{array}\right\}, \quad i = 2, \ldots, N \tag{10.17}$$

where ϕ_i's are the BCOPs. These are normalized by the condition

$$\hat{\phi}_i = \frac{\phi_i}{\|\phi_i\|} \tag{10.18}$$

All the integrals involved in the inner product were evaluated in closed form as discussed in the work of Singh and Chakraverty (1993).

10.5   Some Numerical Results and Discussions for Annular Plates

The computations can be simplified by considering the shape functions to represent symmetric–symmetric, symmetric–antisymmetric, antisymmetric–symmetric, and antisymmetric–antisymmetric groups. First, few natural frequencies required can be arranged in an ascending order by choosing the required number of frequencies from each of the above-mentioned mode groups. Numerical results for natural frequencies have been computed by taking the aspect ratio of the outer boundary, $b/a = m = 0.5$ and 1.0 with the various values of the inner-boundary parameter, k. The results for the nine cases of boundary conditions (CC, CS, CF; SC, SS, SF; FC, FS, FF) at both the inner and outer edges are provided. The C, S, and F designate clamped, simply supported, and free boundary, respectively, and the first and the second letters denote the conditions at the outer and inner edges, respectively. Before going into the details of the computations, we would first discuss about the convergence. Poisson's ratio, ν, is taken as $1/3$ in all the calculations.

10.5.1   Convergence Study

The convergence of the results is studied by computing the results for different values of N, until the first four significant digits converge. It was found that the results converged for the N value from 26 to 30 in all the cases. In particular, for the annular circular plates, we went up to $N = 20$, and for annular elliptic plates, the value of N was taken up to 30. The convergence of the frequency parameters for symmetric–symmetric modes for annular circular plates with $k = 0.8$ and $\nu = 1/3$ is shown in Table 10.1, where convergence studies are shown for CC, CS, CF, and SS boundary conditions. Furthermore, the convergence study for annular elliptic plates ($m = 0.5$, $\nu = 1/3$) is shown in Table 10.2 for $k = 0.2$ and 0.5, showing the results for CS, CF, SS, and FC boundary conditions.

10.5.2   Comparison between Exact and BCOPs Results

The exact results for annular circular plate in terms of Bessel functions may be obtained. Accordingly, the fundamental frequencies were computed by using BCOPs for various values of k, viz., $k = 0.2$–0.8 at the interval of 0.2, and were compared with the exact results, as shown in Table 10.3. All nine boundary conditions were considered in this comparison analysis for the annular circular plates. It can be observed from Table 10.3 that the results from BCOPs method are in accordance with the exact method. Table 10.3

TABLE 10.1

Convergence of Frequency Parameters for Symmetric–Symmetric Modes for Annular Circular Plates ($k = 0.8$, $\nu = 1/3$)

N	CC Boundary	CS Boundary	CF Boundary	SS Boundary
2	570.53	404.80	94.144	277.92
3	559.66	395.96	92.863	267.85
4	559.64	395.52	92.862	265.88
10	559.16	389.54	92.819	247.08
11	559.16	389.54	92.819	247.08
12	559.16	389.54	92.819	247.08
16	559.16	389.51	92.815	247.07
17	559.16	389.51	92.815	247.07
18	559.16	389.51	92.815	247.07
19	559.16	389.51	92.815	247.07
20	559.16	389.51	92.815	247.07

Note: Convergence obtained for circular annular plate in 20 approximations.

also reveals that the comparison is excellent for large values of k, such as $k = 0.6$ and 0.8. Moreover, as expected, the fundamental frequencies increased as we increased the value of k (i.e., the whole size of the annular circular plate) for all the boundary conditions, apart from for the exceptional case of FF boundary. For the FF boundary condition, the frequencies decreased as k increased.

10.5.3 Annular Circular Plate

As discussed earlier, $m = 1.0$ gives the results for annular circular plates. Tables 10.4 through 10.12 show the results of the frequency parameters for

TABLE 10.2

Convergence of Frequency Parameters for Symmetric–Symmetric Modes for Annular Elliptic Plates ($m = 0.5$, $\nu = 1/3$)

N	CS Boundary $k = 0.2$	CF Boundary $k = 0.2$	SS Boundary $k = 0.6$	FC Boundary $k = 0.6$
3	74.155	29.122	152.94	60.307
10	61.640	28.763	99.765	32.572
15	59.212	28.517	95.197	28.659
20	58.401	28.386	93.063	26.737
25	57.493	28.209	92.784	26.566
27	56.669	28.034	92.083	25.676
28	56.668	28.029	92.075	25.675
29	56.668	28.028	92.074	25.675
30	56.668	28.028	92.074	25.675

Note: Convergence obtained for elliptic annular plate in 30 approximations.

TABLE 10.3

Comparison of Fundamental Frequency Results between Exact and BCOPs for Annular Circular Plates with All Boundary Conditions, $\nu = 1/3$

B.C.	Method Used	$k \rightarrow 0.2$	0.4	0.5	0.6	0.8
CC	BCOPs	35.12	61.88	89.25	139.6	559.1
	Exact	34.61	61.87	89.25	139.6	559.1
CS	BCOPs	26.68	44.93	63.85	98.79	389.5
	Exact	26.61	44.93	63.85	98.79	389.5
CF	BCOPs	10.46	13.50	17.60	25.24	92.81
	Exact	10.34	13.50	17.59	25.24	92.81
SC	BCOPs	23.34	41.27	59.91	94.26	381.6
	Exact	22.76	41.26	59.90	94.26	381.6
SS	BCOPs	16.86	28.08	40.01	62.12	247.0
	Exact	16.72	28.08	40.01	62.12	247.0
SF	BCOPs	4.851	4.748	5.051	5.663	9.455
	Exact	4.732	4.743	5.043	5.663	9.455
FC	BCOPs	5.384	9.082	13.10	20.60	84.67
	Exact	5.214	9.071	13.09	20.60	84.67
FS	BCOPs	3.466	3.634	4.074	4.809	8.782
	Exact	3.313	3.629	4.070	4.809	8.782
FF	BCOPs	5.056	4.533	4.214	3.865	3.197
	Exact	5.049	4.532	4.207	3.864	3.197

TABLE 10.4

First Three Frequency Parameters for Symmetric–Symmetric, Symmetric–Antisymmetric, Antisymmetric–Symmetric, and Antisymmetric–Antisymmetric Modes for Annular Circular Plates (CC Boundary, $\nu = 1/3$)

	Symmetric–Symmetric			Symmetric–Antisymmetric		
k	First	Second	Third	First	Second	Third
0.0	27.880	42.039	73.636	32.310	55.524	89.472
0.2	35.116	46.035	75.706	37.909	58.182	95.257
0.4	61.876	67.166	88.548	63.038	74.970	105.29
0.6	139.61	143.13	155.04	140.48	147.85	166.07
0.8	559.16	561.89	570.24	559.84	565.34	576.68

	Antisymmetric–Symmetric			Antisymmetric–Antisymmetric		
k	First	Second	Third	First	Second	Third
0.0	32.309	55.508	89.469	40.904	72.289	104.87
0.2	37.824	57.957	95.104	44.980	74.248	117.18
0.4	63.038	74.999	105.67	66.869	86.879	125.29
0.6	140.48	147.86	166.44	143.13	154.80	178.34
0.8	559.84	565.34	576.68	561.89	570.24	584.76

TABLE 10.5

First Three Frequency Parameters for Symmetric–Symmetric, Symmetric–
Antisymmetric, Antisymmetric–Symmetric, and Antisymmetric–Antisymmetric
Modes for Annular Circular Plates (CS Boundary, $\nu = 1/3$)

	Symmetric–Symmetric			Symmetric–Antisymmetric		
k	First	Second	Third	First	Second	Third
0.0	23.147	37.087	63.905	26.693	51.921	73.489
0.2	26.680	39.772	71.185	30.389	53.364	85.999
0.4	44.932	52.449	77.904	46.735	62.669	96.344
0.6	98.793	104.01	120.40	100.08	110.72	133.26
0.8	389.51	393.37	405.11	390.47	398.24	414.04

	Antisymmetric–Symmetric			Antisymmetric–Antisymmetric		
k	First	Second	Third	First	Second	Third
0.0	26.796	51.980	73.819	36.746	70.101	91.558
0.2	30.289	53.290	85.827	39.420	70.743	100.35
0.4	46.735	62.600	96.319	52.389	77.229	117.97
0.6	100.08	110.72	133.25	104.01	120.40	149.44
0.8	390.47	398.24	414.04	393.37	405.11	425.12

TABLE 10.6

First Three Frequency Parameters for Symmetric–Symmetric, Symmetric–
Antisymmetric, Antisymmetric–Symmetric, and Antisymmetric–Antisymmetric
Modes for Annular Circular Plates (CF Boundary, $\nu = 1/3$)

	Symmetric–Symmetric			Symmetric–Antisymmetric		
k	First	Second	Third	First	Second	Third
0.0	10.215	34.877	39.771	21.260	51.030	60.828
0.2	10.466	33.927	43.164	21.173	50.591	60.681
0.4	13.501	32.055	66.937	19.477	47.893	72.031
0.6	25.540	36.434	63.672	28.500	48.118	82.734
0.8	92.815	98.781	116.18	94.316	106.11	128.92

	Antisymmetric–Symmetric			Antisymmetric–Antisymmetric		
k	First	Second	Third	First	Second	Third
0.0	21.260	51.030	60.828	34.877	69.665	84.582
0.2	21.143	50.589	60.494	33.913	69.558	82.727
0.4	19.460	47.812	72.000	31.736	66.810	86.608
0.6	28.499	48.108	82.558	36.422	63.180	105.05
0.8	94.316	106.11	128.91	98.781	116.18	144.25

TABLE 10.7

First Three Frequency Parameters for Symmetric–Symmetric, Symmetric–Antisymmetric, Antisymmetric–Symmetric, and Antisymmetric–Antisymmetric Modes for Annular Circular Plates (SC Boundary, $\nu = 1/3$)

	Symmetric–Symmetric			Symmetric–Antisymmetric		
k	First	Second	Third	First	Second	Third
0.0	18.749	31.467	60.680	22.424	43.548	74.425
0.2	23.339	34.071	62.185	26.103	45.543	80.569
0.4	41.270	47.523	70.636	42.634	56.142	87.433
0.6	94.263	98.583	113.30	95.324	104.28	125.47
0.8	381.64	385.17	395.95	382.52	389.63	404.21

	Antisymmetric–Symmetric			Antisymmetric–Antisymmetric		
k	First	Second	Third	First	Second	Third
0.0	22.512	43.708	74.817	30.398	59.283	90.006
0.2	26.045	45.374	80.323	33.015	60.555	99.098
0.4	42.631	56.040	87.111	47.046	68.664	107.12
0.6	95.323	104.39	125.93	98.567	112.62	140.69
0.8	382.52	389.64	404.31	385.17	395.95	414.60

TABLE 10.8

First Three Frequency Parameters for Symmetric–Symmetric, Symmetric–Antisymmetric, Antisymmetric–Symmetric, and Antisymmetric–Antisymmetric Modes for Annular Circular Plates (SS Boundary, $\nu = 1/3$)

	Symmetric–Symmetric			Symmetric–Antisymmetric		
k	First	Second	Third	First	Second	Third
0.0	15.154	27.443	51.313	18.190	40.764	60.139
0.2	16.856	29.165	58.179	20.534	41.709	69.409
0.4	28.085	36.449	62.399	30.087	47.233	80.187
0.6	62.126	68.351	87.185	63.680	76.153	101.54
0.8	247.07	252.00	266.81	248.30	258.17	277.92

	Antisymmetric–Symmetric			Antisymmetric–Antisymmetric		
k	First	Second	Third	First	Second	Third
0.0	18.254	40.819	60.394	27.115	57.294	76.810
0.2	20.466	41.644	69.244	28.833	57.733	83.481
0.4	30.086	47.146	80.114	36.270	61.688	100.83
0.6	63.680	76.153	101.48	68.350	87.107	119.15
0.8	248.30	258.17	277.92	252.00	266.81	291.55

TABLE 10.9

First Three Frequency Parameters for Symmetric–Symmetric, Symmetric–Antisymmetric, Antisymmetric–Symmetric, and Antisymmetric–Antisymmetric Modes for Annular Circular Plates (SF Boundary, $\nu = 1/3$)

	Symmetric–Symmetric			Symmetric–Antisymmetric		
k	First	Second	Third	First	Second	Third
0.0	4.9839	25.652	29.758	13.939	39.995	48.516
0.2	4.8514	25.006	31.753	13.911	39.745	48.552
0.4	4.7481	23.822	47.323	12.095	37.947	53.339
0.6	5.6632	22.214	51.421	11.729	35.019	70.349
0.8	9.4551	29.653	59.940	16.781	44.112	77.285

	Antisymmetric–Symmetric			Antisymmetric–Antisymmetric		
k	First	Second	Third	First	Second	Third
0.0	13.939	39.995	48.516	25.652	56.878	70.153
0.2	13.902	39.744	48.464	24.997	56.823	68.519
0.4	12.064	37.910	53.268	23.561	55.147	69.739
0.6	11.729	34.982	70.165	22.162	50.543	91.888
0.8	16.781	44.111	77.273	29.653	59.900	96.418

TABLE 10.10

First Three Frequency Parameters for Symmetric–Symmetric, Symmetric–Antisymmetric, Antisymmetric–Symmetric, and Antisymmetric–Antisymmetric Modes for Annular Circular Plates (FC Boundary, $\nu = 1/3$)

	Symmetric–Symmetric			Symmetric–Antisymmetric		
k	First	Second	Third	First	Second	Third
0.0	4.5379	6.9729	22.162	4.5239	12.937	33.467
0.2	5.3841	7.6203	22.416	5.4042	13.309	33.668
0.4	9.0818	10.736	23.887	9.1880	15.354	34.725
0.6	20.608	22.430	32.174	20.980	25.774	41.352
0.8	84.678	87.173	95.391	85.289	90.481	102.15

	Antisymmetric–Symmetric			Antisymmetric–Antisymmetric		
k	First	Second	Third	First	Second	Third
0.0	4.5414	12.961	33.491	6.6269	21.860	44.434
0.2	5.3943	13.284	33.630	7.2651	22.051	47.226
0.4	9.1755	15.341	34.669	10.528	23.331	48.012
0.6	20.980	25.775	41.308	22.449	31.752	53.349
0.8	85.289	90.470	102.11	87.177	95.404	116.49

TABLE 10.11

First Three Frequency Parameters for Symmetric–Symmetric, Symmetric–Antisymmetric, Antisymmetric–Symmetric, and Antisymmetric–Antisymmetric Modes for Annular Circular Plates (FS Boundary, $\nu = 1/3$)

	Symmetric–Symmetric			Symmetric–Antisymmetric		
k	First	Second	Third	First	Second	Third
0.0	3.8022	5.7833	21.602	3.1543	12.358	26.684
0.2	3.4664	6.1958	21.672	3.6994	12.504	30.150
0.4	3.6342	7.1596	22.224	4.0090	13.285	33.464
0.6	4.8088	9.7375	24.704	6.0522	15.856	35.719
0.8	8.7829	20.152	40.287	12.482	29.513	52.553

	Antisymmetric–Symmetric			Antisymmetric–Antisymmetric		
k	First	Second	Third	First	Second	Third
0.0	3.1713	12.367	26.772	5.6723	21.546	38.209
0.2	3.6793	12.493	30.074	6.0809	21.597	40.778
0.4	4.0011	13.258	33.448	6.9180	22.064	47.031
0.6	6.0472	15.854	35.665	9.7211	24.395	48.955
0.8	12.481	29.514	52.554	20.152	40.277	66.469

TABLE 10.12

First Three Frequency Parameters for Symmetric–Symmetric, Symmetric–Antisymmetric, Antisymmetric–Symmetric, and Antisymmetric–Antisymmetric Modes for Annular Circular Plates (FF Boundary, $\nu = 1/3$)

	Symmetric–Symmetric			Symmetric–Antisymmetric		
k	First	Second	Third	First	Second	Third
0.0	5.2512	9.0761	21.491	12.221	20.517	33.012
0.2	5.0579	8.8117	21.488	12.190	20.489	33.011
0.4	4.5334	8.5829	21.279	11.772	17.911	32.923
0.6	3.8650	10.546	19.904	10.568	18.210	31.649
0.8	3.1975	16.910	18.264	8.8727	27.293	29.394

	Antisymmetric–Symmetric			Antisymmetric–Antisymmetric		
k	First	Second	Third	First	Second	Third
0.0	12.221	20.517	33.012	5.2512	21.491	35.240
0.2	12.190	20.477	33.011	5.0555	21.487	34.359
0.4	11.772	17.797	32.923	4.5327	21.272	33.093
0.6	10.568	18.209	31.641	3.8656	19.863	32.238
0.8	8.8727	27.289	29.394	3.1973	16.895	40.047

all the nine boundary conditions. In each of these tables, the four types of modes, viz., symmetric–symmetric, symmetric–antisymmetric, antisymmetric–symmetric, and antisymmetric–antisymmetric, are incorporated and in each of the modes, the first three frequencies of that particular mode are given. Results for the special cases, such as full circular plate ($k = 0$, $m = 1.0$) are also shown. It is interesting to note the effect of hole size on the natural frequencies, from Tables 10.4 through 10.12. As k increases for the annular circular plate, the frequencies increase for all the boundary conditions, except for the exceptional case of FF boundary condition, where the frequencies decreased as k increased, as shown in Table 10.12.

Tables 10.4 through 10.12 show that the frequencies for any of the hole size are maximum for CC annular circular plates. The effect of different boundary conditions on the natural frequencies can be well investigated by dividing them into the following three sets:

(i) CC, CS, and CF

(ii) SC, SS, and SF

(iii) FC, FS, and FF

It can be observed from Tables 10.4 through 10.6 (set (i) boundary conditions) that all the frequencies decrease from CC to CF. Similar behavior is also observed for set (ii) boundary conditions, as shown in Tables 10.7 through 10.9. Here, the frequencies decrease from SC to SF. For set (iii) boundary conditions with smaller hole size, i.e., $k = 0$ and 0.2, it can be noticed from Tables 10.10 through 10.12 that the FC condition gives higher frequencies, while the FS condition gives lower frequencies. But, for larger hole-size, viz., for $k \geq 0.4$, the results are smaller for the FF boundary, as in sets (i) and (ii). It may also be noted that for set (iii) with $k \leq 0.2$ (smaller hole-size), the frequencies (particularly the lower modes) with conditions FC and FF are closer. This behavior is analogous to that of beams with CC and FF boundary conditions, where the frequencies are identical.

10.5.4 Annular Elliptic Plate

We considered the annular elliptic plate with aspect ratio $m = 0.5$ as an example of this investigation. Accordingly, Tables 10.13 through 10.21 give the results of all the modes for annular elliptic plates, separately. The results of the computations for various values of k and for different boundary conditions at the outer and inner edges are presented in these tables. Again, Poisson's ratio, ν, was taken as $1/3$ in all the calculations. First few natural frequencies were arranged in an ascending order by choosing them from each of the mode groups. Similar to the observations in the annular circular plates, the results for the special cases, viz., full elliptic ($k = 0$) plates were computed.

TABLE 10.13

First Three Frequency Parameters for Symmetric–Symmetric, Symmetric–Antisymmetric, Antisymmetric–Symmetric, and Antisymmetric–Antisymmetric Modes for Annular Elliptic Plates (CC Boundary, $m = 0.5$, $\nu = 1/3$)

	Symmetric–Symmetric			Symmetric–Antisymmetric		
k	First	Second	Third	First	Second	Third
0.0	59.879	99.894	135.15	103.68	132.90	186.94
0.2	67.305	119.79	151.71	66.092	120.32	172.32
0.4	95.496	179.12	232.71	154.31	219.80	254.68
0.6	181.68	326.84	432.08	265.11	385.96	483.38
0.8	641.71	1019.3	1428.7	787.43	1105.5	1427.3

	Antisymmetric–Symmetric			Antisymmetric–Antisymmetric		
k	First	Second	Third	First	Second	Third
0.0	58.673	106.18	162.29	105.23	153.35	218.01
0.2	115.63	148.08	195.41	114.58	163.33	227.40
0.4	94.543	178.28	233.52	152.68	219.22	272.56
0.6	182.66	329.22	433.32	260.65	382.87	476.89
0.8	632.01	946.29	1261.7	796.74	1138.6	1487.6

TABLE 10.14

First Three Frequency Parameters for Symmetric–Symmetric, Symmetric–Antisymmetric, Antisymmetric–Symmetric, and Antisymmetric–Antisymmetric Modes for Annular Elliptic Plates (CS Boundary, $m = 0.5$, $\nu = 1/3$)

	Symmetric–Symmetric			Symmetric–Antisymmetric		
k	First	Second	Third	First	Second	Third
0.0	50.670	85.011	117.48	87.961	118.87	172.32
0.2	56.668	96.110	126.25	56.222	102.77	151.94
0.4	78.757	146.69	177.92	130.03	171.62	207.61
0.6	143.03	262.57	339.19	214.48	308.71	370.77
0.8	468.55	730.55	967.96	601.70	836.68	1051.3

	Antisymmetric–Symmetric			Antisymmetric–Antisymmetric		
k	First	Second	Third	First	Second	Third
0.0	49.704	91.490	145.04	93.997	141.80	205.73
0.2	97.493	125.44	177.33	101.21	148.28	210.01
0.4	78.409	147.38	191.59	129.63	181.36	236.35
0.6	142.74	261.82	339.20	213.84	308.04	374.51
0.8	467.83	722.87	944.94	600.91	835.93	1044.4

TABLE 10.15

First Three Frequency Parameters for Symmetric–Symmetric, Symmetric–Antisymmetric, Antisymmetric–Symmetric, and Antisymmetric–Antisymmetric Modes for Annular Elliptic Plates (CF Boundary, $m = 0.5$, $\nu = 1/3$)

	Symmetric–Symmetric			Symmetric–Antisymmetric		
k	First	Second	Third	First	Second	Third
0.0	27.377	55.975	102.64	69.857	109.93	165.51
0.2	28.028	56.980	103.47	39.521	77.027	133.21
0.4	36.347	62.686	104.60	54.029	103.35	160.59
0.6	59.745	93.932	130.11	86.300	112.61	166.19
0.8	151.56	263.04	330.64	218.81	301.25	350.54

	Antisymmetric–Symmetric			Antisymmetric–Antisymmetric		
k	First	Second	Third	First	Second	Third
0.0	39.497	76.995	132.95	88.047	135.71	199.41
0.2	69.182	108.54	163.69	86.034	134.24	198.18
0.4	41.041	82.638	133.11	77.093	128.17	194.81
0.6	59.885	106.67	156.49	91.713	135.93	199.20
0.8	151.46	262.46	331.51	218.54	300.61	358.34

TABLE 10.16

First Three Frequency Parameters for Symmetric–Symmetric, Symmetric–Antisymmetric, Antisymmetric–Symmetric, and Antisymmetric–Antisymmetric Modes for Annular Elliptic Plates (SC Boundary, $m = 0.5$, $\nu = 1/3$)

	Symmetric–Symmetric			Symmetric–Antisymmetric		
k	First	Second	Third	First	Second	Third
0.0	39.431	73.042	108.19	73.425	100.96	154.19
0.2	44.522	87.072	117.83	43.458	91.259	140.66
0.4	63.770	133.40	169.92	111.50	160.75	193.19
0.6	123.15	243.81	328.65	190.91	290.91	360.98
0.8	436.14	718.28	988.20	570.09	836.68	1097.1

	Antisymmetric–Symmetric			Antisymmetric–Antisymmetric		
k	First	Second	Third	First	Second	Third
0.0	38.518	81.781	133.06	76.350	122.02	186.03
0.2	82.254	110.05	160.04	82.696	128.45	191.39
0.4	62.899	132.65	180.21	110.06	164.74	219.60
0.6	122.52	241.32	325.70	188.93	287.99	360.77
0.8	438.90	736.88	1007.0	571.40	823.05	1060.8

TABLE 10.17

First Three Frequency Parameters for Symmetric–Symmetric, Symmetric–Antisymmetric, Antisymmetric–Symmetric, and Antisymmetric–Antisymmetric Modes for Annular Elliptic Plates (SS Boundary, $m = 0.5$, $\nu = 1/3$)

	Symmetric–Symmetric			Symmetric–Antisymmetric		
k	First	Second	Third	First	Second	Third
0.0	32.235	61.363	92.804	60.612	90.164	141.47
0.2	35.863	66.889	98.487	35.736	77.202	123.15
0.4	50.220	101.77	128.03	89.659	118.68	162.62
0.6	92.074	186.43	238.51	149.38	219.51	255.79
0.8	303.36	516.71	607.96	415.25	602.39	756.96

	Antisymmetric–Symmetric			Antisymmetric–Antisymmetric		
k	First	Second	Third	First	Second	Third
0.0	31.599	69.830	118.85	67.428	112.32	173.43
0.2	67.043	94.181	144.87	72.154	116.66	176.36
0.4	49.933	105.89	147.99	90.303	135.54	192.70
0.6	91.777	185.91	244.65	148.82	220.24	274.74
0.8	302.77	321.21	512.67	413.96	598.50	755.43

TABLE 10.18

First Three Frequency Parameters for Symmetric–Symmetric, Symmetric–Antisymmetric, Antisymmetric–Symmetric, and Antisymmetric–Antisymmetric Modes for Annular Elliptic Plates (SF Boundary, $m = 0.5$, $\nu = 1/3$)

	Symmetric–Symmetric			Symmetric–Antisymmetric		
k	First	Second	Third	First	Second	Third
0.0	13.271	38.425	81.308	46.189	83.067	135.21
0.2	12.790	38.801	81.617	23.708	57.625	109.68
0.4	12.258	40.935	81.351	31.848	77.613	132.01
0.6	14.564	42.667	91.608	27.151	72.868	130.27
0.8	24.368	59.826	120.28	38.313	92.586	155.69

	Antisymmetric–Symmetric			Antisymmetric–Antisymmetric		
k	First	Second	Third	First	Second	Third
0.0	23.722	57.592	109.59	62.810	107.15	167.34
0.2	45.982	82.365	134.04	61.439	106.01	166.41
0.4	22.532	59.433	108.66	56.221	101.33	163.52
0.6	24.037	65.649	118.09	48.752	99.113	162.38
0.8	36.267	88.808	153.58	64.607	122.65	190.90

TABLE 10.19

First Three Frequency Parameters for Symmetric–Symmetric, Symmetric–Antisymmetric, Antisymmetric–Symmetric, and Antisymmetric–Antisymmetric Modes for Annular Elliptic Plates (FC Boundary, $m = 0.5$, $\nu = 1/3$)

	Symmetric–Symmetric			Symmetric–Antisymmetric		
k	First	Second	Third	First	Second	Third
0.0	7.5349	17.736	37.479	13.355	58.569	26.984
0.2	8.4482	20.783	39.359	8.1673	25.678	55.299
0.4	12.501	33.120	48.410	22.776	39.918	65.889
0.6	25.675	63.430	86.663	42.907	77.811	96.467
0.8	97.016	187.47	270.45	138.25	227.95	299.88

	Antisymmetric–Symmetric			Antisymmetric–Antisymmetric		
k	First	Second	Third	First	Second	Third
0.0	6.9815	23.761	53.408	14.174	39.932	80.077
0.2	15.406	28.972	59.775	15.687	41.312	81.099
0.4	12.194	34.898	63.249	22.346	48.655	86.369
0.6	25.383	63.047	94.357	42.444	79.388	112.31
0.8	96.510	184.77	266.14	137.24	225.51	296.54

TABLE 10.20

First Three Frequency Parameters for Symmetric–Symmetric, Symmetric–Antisymmetric, Antisymmetric–Symmetric, and Antisymmetric–Antisymmetric Modes for Annular Elliptic Plates (FS Boundary, $m = 0.5$, $\nu = 1/3$)

	Symmetric–Symmetric			Symmetric–Antisymmetric		
k	First	Second	Third	First	Second	Third
0.0	5.6839	15.094	33.279	9.5556	24.190	55.763
0.2	6.2539	13.752	34.555	5.8945	22.090	50.148
0.4	7.6656	14.687	38.398	10.677	25.684	59.040
0.6	10.841	20.225	47.929	14.146	32.354	67.315
0.8	22.224	39.557	81.684	28.325	59.240	106.04

	Antisymmetric–Symmetric			Antisymmetric–Antisymmetric		
k	First	Second	Third	First	Second	Third
0.0	5.0096	20.793	48.420	11.637	37.561	77.559
0.2	10.924	24.921	56.370	12.743	38.526	78.038
0.4	7.7702	24.820	55.166	15.048	40.470	80.409
0.6	11.885	32.452	66.263	20.323	48.158	89.324
0.8	25.663	44.913	59.333	41.332	80.918	133.29

TABLE 10.21

First Three Frequency Parameters for Symmetric–Symmetric, Symmetric–Antisymmetric, Antisymmetric–Symmetric, and Antisymmetric–Antisymmetric Modes for Annular Elliptic Plates (FF Boundary, $m = 0.5$, $\nu = 1/3$)

	Symmetric–Symmetric			Symmetric–Antisymmetric		
k	First	Second	Third	First	Second	Third
0.0	6.5886	27.763	31.189	21.621	54.518	70.532
0.2	6.4036	26.342	31.154	16.705	42.792	49.690
0.4	5.8762	23.859	31.779	19.910	50.888	54.570
0.6	5.2014	25.206	32.966	16.318	45.959	50.639
0.8	5.1562	25.402	46.980	13.700	43.352	69.572

	Antisymmetric–Symmetric			Antisymmetric–Antisymmetric		
k	First	Second	Third	First	Second	Third
0.0	16.735	42.839	49.780	10.323	36.261	76.480
0.2	21.551	54.334	70.252	9.9101	35.901	76.248
0.4	15.978	40.823	49.454	8.8065	35.060	75.018
0.6	14.732	40.728	51.718	7.4034	31.325	78.998
0.8	13.031	41.867	66.870	6.9624	26.857	64.233

The results of annular elliptic plates are also shown in Figures 10.2 through 10.10. From Tables 10.13 through 10.21, one can note the effect of hole size on the natural frequencies. As k increases, the frequencies increase for all the boundary conditions, except for the exceptional case of FF boundary. For FF boundary condition, the frequencies decreased as k increased for a particular value of m, which is shown in Table 10.21.

Tables 10.2 through 10.10 show the variation of first 12 frequency parameters chosen from the four types of modes with various boundary conditions at the outer and inner edges of the annular plate. These figures also show that the frequencies for any of the hole size are maximum for CC annular plates. Similar to the investigation carried out for annular circular plates, the effect of different boundary conditions on the natural frequencies can be well investigated for annular elliptic plates also, by dividing them into the following three sets:

(i) CC, CS, and CF

(ii) SC, SS, and SF

(iii) FC, FS, and FF

From Figures 10.2 through 10.4 and Tables 10.13 through 10.15 (set (i) boundary conditions), it can be observed that all the frequencies decrease from CC to CF. Similar behavior can also be observed for set (ii) boundary conditions, as shown in Figures 10.5 through 10.7 and Tables 10.16 through

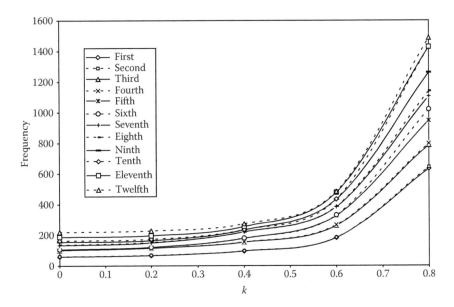

FIGURE 10.2
Effect of first twelve frequency parameters on *k* value of annular elliptic plate (*m* = 0.5, CC boundary).

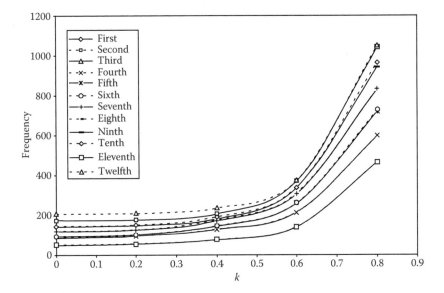

FIGURE 10.3
Effect of first twelve frequency parameters on *k* value of annular elliptic plate (*m* = 0.5, CS boundary).

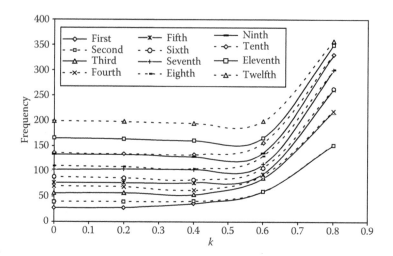

FIGURE 10.4
Effect of first twelve frequency parameters on k value of annular elliptic plate ($m = 0.5$, CF boundary).

10.18, where frequencies decrease from SC to SF. However, Figures 10.8 through 10.10 and Tables 10.19 through 10.21 show that the FC condition gives higher frequencies, while the FS condition gives lower frequencies for set (iii) boundary conditions with smaller hole-size (i.e., $k = 0$ and 0.2). But, for larger hole–size, viz., for $k \geq 0.4$, the results are smaller for the FF boundary, as in sets (i) and (ii). It may also be noted that for set (iii) with

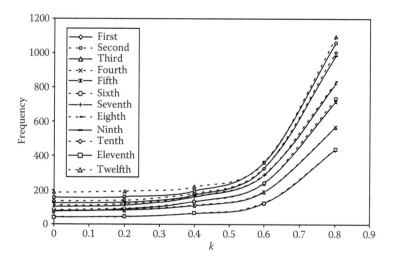

FIGURE 10.5
Effect of first twelve frequency parameters on k value of annular elliptic plate ($m = 0.5$, SC boundary).

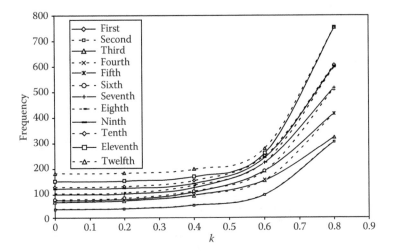

FIGURE 10.6
Effect of first twelve frequency parameters on k value of annular elliptic plate ($m=0.5$, SS boundary).

$k \leq 0.2$ (smaller hole-size), the frequencies (particularly the lower modes) for conditions FC and FF are closer. The above behaviors are the same as those for annular circular plates.

Thus, in this chapter, the 2D BCOPs as shape functions in the Rayleigh–Ritz method have been used to study the annular plate with a curved boundary (especially for elliptical). Also, the effects of boundary conditions

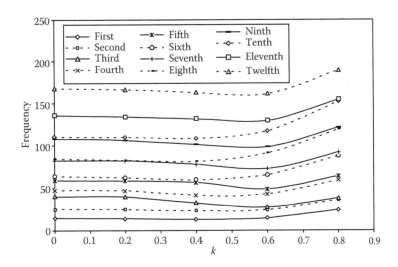

FIGURE 10.7
Effect of first twelve frequency parameters on k value of annular elliptic plate ($m=0.5$, SF boundary).

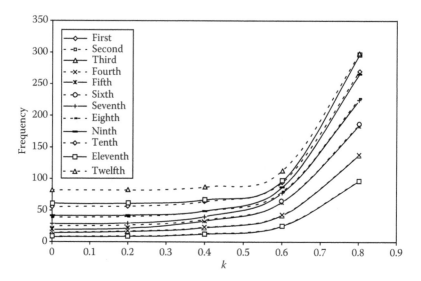

FIGURE 10.8
Effect of first twelve frequency parameters on k value of annular elliptic plate ($m=0.5$, FC boundary).

and hole size on different modes of vibrations have been fully investigated. The use of 2D BCOPs in the Rayleigh–Ritz method makes the said problem a computationally efficient and simple numerical technique for finding the vibration characteristics. It is important to mention (as reported in earlier chapters also) that the generation of orthogonal polynomials is very much

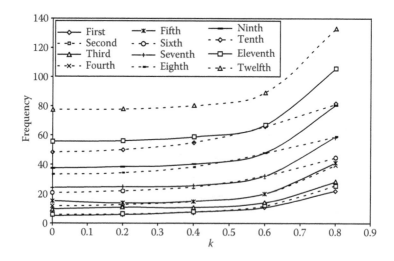

FIGURE 10.9
Effect of first twelve frequency parameters on k value of annular elliptic plate ($m=0.5$, FS boundary).

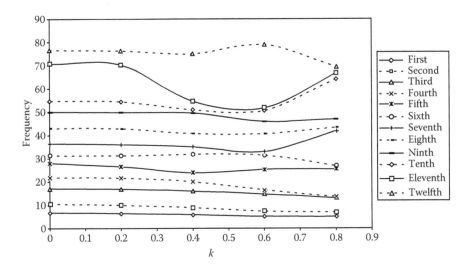

FIGURE 10.10
Effect of first twelve frequency parameters on k value of annular elliptic plate ($m = 0.5$, FF boundary).

sensitive to the numerical errors, as the approximations are increased owing to the rounding errors, which grow to an extent of rendering the results to diverge. Therefore, all the computations were carried out with double-precision arithmetic. This study can be generalized to other types of plate geometries and material properties considering various complicating effects, such as plates with variable thickness, orthotropy, and non-homogeneity. These complicating effects are investigated in the subsequent chapters.

References

Bhat, R.B. 1985. Natural frequencies of rectangular plates using characteristic orthogonal polynomials in Rayleigh-Ritz method. *Journal of Sound and Vibration*, 102: 493–499.

Bhat, R.B. 1987. Flexural vibration of polygonal plates using characteristic orthogonal polynomials in two variables. *Journal of Sound and Vibration*, 114: 65–71.

Chakraverty, S. 1992. Numerical solution of vibration of plates. PhD Thesis, Department of Applied Mathematics, University of Roorkee, Roorkee - 247 667, Uttar Pradesh, India.

Chakraverty, S., Bhat, R.B., and Stiharu I. 2001. Free vibration of annular elliptic plates using boundary characteristic orthogonal polynomials as shape functions in the Rayleigh-Ritz method. *Journal of Sound and Vibration*, 241(3): 524–539.

Leissa, A.W. 1969. *Vibration of Plates*, NASA SP 160, U.S. Government Printing Office, Washington, DC.

Rajalingham, C. and Bhat, R.B. 1991. Vibration of elliptic plates using boundary characteristic orthogonal polynomials. *International Journal of Mechanical Sciences,* 33: 705–716.

Rajalingham, C. and Bhat, R.B. 1992. Axisymmetric vibration of circular plates and its analog in elliptic plates using characteristic orthogonal polynomials. *Journal of Sound and Vibration,* 161: 109–117.

Rajalingham, C., Bhat, R.B., and Xistris, G.D. 1993. Natural frequencies and mode shapes of elliptic plates with boundary characteristic orthogonal polynomials as assumed modes. *ASME Journal of Vibration and Acoustics,* 115: 353–358.

Rajalingham, C., Bhat, R.B., and Xistris, G.D. 1994. Vibration of clamped elliptic plate using exact circular plate modes as shape functions in Rayleigh-Ritz method. *International Journal of Mechanical Sciences,* 36:231–246.

Singh, B. and Chakraverty, S. 1991. Transverse vibration of completely free elliptic and circular plates using orthogonal polynomials in Rayleigh-Ritz method. *International Journal of Mechanical Sciences,* 33: 741–751.

Singh, B. and Chakraverty, S. 1992a. On the use of orthogonal polynomials in Rayleigh-Ritz method for the study of transverse vibration of elliptic plates. *Computers and Structures,* 43: 439–443.

Singh, B. and Chakraverty, S. 1992b. Transverse vibration of simply-supported elliptic and circular plates using orthogonal polynomials in two variables. *Journal of Sound and Vibration,* 152: 149–155.

Singh, B. and Chakraverty, S. 1993. Transverse vibration of annular circular and elliptic plates using the characteristic orthogonal polynomials in two dimensions. *Journal of Sound and Vibration,* 162: 537–546.

11

Plates with Nonhomogeneous Material Properties

11.1 Introduction

Nonhomogeneous elastic plates are widely used now-a-days in the design of space vehicles, modern missiles, aircraft wings, etc. The nonhomogeneity that occurs in the bodies is especially due to the imperfections of the materials. A few works were done on the free vibration of nonhomogeneous plates. The excellent surveys by Leissa (1978, 1981, 1987) cover some of the papers on nonhomogeneous circular, rectangular, and square plates. Nonhomogeneous rectangular and square plates have also been studied by Tomar et al. (1982, 1983), Laura and Gutierrez (1984), and Rao et al. (1974, 1976). Tomar et al. (1983) considered a nonhomogeneous infinite plate of linearly varying thickness. Furthermore, circular plates with nonhomogeneity have also been studied by Tomar et al. (1982), Pan (1976), and Mishra and Das (1971). Tomar et al. (1982, 1983) have used a series method of solution. Explicit closed-form expressions were derived for the frequencies of a nonhomogeneous free circular plate by Mishra and Das (1971), whereas Rao et al. (1974) used a high-precision triangular finite element for a simply supported square plate.

The investigation presented in this chapter shows the use of powerful two-dimensional boundary characteristic orthogonal polynomials (2D BCOPs) in the Rayleigh–Ritz method for the free vibration of nonhomogeneous plates. In particular, the example of circular and elliptic plate with nonhomogeneous density and Young's modulus is taken into consideration. Extensive and wide varieties of results are given for natural frequencies of the plates with different boundary conditions, viz., clamped, simply supported, or free at the edges. However, the results for a circular plate are obtained as a special case. As pointed out in earlier chapters, the orthogonal polynomials have been used extensively to determine the vibration characteristics of different types of plate geometries with various boundary conditions at the edges. In this study, the same procedure has been used to generate the polynomials and to use them in the investigation of transverse vibration of nonhomogeneous elliptic plates with different boundary conditions. To apply this method, three steps as reported earlier have to be followed.

11.2   Basic Equations and Method of Solution

Let us consider that the domain occupied by the elliptic plate is

$$S = \{(x,y),\ x^2/a^2 + y^2/b^2 \leq 1,\quad x, y \in R\} \tag{11.1}$$

where a and b are the semimajor and semiminor axes of the ellipse, respectively.

By following the standard procedure, we can assume the displacement to be of the form

$$w(x,y,t) = W(x,y)\exp(i\omega t) \tag{11.2}$$

where ω is the natural frequency and x, y, and t are the space and time coordinates, and equating the maximum strain and kinetic energies gives the Rayleigh quotient

$$\omega^2 = \frac{\iint\limits_{S} D\left[W_{xx}^2 + W_{yy}^2 + 2\nu W_{xx}W_{yy} + 2(1-\nu)W_{xy}^2\right]dxdy}{h\iint\limits_{S} \rho W^2 dxdy} \tag{11.3}$$

where $D = Eh^3/(12(1-\nu^2))$ is flexural rigidity, while E, ρ, ν, and h are Young's modulus, density, Poisson's ratio, and plate thickness, respectively. W_{xx} is the second derivative of W with respect to x. In this chapter, two types of nonhomogeneity are included for the purpose of illustration.

11.2.1   Type 1 Nonhomogeneity

If one assumes that Young's modulus varies linearly and the density varies parabolically, then the nonhomogeneity of the elliptic plate can be characterized by taking

$$E = E_0(1 + \alpha X) \tag{11.4}$$

$$\rho = \rho_0(1 + \beta X^2) \tag{11.5}$$

where E_0, ρ_0 are constants, and α, β are the parameters designating the nonhomogeneity, $X = x/a$ and $Y = y/a$.

$$\text{Then } D = D_0(1 + \alpha X) \tag{11.6}$$

where $D_0 = E_0 h^3/(12(1-\nu^2))$.

By substituting Equations 11.5 and 11.6 in Equation 11.3, assuming the N-term approximation as

$$W(x,y) = \sum_{j=1}^{N} c_j \phi_j$$

and minimizing ω^2 as a function of the coefficients c_j's, we have

$$\sum_{j=1}^{N} (a_{ij} - \lambda^2 b_{ij}) c_j = 0, \quad i = 1, 2, \ldots, N \tag{11.7}$$

where

$$a_{ij} = \iint_{S'} (1 + \alpha X) \left[\phi_i^{XX} \phi_j^{XX} + \phi_i^{YY} \phi_j^{YY} + \nu \left(\phi_i^{XX} \phi_j^{YY} + \phi_i^{YY} \phi_j^{XX} \right) \right.$$

$$\left. + 2(1 - \nu) \phi_i^{XY} \phi_j^{XY} \right] dX dY \tag{11.8}$$

and

$$b_{ij} = \iint_{S'} (1 + \beta X^2) \phi_i \phi_j dX dY \tag{11.9}$$

11.2.2 Type 2 Nonhomogeneity

In this case, let the nonhomogeneity of the plate be characterized by taking

$$E = E_0 (1 - X^2 - Y^2/m^2)^{\alpha} = E_0 P^{\alpha} \tag{11.10}$$

$$\rho = \rho_0 (1 - X^2 - Y^2/m^2)^{\beta} = \rho_0 P^{\beta} \tag{11.11}$$

where $P = (1 - X^2 - Y^2/m^2)$; E_0, ρ_0 are constants; α, β (integers) are the indices of nonhomogeneity; $X = x/a$, $Y = y/a$, and $m(=b/a)$ are the aspect ratios of the ellipse.

$$\text{Then } D = E_0 P^{\alpha} h^3 / (12(1 - \nu^2)) = D_0 P^{\alpha} \tag{11.12}$$

where again $D_0 = E_0 h^3 / (12(1 - \nu^2))$.

Now, putting Equations 11.11 and 11.12 in Equation 11.3, substituting the N-term approximation

$$W(x,y) = \sum_{j=1}^{N} c_j \phi_j$$

and minimizing ω^2 as a function of the coefficients c_i's, we again obtain

$$\sum_{j=1}^{N}(a_{ij} - \lambda^2 b_{ij})c_j = 0, \quad i = 1, 2, \ldots, N \tag{11.13}$$

where

$$a_{ij} = \iint_{S'} P^\alpha \left[\phi_i^{XX}\phi_j^{XX} + \phi_i^{YY}\phi_j^{YY} + \nu\left(\phi_i^{XX}\phi_j^{YY} + \phi_i^{YY}\phi_j^{XX}\right) \right.$$

$$\left. + 2(1-\nu)\phi_i^{XY}\phi_j^{XY} \right] dXdY \tag{11.14}$$

$$b_{ij} = \iint_{S'} P^\beta \phi_i\phi_j dXdY \tag{11.15}$$

The frequency parameter in both the above cases may be written as

$$\lambda^2 = a^4 \rho_0 h\omega^2 / D_0 \tag{11.16}$$

The ϕ_j's are the orthogonal polynomials and are described in the following section. ϕ_i^{XX} is the second derivative of ϕ_i with respect to X and the new domain S' is defined by

$$S' = \{(X,Y), X^2 + Y^2/m^2 \leq 1, \quad X,Y \in R\}$$

where $m(=b/a)$ is the aspect ratio of the ellipse.
Since, the ϕ_i's are orthogonal, Equation 11.7 or Equation 11.3 reduces to

$$\sum_{j=1}^{N}(a_{ij} - \lambda^2 \delta_{ij})c_j = 0, \quad i = 1, 2, \ldots, N \tag{11.17}$$

where

$$\delta_{ij} = 0, \quad \text{if} \quad i \neq j$$
$$\delta_{ij} = 1, \quad \text{if} \quad i = j$$

The type of nonhomogeneity considered here is mainly for the illustration purpose. The present approach shows how the nonhomogeneity, if it occurs on a plate structure owing to the imperfections of the material, can be theoretically investigated for its vibration characteristics.

11.3 Orthogonal Polynomials Generation

The present polynomials have been generated in the same way as described in the earlier chapters, but for clarity the method is again described as follows.

We start with a suitable set of linearly independent functions

$$F_i(x,y) = f(x,y)\{f_i(x,y)\}, \quad i = 1, 2, \ldots, N \tag{11.18}$$

where $f(x,y)$ satisfies the essential boundary conditions and $f_i(x,y)$ are the linearly independent functions involving products of non-negative integral powers of x and y. The function $f(x,y)$ is defined by

$$f(x,y) = (1 - x^2 - y^2/m^2)^p \tag{11.19}$$

where p takes the value of 0, 1, or 2 representing the boundary of the plate as free, simply supported, or clamped, respectively.

We can now generate the orthogonal set from $F_i(x,y)$ by the well-known Gram–Schmidt process. For this, we define the inner product of the two functions $f(x,y)$ and $g(x,y)$ over a domain S by

$$<f,g> = \iint\limits_S \psi(x,y)f(x,y)g(x,y)dxdy \tag{11.20}$$

where $\psi(x,y)$ is the weight function.

Then, the norm of f is given by

$$\| f \| = \langle f, f \rangle^{1/2} \tag{11.21}$$

Now, proceeding as in the previous chapters, the Gram–Schmidt process can be written as

$$\phi_1 = F_1$$

$$\left.\begin{array}{l} \phi_i = F_i - \displaystyle\sum_{j=1}^{i-1} \alpha_{ij}\phi_j \\[2mm] \alpha_{ij} = \langle F_i, \phi_j \rangle / \langle \phi_j, \phi_j \rangle, \quad j = 1, 2, \ldots, (i-1) \end{array}\right\}, \quad i = 2, \ldots, N \tag{11.22}$$

where the ϕ_i's are the orthogonal polynomials. The orthonormal polynomials are then generated by

$$\widehat{\phi}_i = \phi_i / \| \phi_i \|$$

11.4   Some Numerical Results and Discussions

The BCOPs were used by choosing the functions $f_i(x,y)$ appropriately and then generating the corresponding orthogonal functions, thus performing an exhaustive study of various modes of vibrations as described earlier. By taking these functions in both x and y as even, one can discuss all those modes that are symmetric about both the major and minor axes. When the functions are even in x and odd in y or vice versa, we can have symmetric–antisymmetric or antisymmetric–symmetric modes. Similarly, if we take the function in both x and y as odd, then we may have the case of antisymmetric–antisymmetric mode of vibration. One can use various types of combinations of these functions and obtain the corresponding results for various values of $m = b/a$, i.e., the aspect ratios of the elliptic plate along with the variety of nonhomogeneity parameters, as defined earlier in this chapter. In the following section, the results for the two types of nonhomogeneity will be discussed.

11.4.1   Results for Type 1 Nonhomogeneity

Numerical results were computed for type 1 nonhomogeneity for different values of the nonhomogeneity parameters α and β with different boundary conditions (clamped, simply supported, and free), for various values of the aspect ratios of the ellipse. As mentioned earlier, the first three frequencies for all the modes of symmetric–symmetric, symmetric–antisymmetric, antisymmetric–symmetric, and antisymmetric–antisymmetric are reported. Although the results can be worked out for various combinations of α and β, only a few of them are reported here in Tables 11.1 through 11.32. The value of ν has been taken as 0.3 in all the calculations. In each of the tables, the three boundary conditions, viz., clamped, simply supported, and free, are incorporated for different m values. The last row in all the tables gives the results for a circular plate (i.e. $m = 1.0$). The first few frequencies can be obtained from the sets of various combinations of symmetric–symmetric,

TABLE 11.1

First Three Frequency Parameters for Symmetric–Symmetric Modes (Clamped, Simply Supported, and Free) for $\alpha = 0.2$, $\beta = 0$ (Type 1 Nonhomogeneity)

	Clamped			Simply Supported			Free		
$m = b/a$	First	Second	Third	First	Second	Third	First	Second	Third
0.2	150.17	199.18	274.65	70.164	113.75	181.69	6.8324	33.697	93.728
0.4	40.976	73.068	123.37	19.779	47.980	95.021	6.7918	33.152	42.445
0.5	27.670	57.452	106.05	13.419	39.340	84.414	6.7314	28.171	32.412
0.6	20.466	49.166	94.475	9.9238	34.659	70.676	6.6398	20.246	31.230
0.8	13.484	41.173	56.212	6.4985	29.924	42.044	6.2717	12.334	27.271
1.0	10.469	35.617	40.555	5.0112	26.153	30.223	5.4679	9.1226	22.522

TABLE 11.2

First Three Frequency Parameters for Symmetric–Antisymmetric Modes (Clamped, Simply Supported, and Free) for $\alpha = 0.2$, $\beta = 0$ (Type 1 Nonhomogeneity)

	Clamped			Simply Supported			Free		
$m = b/a$	First	Second	Third	First	Second	Third	First	Second	Third
0.2	404.37	490.91	635.73	70.164	113.75	181.69	105.99	49.469	410.61
0.4	106.60	154.58	220.79	19.779	47.980	95.021	26.965	65.364	108.30
0.5	70.294	112.08	170.86	13.419	39.340	84.414	22.526	57.173	71.128
0.6	50.443	88.497	143.25	9.9238	34.659	70.676	19.528	50.642	51.281
0.8	30.643	64.564	89.372	6.4985	29.924	42.044	15.575	30.085	42.185
1.0	21.552	52.197	61.471	5.0112	26.153	30.223	12.783	20.738	34.632

TABLE 11.3

First Three Frequency Parameters for Antisymmetric–Symmetric Modes (Clamped, Simply Supported, and Free) for $\alpha = 0.2$, $\beta = 0$ (Type 1 Nonhomogeneity)

	Clamped			Simply Supported			Free		
$m = b/a$	First	Second	Third	First	Second	Third	First	Second	Third
0.2	172.59	229.72	319.83	90.085	141.36	224.30	17.727	54.861	138.29
0.4	54.931	95.631	156.09	31.829	68.579	126.76	17.549	53.592	61.430
0.5	40.324	79.242	137.87	24.178	59.130	114.95	17.284	43.877	52.001
0.6	32.489	70.555	117.27	20.011	53.941	91.849	16.857	33.935	49.491
0.8	25.063	61.452	76.372	15.990	47.975	60.406	15.264	24.252	42.144
1.0	21.910	52.199	62.457	14.220	40.898	49.658	12.783	20.852	34.632

TABLE 11.4

First Three Frequency Parameters for Antisymmetric–Antisymmetric Modes (Clamped, Simply Supported, and Free) for $\alpha = 0.2$, $\beta = 0$ (Type 1 Nonhomogeneity)

	Clamped			Simply Supported			Free		
$m = b/a$	First	Second	Third	First	Second	Third	First	Second	Third
0.2	443.88	542.42	715.50	300.08	392.39	567.64	25.959	75.307	141.36
0.4	128.75	184.29	261.11	90.656	142.60	220.53	13.230	44.059	91.018
0.5	89.278	138.89	207.17	63.818	109.87	177.51	10.683	37.850	80.565
0.6	67.498	113.57	177.22	48.803	91.218	148.62	8.9740	33.533	70.570
0.8	45.670	87.168	114.02	33.443	71.019	94.303	6.8037	27.384	46.592
1.0	35.658	71.340	86.197	26.168	58.279	71.418	5.4688	22.518	35.925

TABLE 11.5

First Three Frequency Parameters for Symmetric–Symmetric Modes (Clamped, Simply Supported, and Free) for $\alpha = 0$, $\beta = 0.2$ (Type 1 Nonhomogeneity)

	Clamped			Simply Supported			Free		
$m = b/a$	First	Second	Third	First	Second	Third	First	Second	Third
0.2	149.17	193.40	259.03	69.311	109.17	169.17	6.4931	31.696	85.214
0.4	40.423	69.890	116.29	19.344	45.541	88.988	6.4486	31.131	41.624
0.5	27.205	54.727	99.909	13.085	37.294	79.140	6.3893	27.529	30.393
0.6	20.056	46.714	91.166	9.6623	32.837	69.859	6.2944	19.729	29.237
0.8	13.130	39.033	55.491	6.3237	28.336	41.419	5.9337	11.976	25.467
1.0	10.136	34.291	39.262	4.8800	25.084	29.298	5.1747	8.8204	21.028

TABLE 11.6

First Three Frequency Parameters for Symmetric–Antisymmetric Modes (Clamped, Simply Supported, and Free) for $\alpha = 0$, $\beta = 0.2$ (Type 1 Nonhomogeneity)

	Clamped			Simply Supported			Free		
$m = b/a$	First	Second	Third	First	Second	Third	First	Second	Third
0.2	402.56	479.63	602.12	261.87	333.81	455.90	47.215	99.684	202.48
0.4	105.67	149.58	209.43	69.492	110.23	168.75	25.571	61.235	107.08
0.5	69.538	107.97	161.93	45.871	81.256	132.97	21.298	53.465	70.160
0.6	49.802	84.964	135.64	32.898	64.940	112.74	18.415	47.856	49.870
0.8	30.140	61.735	88.470	19.855	47.990	70.577	14.630	29.514	39.286
1.0	21.123	50.052	60.386	13.780	39.046	48.096	11.991	20.268	32.234

TABLE 11.7

First Three Frequency Parameters for Antisymmetric–Symmetric Modes (Clamped, Simply Supported, and Free) for $\alpha = 0$, $\beta = 0.2$ (Type 1 Nonhomogeneity)

	Clamped			Simply Supported			Free		
$m = b/a$	First	Second	Third	First	Second	Third	First	Second	Third
0.2	169.48	220.94	298.60	87.519	134.47	206.85	16.744	51.420	124.55
0.4	53.188	90.786	146.44	30.497	64.780	118.05	16.551	50.150	59.163
0.5	38.851	75.020	129.42	23.110	55.824	107.23	16.282	42.118	48.586
0.6	31.184	66.700	114.55	19.109	50.905	89.379	15.857	32.507	46.174
0.8	23.932	58.109	74.113	15.255	45.291	58.452	14.318	23.185	39.242
1.0	20.857	50.052	59.508	13.555	39.046	47.340	11.991	19.873	32.234

TABLE 11.8

First Three Frequency Parameters for Antisymmetric–Antisymmetric Modes
(Clamped, Simply Supported, and Free) for $\alpha = 0$, $\beta = 0.2$ (Type 1 Nonhomogeneity)

$m = b/a$	Clamped			Simply Supported			Free		
	First	Second	Third	First	Second	Third	First	Second	Third
0.2	438.30	524.93	670.30	295.05	376.58	524.22	24.968	71.557	131.90
0.4	125.90	176.88	245.82	88.137	135.98	205.12	12.672	41.551	84.847
0.5	86.957	132.77	195.20	61.786	104.46	165.54	10.209	35.590	75.041
0.6	65.521	108.25	167.03	47.100	86.552	142.93	8.5554	31.452	67.640
0.8	44.103	82.882	111.25	32.145	67.266	91.803	6.4579	25.593	44.921
1.0	34.307	68.186	83.227	25.095	55.440	68.910	5.1721	21.026	34.505

TABLE 11.9

First Three Frequency Parameters for Symmetric–Symmetric Modes (Clamped,
Simply Supported, and Free) for $\alpha = 0.2$, $\beta = 0.2$ (Type 1 Nonhomogeneity)

$m = b/a$	Clamped			Simply Supported			Free		
	First	Second	Third	First	Second	Third	First	Second	Third
0.2	149.73	196.40	267.16	69.806	111.51	175.64	6.5467	32.519	89.532
0.4	40.759	71.580	119.95	19.612	46.753	91.998	6.5072	31.975	42.138
0.5	27.501	56.198	103.10	13.291	38.298	81.761	6.4515	27.936	31.246
0.6	20.330	48.056	94.100	9.8231	33.726	70.335	6.3615	20.056	30.091
0.8	13.386	40.228	55.925	6.4284	29.112	41.790	6.0166	12.183	26.266
1.0	10.390	35.085	39.983	4.9558	25.652	29.767	5.2798	8.9432	21.699

TABLE 11.10

First Three Frequency Parameters for Symmetric–Antisymmetric Modes (Clamped,
Simply Supported, and Free) for $\alpha = 0.2$, $\beta = 0.2$ (Type 1 Nonhomogeneity)

$m = b/a$	Clamped			Simply Supported			Free		
	First	Second	Third	First	Second	Third	First	Second	Third
0.2	403.54	485.47	619.70	262.78	339.10	472.77	48.103	102.64	214.30
0.4	106.21	152.18	215.27	69.979	112.50	174.46	26.139	63.155	107.79
0.5	69.984	110.13	166.51	46.271	83.090	137.34	21.804	55.183	70.737
0.6	50.192	86.834	139.56	33.237	66.495	116.38	18.877	49.424	50.353
0.8	30.465	63.260	88.996	20.113	49.209	71.049	15.025	29.859	40.620
1.0	21.416	51.246	61.000	13.986	39.999	48.575	12.329	20.532	33.338

TABLE 11.11

First Three Frequency Parameters for Antisymmetric–Symmetric Modes (Clamped, Simply Supported, and Free) for $\alpha = 0.2$, $\beta = 0.2$ (Type 1 Nonhomogeneity)

	Clamped			Simply Supported			Free		
$m = b/a$	First	Second	Third	First	Second	Third	First	Second	Third
0.2	171.12	225.45	309.62	88.865	137.95	215.86	17.078	52.953	131.65
0.4	54.146	93.315	151.40	31.181	66.653	122.42	16.900	51.697	60.337
0.5	39.681	77.245	133.77	23.645	57.446	111.10	16.640	43.007	50.130
0.6	31.937	68.748	116.01	19.551	52.394	90.673	16.223	33.204	47.683
0.8	24.610	59.909	75.338	15.608	46.619	59.465	14.695	23.646	40.576
1.0	21.504	51.246	61.094	13.874	39.999	48.489	12.328	20.247	33.338

TABLE 11.12

First Three Frequency Parameters for Antisymmetric–Antisymmetric Modes (Clamped, Simply Supported, and Free) for $\alpha = 0.2$, $\beta = 0.2$ (Type 1 Nonhomogeneity)

	Clamped			Simply Supported			Free		
$m = b/a$	First	Second	Third	First	Second	Third	First	Second	Third
0.2	441.20	533.94	693.84	297.66	384.73	546.93	25.181	73.206	136.46
0.4	127.41	180.69	253.65	89.458	139.32	213.00	12.817	42.667	87.783
0.5	88.208	135.93	201.31	62.852	107.18	171.61	10.343	36.597	77.655
0.6	66.601	111.02	172.24	47.992	88.893	147.06	8.6828	32.379	69.457
0.8	44.986	85.140	112.73	32.820	69.147	93.104	6.5752	26.391	45.747
1.0	35.089	69.885	84.808	25.649	56.886	70.179	5.2801	21.696	35.167

TABLE 11.13

First Three Frequency Parameters for Symmetric–Symmetric Modes (Clamped, Simply Supported, and Free) for $\alpha = 0.4$, $\beta = 0.2$ (Type 1 Nonhomogeneity)

	Clamped			Simply Supported			Free		
$m = b/a$	First	Second	Third	First	Second	Third	First	Second	Third
0.2	150.26	199.27	275.06	70.278	113.73	181.90	6.5973	33.301	93.654
0.4	41.080	73.161	123.41	19.867	47.887	94.876	6.5622	32.773	42.628
0.5	27.783	57.569	106.09	13.488	39.237	84.257	6.5098	28.324	32.050
0.6	20.590	49.301	94.597	9.9771	34.556	70.789	6.4241	20.368	30.894
0.8	13.628	41.329	56.343	6.5286	29.837	42.145	6.0934	12.384	27.013
1.0	10.628	35.763	40.716	5.0281	26.162	30.232	5.3780	9.0610	22.326

TABLE 11.14

First Three Frequency Parameters for Symmetric–Antisymmetric Modes (Clamped, Simply Supported, and Free) for $\alpha = 0.4$, $\beta = 0.2$ (Type 1 Nonhomogeneity)

	Clamped			Simply Supported			Free		
$m=b/a$	First	Second	Third	First	Second	Third	First	Second	Third
0.2	404.48	491.18	636.82	263.65	344.25	489.08	48.940	105.47	225.51
0.4	106.71	154.64	220.89	70.443	114.64	179.97	26.674	64.972	108.48
0.5	70.410	112.15	170.88	46.652	84.813	141.54	22.279	56.800	71.289
0.6	50.564	88.586	143.26	33.560	67.952	119.86	19.310	50.742	50.967
0.8	30.775	64.682	89.502	20.359	50.349	71.502	15.396	30.190	41.865
1.0	21.695	52.327	61.613	14.182	40.876	49.045	12.646	20.786	34.365

TABLE 11.15

First Three Frequency Parameters for Antisymmetric–Symmetric Modes (Clamped, Simply Supported, and Free) for $\alpha = 0.4$, $\beta = 0.2$ (Type 1 Nonhomogeneity)

	Clamped			Simply Supported			Free		
$m=b/a$	First	Second	Third	First	Second	Third	First	Second	Third
0.2	172.68	229.78	320.29	90.133	141.27	224.54	17.393	54.414	138.39
0.4	55.050	95.670	156.11	31.825	68.401	126.62	17.228	53.162	61.442
0.5	40.461	79.308	137.87	24.149	58.957	114.79	16.975	43.845	51.587
0.6	32.643	70.642	117.39	19.969	53.779	91.898	16.567	33.862	49.101
0.8	25.241	61.561	76.501	15.940	47.850	60.426	15.048	24.083	41.820
1.0	22.102	52.333	62.590	14.173	40.877	49.575	12.646	20.602	34.365

TABLE 11.16

First Three Frequency Parameters for Antisymmetric–Antisymmetric Modes (Clamped, Simply Supported, and Free) for $\alpha = 0.4$, $\beta = 0.2$ (Type 1 Nonhomogeneity)

	Clamped			Simply Supported			Free		
$m=b/a$	First	Second	Third	First	Second	Third	First	Second	Third
0.2	443.98	542.78	716.66	300.16	392.68	568.77	25.382	74.752	140.86
0.4	128.84	184.28	261.21	90.704	142.48	220.61	12.955	43.708	90.573
0.5	89.390	138.90	207.16	63.857	109.74	177.45	10.470	37.536	80.123
0.6	67.621	113.60	177.18	48.833	91.082	148.68	8.8036	33.242	70.565
0.8	45.816	87.233	114.13	33.457	70.898	94.342	6.6865	27.134	46.530
1.0	35.821	71.438	86.320	26.171	58.221	71.392	5.3828	22.321	35.795

TABLE 11.17

First Three Frequency Parameters for Symmetric–Symmetric Modes (Clamped, Simply Supported, and Free) for $\alpha = 0.8$, $\beta = 0.2$ (Type 1 Nonhomogeneity)

	Clamped			Simply Supported			Free		
$m = b/a$	First	Second	Third	First	Second	Third	First	Second	Third
0.2	151.26	204.76	290.26	71.163	117.91	193.82	6.6913	34.771	101.40
0.4	41.684	76.069	129.89	20.347	49.977	100.32	6.6640	34.266	43.550
0.5	28.314	60.076	111.65	13.860	40.964	88.963	6.6171	29.054	33.548
0.6	21.078	51.570	95.531	10.267	36.082	71.646	6.5387	20.959	32.382
0.8	14.077	43.320	57.136	6.7178	31.164	42.818	6.2328	12.767	28.388
1.0	11.068	36.870	42.186	5.1641	27.052	31.144	5.5573	9.2905	23.479

TABLE 11.18

First Three Frequency Parameters for Symmetric–Antisymmetric Modes (Clamped, Simply Supported, and Free) for $\alpha = 0.8$, $\beta = 0.2$ (Type 1 Nonhomogeneity)

	Clamped			Simply Supported			Free		
$m = b/a$	First	Second	Third	First	Second	Third	First	Second	Third
0.2	406.26	502.31	669.83	265.30	354.25	520.21	50.495	110.88	246.42
0.4	107.67	159.22	231.66	71.314	118.64	190.51	27.661	68.375	109.78
0.5	71.212	115.90	179.14	47.369	88.005	149.54	23.156	59.810	72.333
0.6	51.265	91.816	150.17	34.170	70.643	122.81	20.107	51.659	53.645
0.8	31.358	67.282	90.463	20.824	52.444	72.362	16.078	30.818	44.152
1.0	22.219	54.218	62.844	14.554	42.450	49.970	13.232	21.269	36.247

TABLE 11.19

First Three Frequency Parameters for Antisymmetric–Symmetric Modes (Clamped, Simply Supported, and Free) for $\alpha = 0.8$, $\beta = 0.2$ (Type 1 Nonhomogeneity)

	Clamped			Simply Supported			Free		
$m = b/a$	First	Second	Third	First	Second	Third	First	Second	Third
0.2	175.60	238.07	340.70	92.484	147.51	241.04	17.980	57.172	150.98
0.4	56.725	99.983	164.98	33.016	71.616	134.59	17.836	55.909	63.493
0.5	41.901	83.067	145.53	25.082	61.728	121.78	17.595	45.400	54.309
0.6	33.940	74.080	119.96	20.743	56.314	94.190	17.201	35.088	51.735
0.8	26.392	64.535	78.679	16.556	50.094	62.220	15.696	24.903	44.107
1.0	23.190	54.274	65.358	14.726	42.469	51.597	13.232	21.266	36.247

TABLE 11.20

First Three Frequency Parameters for Antisymmetric–Antisymmetric Modes
(Clamped, Simply Supported, and Free) for $\alpha = 0.8$, $\beta = 0.2$ (Type 1 Nonhomogeneity)

	Clamped			Simply Supported			Free		
$m = b/a$	First	Second	Third	First	Second	Third	First	Second	Third
0.2	449.26	560.01	760.32	304.89	408.11	610.15	25.757	77.609	149.24
0.4	131.51	190.98	275.69	93.016	148.39	235.12	13.211	45.616	95.824
0.5	91.588	144.37	218.25	65.720	114.47	188.55	10.708	39.252	84.734
0.6	69.515	118.33	184.15	50.393	95.117	151.73	9.0293	34.819	72.616
0.8	47.352	91.041	116.78	34.638	74.104	96.668	6.8948	28.492	47.991
1.0	37.172	74.211	89.178	27.138	60.631	73.685	5.5754	23.464	36.965

TABLE 11.21

First Three Frequency Parameters for Symmetric–Symmetric Modes (Clamped,
Simply Supported, and Free) for $\alpha = 0.3$, $\beta = 0.5$ (Type 1 Nonhomogeneity)

	Clamped			Simply Supported			Free		
$m = b/a$	First	Second	Third	First	Second	Third	First	Second	Third
0.2	149.33	193.84	260.93	69.512	109.50	170.78	6.2225	31.422	86.360
0.4	40.599	70.303	117.11	19.496	45.660	89.483	6.1864	30.889	41.934
0.5	27.395	55.157	100.66	13.205	37.368	79.542	6.1350	27.788	30.176
0.6	20.261	47.157	91.893	9.7545	32.894	70.056	6.0519	19.937	29.053
0.8	13.364	39.487	55.716	6.3774	28.392	41.597	5.7395	12.075	25.358
1.0	10.395	34.653	39.598	4.9122	25.188	29.421	5.0860	8.7751	20.953

TABLE 11.22

First Three Frequency Parameters for Symmetric–Antisymmetric Modes (Clamped,
Simply Supported, and Free) for $\alpha = 0.3$, $\beta = 0.5$ (Type 1 Nonhomogeneity)

	Clamped			Simply Supported			Free		
$m = b/a$	First	Second	Third	First	Second	Third	First	Second	Third
0.2	402.75	480.42	606.19	262.09	334.62	460.42	46.734	99.724	206.16
0.4	105.87	150.00	210.56	69.691	110.53	169.97	25.337	61.227	107.38
0.5	69.736	108.39	162.82	46.060	81.503	133.80	21.110	53.443	70.435
0.6	50.006	85.385	136.43	33.076	65.149	113.37	18.256	47.825	50.115
0.8	30.360	62.163	88.694	20.011	48.150	70.769	14.505	29.697	39.244
1.0	21.358	50.449	60.660	13.914	39.192	48.265	11.902	20.375	32.192

TABLE 11.23

First Three Frequency Parameters for Antisymmetric–Symmetric Modes (Clamped, Simply Supported, and Free) for $\alpha = 0.3$, $\beta = 0.5$ (Type 1 Nonhomogeneity)

$m = b/a$	Clamped			Simply Supported			Free		
	First	Second	Third	First	Second	Third	First	Second	Third
0.2	169.72	221.62	301.31	87.739	134.97	209.22	16.421	51.265	126.78
0.4	53.473	91.343	147.54	30.593	64.970	118.90	16.250	50.029	59.340
0.5	39.157	75.581	130.41	23.155	55.975	107.95	15.999	42.205	48.488
0.6	31.510	67.262	114.85	19.122	51.046	89.586	15.600	32.514	46.100
0.8	24.289	58.652	74.452	15.246	45.437	58.618	14.149	23.046	39.198
1.0	21.233	50.448	60.010	13.546	39.191	47.484	11.898	19.636	32.192

TABLE 11.24

First Three Frequency Parameters for Antisymmetric–Antisymmetric Modes (Clamped, Simply Supported, and Free) for $\alpha = 0.3$, $\beta = 0.5$ (Type 1 Nonhomogeneity)

$m = b/a$	Clamped			Simply Supported			Free		
	First	Second	Third	First	Second	Third	First	Second	Third
0.2	438.59	526.30	675.96	295.34	377.97	530.46	24.230	71.235	132.38
0.4	126.19	177.47	247.43	88.350	136.42	206.93	12.331	41.390	85.043
0.5	87.248	133.33	196.42	61.972	104.81	166.81	9.9506	35.455	75.176
0.6	65.821	108.81	168.07	47.259	86.850	143.90	8.3531	31.332	67.733
0.8	44.425	83.420	111.58	32.261	67.507	92.005	6.3258	25.493	44.952
1.0	34.647	68.639	83.651	25.182	55.647	69.092	5.0800	20.950	34.447

TABLE 11.25

First Three Frequency Parameters for Symmetric–Symmetric Modes (Clamped, Simply Supported, and Free) for $\alpha = 0.5$, $\beta = 0.3$ (Type 1 Nonhomogeneity)

$m = b/a$	Clamped			Simply Supported			Free		
	First	Second	Third	First	Second	Third	First	Second	Third
0.2	150.3	199.32	275.26	70.333	113.73	182.00	6.4954	33.138	93.665
0.4	41.13	73.202	123.43	19.909	47.846	94.824	6.4627	32.618	42.717
0.5	27.838	57.620	106.11	13.521	39.191	84.199	6.4127	28.397	31.902
0.6	20.649	49.360	94.656	10.002	34.511	70.844	6.3306	20.427	30.756
0.8	13.696	41.398	56.406	6.5462	29.799	42.194	6.0154	12.41	26.905
1.0	10.703	35.830	40.789	5.0358	26.167	30.239	5.3366	9.0399	22.243

TABLE 11.26

First Three Frequency Parameters for Symmetric–Antisymmetric Modes (Clamped, Simply Supported, and Free) for $\alpha = 0.5$, $\beta = 0.3$ (Type 1 Nonhomogeneity)

	Clamped			Simply Supported			Free		
$m = b/a$	First	Second	Third	First	Second	Third	First	Second	Third
0.2	404.53	491.31	637.33	263.71	344.36	489.58	48.713	105.27	411.01
0.4	106.77	154.67	220.95	70.498	114.63	180.02	26.549	64.813	108.57
0.5	70.466	112.18	170.90	46.705	84.793	141.53	22.175	56.648	71.368
0.6	50.623	88.626	143.27	33.610	67.929	119.82	19.217	50.703	50.929
0.8	30.839	64.734	89.565	20.403	50.324	71.555	15.319	30.242	41.731
1.0	21.764	52.386	61.681	14.220	40.866	49.081	12.588	34.253	20.813

TABLE 11.27

First Three Frequency Parameters for Antisymmetric–Symmetric Modes (Clamped, Simply Supported, and Free) for $\alpha = 0.5$, $\beta = 0.3$ (Type 1 Nonhomogeneity)

	Clamped			Simply Supported			Free		
$m = b/a$	First	Second	Third	First	Second	Third	First	Second	Third
0.2	172.72	229.82	320.51	90.155	141.23	224.67	17.252	54.237	138.49
0.4	55.104	95.687	156.12	31.821	68.325	126.58	17.092	52.991	61.446
0.5	40.524	79.337	137.88	24.134	58.883	114.74	16.845	43.830	51.421
0.6	32.712	70.679	117.44	19.948	53.711	91.921	16.445	33.828	48.944
0.8	25.321	61.609	76.561	15.916	47.797	60.435	14.956	24.009	41.686
1.0	22.188	52.393	62.648	14.149	40.867	49.541	12.587	20.494	34.253

TABLE 11.28

First Three Frequency Parameters for Antisymmetric–Antisymmetric Modes (Clamped, Simply Supported, and Free) for $\alpha = 0.5$, $\beta = 0.3$ (Type 1 Nonhomogeneity)

	Clamped			Simply Supported			Free		
$m = b/a$	First	Second	Third	First	Second	Third	First	Second	Third
0.2	444.03	542.95	717.20	300.20	392.82	569.30	25.114	74.519	140.67
0.4	128.89	184.28	261.27	90.725	142.43	220.67	12.827	43.561	90.398
0.5	89.442	138.90	207.17	63.874	109.68	177.44	10.372	37.404	79.946
0.6	67.678	113.61	177.17	48.846	91.022	148.71	8.7248	33.120	70.562
0.8	45.883	87.261	114.19	33.462	70.845	94.360	6.6325	27.029	46.502
1.0	35.896	71.482	86.377	26.170	58.194	71.382	5.3432	22.237	35.738

TABLE 11.29

First Three Frequency Parameters for Symmetric–Symmetric Modes (Clamped, Simply Supported, and Free) for $\alpha = 0.1$, $\beta = 0.6$ (Type 1 Nonhomogeneity)

	Clamped			Simply Supported			Free		
$m = b/a$	First	Second	Third	First	Second	Third	First	Second	Third
0.2	148.52	189.59	250.27	68.827	106.24	162.49	6.0718	30.218	80.994
0.4	40.153	68.022	112.31	19.151	43.999	85.541	6.0317	29.667	41.270
0.5	27.017	53.212	96.497	12.942	35.990	76.094	5.9777	27.271	28.950
0.6	19.924	45.409	88.065	9.5499	31.673	69.401	5.8918	19.525	27.837
0.8	13.068	37.956	55.143	6.2428	27.328	41.104	5.5729	11.803	24.239
1.0	10.112	33.577	38.749	4.8130	24.380	28.835	8.5844	4.9194	20.014

TABLE 11.30

First Three Frequency Parameters for Symmetric–Antisymmetric Modes (Clamped, Simply Supported, and Free) for $\alpha = 0.1$, $\beta = 0.6$ (Type 1 Nonhomogeneity)

	Clamped			Simply Supported			Free		
$m = b/a$	First	Second	Third	First	Second	Third	First	Second	Third
0.2	401.28	472.10	582.94	260.75	327.20	438.64	45.335	95.709	191.75
0.4	105.12	146.33	202.80	69.020	107.36	162.54	24.473	58.614	106.39
0.5	69.127	105.40	156.72	45.522	78.962	128.07	20.352	51.104	69.656
0.6	49.489	82.835	131.23	32.629	63.008	108.58	17.571	45.684	49.480
0.8	29.950	60.127	87.978	19.682	46.481	70.124	13.925	29.245	37.425
1.0	21.004	48.847	59.886	13.657	37.874	47.643	11.412	20.024	30.680

TABLE 11.31

First Three Frequency Parameters for Antisymmetric–Symmetric Modes (Clamped, Simply Supported, and Free) for $\alpha = 0.1$, $\beta = 0.6$ (Type 1 Nonhomogeneity)

	Clamped			Simply Supported			Free		
$m = b/a$	First	Second	Third	First	Second	Third	First	Second	Third
0.2	167.32	215.38	287.12	85.817	130.23	197.87	15.867	49.139	118.12
0.4	52.163	87.967	141.12	29.642	62.435	113.27	15.683	47.896	57.692
0.5	38.054	72.650	124.77	22.407	53.779	102.95	15.426	40.947	46.370
0.6	30.532	64.588	112.78	18.498	49.031	87.740	15.023	31.516	44.036
0.8	23.438	56.312	72.781	14.745	43.642	57.202	13.584	22.348	37.378
1.0	20.437	48.843	57.980	13.094	37.870	45.943	11.405	19.040	30.680

TABLE 11.32

First Three Frequency Parameters for Antisymmetric–Antisymmetric Modes
(Clamped, Simply Supported, and Free) for $\alpha = 0.1, \beta = 0.6$ (Type 1 Nonhomogeneity)

	Clamped			Simply Supported			Free		
$m = b/a$	First	Second	Third	First	Second	Third	First	Second	Third
0.2	434.25	513.78	645.30	291.47	366.86	501.57	23.700	68.834	126.39
0.4	124.01	172.20	237.20	86.464	131.84	196.84	12.022	39.802	81.147
0.5	85.490	129.01	188.39	60.470	101.10	158.99	9.6836	34.031	71.694
0.6	64.332	105.08	161.22	46.013	83.666	137.33	8.1136	30.026	64.558
0.8	43.248	80.409	109.50	31.326	64.955	90.159	6.1229	24.374	43.765
1.0	33.631	66.343	81.556	24.417	53.657	67.356	4.9029	20.011	33.484

symmetric–antisymmetric, antisymmetric–symmetric, and antisymmetric–antisymmetric modes about the two axes of the ellipse. The first few natural frequencies can be chosen from them to ensure that none are left out. To fix N, i.e., the number of approximations needed, calculations were carried out for different values of N until the first five significant digits converged. It was found that the results converged in about 5–6 approximations for a nonhomogeneous circular plate (i.e., $m = 1$) and in 11–13 approximations for a nonhomogeneous elliptic plate. However, to be on the safer side, all the results were obtained by carrying out up to 15 approximations. To get a feel of the convergence, i.e., how fast the results converge, the results of the first frequency for $\alpha = 0.4, \beta = 0.2$, and $m = 0.5$, increasing N from 2 to 15 for the clamped, simply supported, and free boundary conditions of the symmetric–symmetric mode are given in Table 11.33.

TABLE 11.33

Convergence of Fundamental Mode Obtained from Symmetric–Symmetric
Modes for $\alpha = 0.4, \beta = 0.2, m = b/a = 0.5$ (Type 1 Nonhomogeneity)

N	Clamped	Simply Supported	Free
2	8.0086	14.604	27.885
3	7.4048	13.570	27.823
5	6.5605	13.493	27.785
7	6.5130	13.489	27.783
10	6.5103	13.488	27.783
12	6.5098	13.488	27.783
13	6.5098	13.488	27.783
14	6.5098	13.488	27.783
15	6.5098	13.488	27.783
17	6.5098	13.488	27.783
18	6.5098	13.488	27.783
19	6.5098	13.488	27.783
20	6.5098	13.488	27.783

It is interesting to note from Tables 11.1 through 11.32 that as m increases, the frequencies decrease for all the boundary conditions. In particular, the symmetric–symmetric mode, as given in Tables 11.5, 11.9, 11.3, and 11.17 shows that when β, the nonhomogeneity parameter for density, is fixed ($\beta = 0.2$) and α, the nonhomogeneity parameter for Young's modulus, is increased ($\alpha = 0$, 0.2, 0.4, and 0.8), the frequencies increase. Although it is clear that the differences in the values of frequencies are not so significant for the fundamental mode, in the case of higher modes there is, however, a significant increase in the value of the frequencies as α is increased. This is true for all the boundary conditions and for all the mode groups. These can be clearly understood from Tables 11.5 through 11.20.

Similarly, we can observe that when α is fixed ($\alpha = 0.2$) and β increases from 0 to 0.2, the frequencies decrease, as shown in Tables 11.1 through 11.8. Also, we can notice that for modes higher than the first one, there is a significant increasing tendency. Again, this is true for all the boundary conditions and for all the mode groups about the two axes of the ellipse. Hence, the effect of nonhomogeneity cannot be neglected in the vibration of plates.

Tables 11.21 through 11.32 give some results related to other nonhomogeneity parameters, to have the feel of the results for their comparison as a benchmark and to prove that the nonhomogeneity parameter, as mentioned earlier, plays a great role in the study of vibration of plates.

Figures 11.1 through 11.4 depict the variation of the first three frequency parameters with all the boundary conditions for symmetric–symmetric,

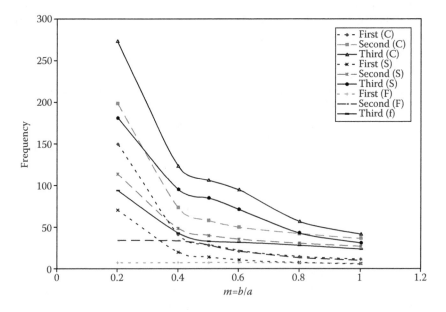

FIGURE 11.1
Variation of frequency versus aspect ratio [symmetric–symmetric modes, (0.2,0)] (type 1 nonhomogeneity).

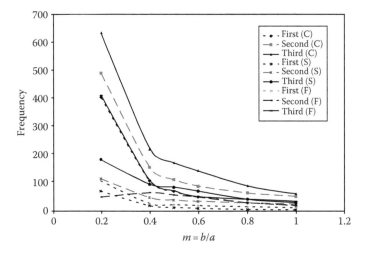

FIGURE 11.2
Variation of frequency versus aspect ratio [symmetric–antisymmetric modes, (0.2,0)] (type 1 nonhomogeneity).

symmetric–antisymmetric, antisymmetric–symmetric, and antisymmetric–antisymmetric modes, respectively, for the nonhomogeneity parameter (0.2, 0). These figures show that the frequencies increase as the aspect ratios are increased. Similar results are shown in Figures 11.5 through 11.8 and in Figures 11.9 through 11.12 for the nonhomogeneity parameters (0, 0.2) and (0.5, 0.3), respectively. An example of convergence pattern for the fundamental

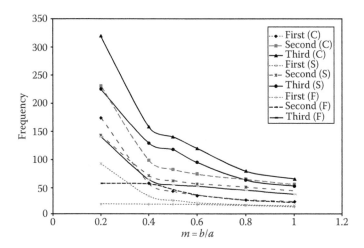

FIGURE 11.3
Variation of frequency versus aspect ratio [antisymmetric–symmetric modes, (0.2,0)] (type 1 nonhomogeneity).

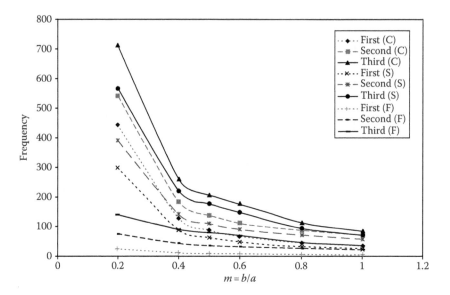

FIGURE 11.4
Variation of frequency versus aspect ratio [antisymmetric–antisymmetric modes, (0.2,0)] (type 1 nonhomogeneity).

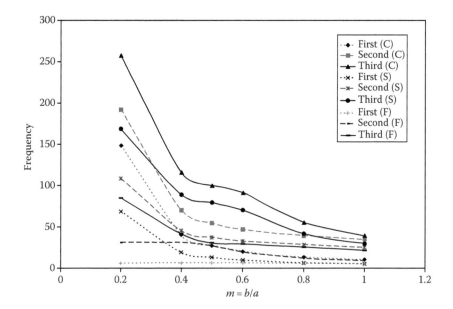

FIGURE 11.5
Variation of frequency versus aspect ratio [symmetric–symmetric modes, (0,0.2)] (type 1 nonhomogeneity).

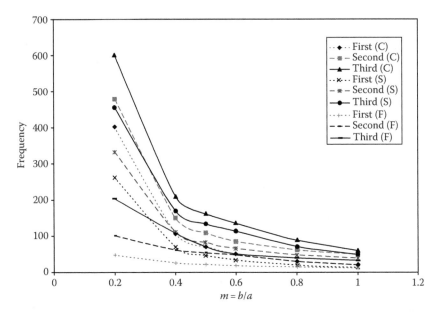

FIGURE 11.6
Variation of frequency versus aspect ratio [symmetric–antisymmetric modes, (0,0.2)] (type 1 nonhomogeneity).

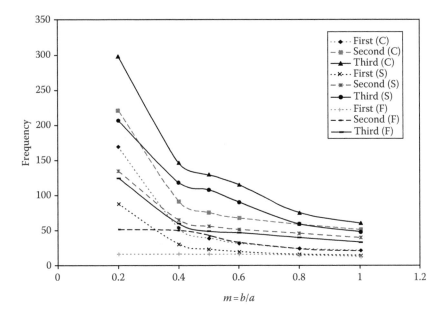

FIGURE 11.7
Variation of frequency versus aspect ratio [antisymmetric–symmetric modes, (0,0.2)] (type 1 nonhomogeneity).

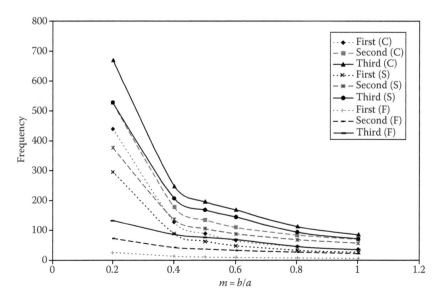

FIGURE 11.8
Variation of frequency versus aspect ratio [antisymmetric–antisymmetric modes, (0,0.2)] (type 1 nonhomogeneity).

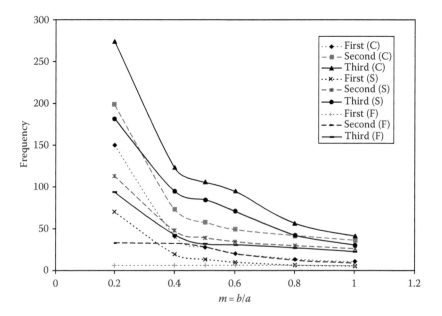

FIGURE 11.9
Variation of frequency versus aspect ratio [symmetric–symmetric modes, (0.5,0.3)], (type 1 nonhomogeneity).

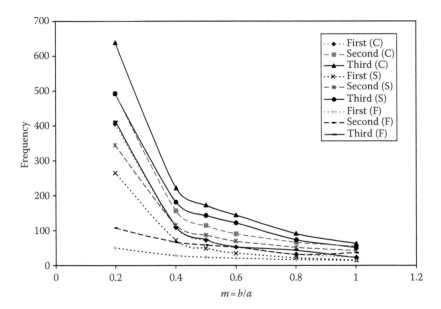

FIGURE 11.10
Variation of frequency versus aspect ratio [symmetric–antisymmetric modes, (0.5,0.3)], (type 1 nonhomogeneity).

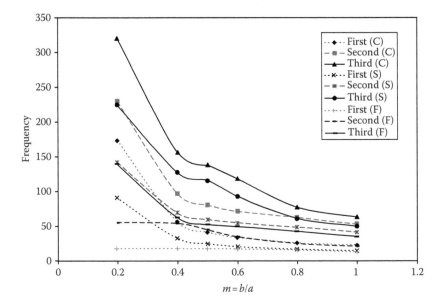

FIGURE 11.11
Variation of frequency versus aspect ratio [antisymmetric–symmetric modes, (0.5,0.3)], (type 1 nonhomogeneity).

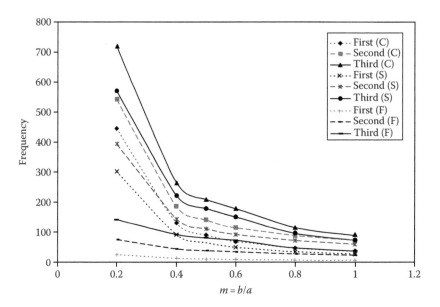

FIGURE 11.12
Variation of frequency versus aspect ratio [antisymmetric–antisymmetric modes, (0.5,0.3)] (type 1 nonhomogeneity).

mode from symmetric–symmetric modes is illustrated in Figure 11.13 with clamped, simply supported, and free boundary conditions, as the number of approximations, N, increased. The nonhomogeneity parameter in this figure has been taken as (0.4, 0.2) for the aspect ratio of $m = 0.5$.

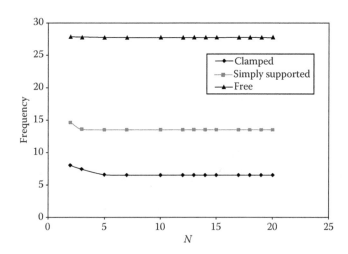

FIGURE 11.13
Convergence of fundamental mode obtained from symmetric–symmetric modes for (0.4,0.2), $m = b/a = 0.5$, (type 1 nonhomogeneity).

11.4.2 Special Case (when $\alpha = \beta = 0$)

This special case reduces the problem to that of a homogeneous elliptic or circular plate, for which results were already reported in the earlier chapters. Using this present computer program taking $\alpha = \beta = 0$, we have again calculated all the results for various values of m as given in Table 11.34, which are exactly the same as given in the papers of Singh and Chakraverty (1991; 1992a,b).

11.4.3 Results for Type 2 Nonhomogeneity

Although results can be worked out for various combinations of α and β (i.e., the degree of nonhomogeneity), only a few are reported here. In Tables 11.35 through 11.39, the first three frequencies are reported for various values of m from 0.1 to 1.0 in steps of 0.1, with all the modes of symmetric–symmetric, symmetric–antisymmetric, antisymmetric–symmetric, and antisymmetric–antisymmetric cases. These tables give the results for five chosen sets of values of the degree of nonhomogeneity of (α,β), viz., (4,2), (4,3), (5,3), (5,4), and (10,8), respectively. The computations have been carried out only for the free boundary condition. Poisson's ratio, ν, was taken as 0.3 in all the calculations. If one wants to find all the first

TABLE 11.34

First Five Frequencies for Homogeneous Case ($\alpha = \beta = 0$) (Type 1 Nonhomogeneity)

m	B.C.	λ_1	λ_2	λ_3	λ_4	λ_5
	C	149.66	171.10	198.55	229.81	266.25
0.2	S	69.684	88.792	112.92	141.23	174.94
	F	6.7778	17.389	25.747	32.817	48.582
	C	40.646	53.982	71.414	93.202	106.08
0.4	S	19.514	31.146	46.823	66.937	89.356
	F	6.7321	13.084	17.195	26.393	32.287
	C	27.377	39.497	55.985	69.858	76.995
0.5	S	13.213	23.641	38.354	46.151	62.764
	F	6.6705	10.548	16.921	22.015	27.768
	C	20.195	31.736	47.820	50.060	66.429
0.6	S	9.7629	19.566	33.122	33.777	47.916
	F	6.5712	8.8447	16.484	19.061	19.921
	C	13.229	24.383	30.322	39.972	44.792
0.8	S	6.3935	15.634	20.012	29.139	32.768
	F	6.1861	6.6841	12.128	14.880	15.174
	C	10.216	21.260	34.878	39.773	51.030
1.0	S	4.9351	13.898	25.613	29.720	39.957
	F	5.3583	9.0031	12.439	20.475	21.835

B.C., Boundary Condition

TABLE 11.35

First Three Frequency Parameters for Symmetric–Symmetric, Symmetric–Antisymmetric, Antisymmetric–Symmetric, and Antisymmetric–Antisymmetric Modes for $\alpha = 4$, $\beta = 2$ (Type 2 Nonhomogeneity)

	Symmetric–Symmetric			Symmetric–Antisymmetric		
m	First	Second	Third	First	Second	Third
0.1	9.9876	30.227	56.883	122.54	191.19	255.19
0.2	9.9704	30.139	56.568	61.783	98.318	134.26
0.3	9.9490	29.953	55.841	41.739	68.351	96.014
0.4	9.9087	29.613	54.423	31.858	53.889	77.605
0.5	9.8462	29.024	42.958	26.019	45.405	66.426
0.6	9.7489	28.031	30.130	22.177	39.692	58.356
0.7	9.5935	22.495	26.473	19.446	35.374	43.724
0.8	9.3347	17.700	24.409	17.369	31.797	33.984
0.9	8.8941	14.678	22.185	15.679	27.453	28.680
1.0	8.1975	12.899	20.079	14.198	22.978	25.922

	Anitsymmetric–Symmetric			Antisymmetric–Antisymmetric		
m	First	Second	Third	First	Second	Third
0.1	19.337	42.736	72.671	81.975	157.93	223.43
0.2	19.298	42.562	72.138	40.987	80.377	116.18
0.3	19.221	42.176	70.869	27.325	55.074	81.940
0.4	19.087	41.442	68.338	20.493	42.753	65.486
0.5	18.864	40.139	51.093	16.395	35.533	55.715
0.6	18.495	37.993	38.336	13.662	30.768	48.919
0.7	17.887	30.939	35.024	11.710	27.304	43.577
0.8	16.930	26.623	31.761	10.246	24.550	39.064
0.9	15.631	24.238	28.679	9.1084	22.192	35.153
1.0	14.198	22.978	25.922	8.1975	20.079	31.749

TABLE 11.36

First Three Frequency Parameters for Symmetric–Symmetric, Symmetric–Antisymmetric, Antisymmetric–Symmetric, and Antisymmetric–Antisymmetric Modes for $\alpha = 4$, $\beta = 3$ (Type 2 Nonhomogeneity)

	Symmetric–Symmetric			Symmetric–Antisymmetric		
m	First	Second	Third	First	Second	Third
0.1	13.831	45.196	89.268	181.01	301.52	427.75
0.2	13.818	45.180	89.254	91.456	156.32	227.34
0.3	13.793	45.074	88.854	61.965	109.75	164.64
0.4	13.751	44.805	87.803	47.445	87.385	134.99
0.5	13.681	44.274	58.689	38.871	74.350	117.44
0.6	13.568	41.155	43.304	33.230	65.653	83.404
0.7	13.383	30.685	41.583	29.222	59.103	61.888

TABLE 11.36 (continued)

First Three Frequency Parameters for Symmetric–Symmetric, Symmetric–Antisymmetric, Antisymmetric–Symmetric, and Antisymmetric–Antisymmetric Modes for $\alpha = 4$, $\beta = 3$ (Type 2 Nonhomogeneity)

	Symmetric–Symmetric			Symmetric–Antisymmetric		
m	First	Second	Third	First	Second	Third
0.8	13.065	24.080	38.905	26.174	48.046	53.592
0.9	12.495	19.898	35.616	23.684	38.758	48.605
1.0	11.539	17.455	32.293	21.471	32.420	44.007

	Antisymmetric–Symmetric			Antisymmetric–Antisymmetric		
m	First	Second	Third	First	Second	Third
0.1	27.981	65.543	116.28	115.00	242.06	360.67
0.2	27.963	65.535	116.24	57.524	123.78	189.78
0.3	27.910	65.329	115.52	38.796	85.346	135.66
0.4	27.794	64.777	108.42	28.796	66.677	109.88
0.5	27.578	63.662	74.131	23.050	55.766	94.784
0.6	27.195	55.479	61.588	19.218	48.589	84.424
0.7	26.524	44.479	58.107	16.478	43.389	76.240
0.8	25.362	37.917	53.477	14.421	39.246	63.769
0.9	23.592	34.273	48.602	12.821	35.633	54.534
1.0	21.471	32.420	44.007	11.539	32.293	48.497

TABLE 11.37

First Three Frequency Parameters for Symmetric–Symmetric, Symmetric–Antisymmetric, Antisymmetric–Symmetric, and Antisymmetric–Antisymmetric Modes for $\alpha = 5$, $\beta = 3$ (Type 2 Nonhomogeneity)

	Symmetric–Symmetric			Symmetric–Antisymmetric		
m	First	Second	Third	First	Second	Third
0.1	12.714	37.090	68.040	156.776	241.178	318.52
0.2	12.701	37.038	67.874	79.062	124.06	167.59
0.3	12.675	36.904	67.374	53.437	86.327	120.01
0.4	12.633	36.630	66.289	40.814	68.183	97.366
0.5	12.567	36.121	54.314	33.365	57.618	83.860
0.6	12.461	35.212	38.038	28.472	50.579	72.539
0.7	12.288	28.340	33.653	25.001	45.296	53.772
0.8	11.991	22.235	31.335	22.369	40.898	41.724
0.9	11.463	18.378	28.609	20.225	33.651	36.994
1.0	10.583	16.124	25.922	18.330	28.142	33.466

(continued)

TABLE 11.37 (continued)

First Three Frequency Parameters for Symmetric–Symmetric, Symmetric–Antisymmetric, Antisymmetric–Symmetric, and Antisymmetric–Antisymmetric Modes for $\alpha = 5$, $\beta = 3$ (Type 2 Nonhomogeneity)

	Antisymmetric–Symmetric			Antisymmetric–Antisymmetric		
m	First	Second	Third	First	Second	Third
0.1	24.125	51.719	86.060	105.83	200.51	280.26
0.2	24.097	51.625	85.779	52.915	102.08	145.77
0.3	24.035	51.357	84.910	35.276	70.001	102.91
0.4	23.919	50.784	82.999	26.457	54.405	82.465
0.5	23.714	49.695	64.262	21.166	45.300	70.472
0.6	23.360	47.737	47.991	17.638	39.323	62.247
0.7	23.746	38.471	44.647	15.118	35.005	55.796
0.8	21.702	32.840	40.827	13.228	31.581	50.259
0.9	20.150	29.729	36.992	11.758	28.622	44.911
1.0	18.330	28.142	33.466	10.583	25.922	39.949

TABLE 11.38

First Three Frequency Parameters for Symmetric–Symmetric, Symmetric–Antisymmetric, Antisymmetric–Symmetric, and Antisymmetric–Antisymmetric Modes for $\alpha = 5$, $\beta = 4$ (Type 2 Nonhomogeneity)

	Symmetric–Symmetric			Symmetric–Antisymmetric		
m	First	Second	Third	First	Second	Third
0.1	16.580	52.178	100.63	216.16	354.29	495.51
0.2	16.567	52.194	100.74	109.18	183.40	262.78
0.3	16.542	52.135	100.52	73.960	128.57	189.99
0.4	16.497	51.922	99.720	56.621	102.30	155.77
0.5	16.423	51.457	70.203	46.394	87.082	135.78
0.6	16.302	49.164	50.571	39.677	77.016	97.358
0.7	16.101	36.608	48.891	34.914	69.512	72.157
0.8	15.745	28.673	46.067	31.303	55.945	63.216
0.9	15.090	23.642	42.329	28.355	45.072	57.456
1.0	13.951	20.716	38.414	25.721	37.675	52.057

	Antisymmetric–Symmetric			Antisymmetric–Antisymmetric		
m	First	Second	Third	First	Second	Third
0.1	32.833	74.707	129.89	139.15	286.42	421.37
0.2	32.825	74.768	130.05	69.598	146.36	221.19
0.3	32.787	74.660	129.63	46.419	100.83	157.79
0.4	32.687	74.233	128.21	34.830	78.745	127.71

TABLE 11.38 (continued)

First Three Frequency Parameters for Symmetric–Symmetric, Symmetric–Antisymmetric, Antisymmetric–Symmetric, and Antisymmetric–Antisymmetric Modes for $\alpha = 5$, $\beta = 4$ (Type 2 Nonhomogeneity)

	Antisymmetric–Symmetric			Antisymmetric–Antisymmetric		
m	First	Second	Third	First	Second	Third
0.5	32.488	73.284	87.697	27.876	65.870	110.25
0.6	32.119	65.422	71.423	23.239	57.442	98.449
0.7	31.445	52.235	68.022	19.924	51.374	89.226
0.8	30.222	44.296	63.033	17.437	46.564	73.430
0.9	28.228	39.871	57.451	15.501	42.356	62.661
1.0	25.721	37.675	52.057	13.951	38.414	55.678

TABLE 11.39

First Three Frequency Parameters for Symmetric–Symmetric, Symmetric–Antisymmetric, Antisymmetric–Symmetric, and Antisymmetric–Antisymmetric Modes for $\alpha = 10$, $\beta = 8$ (Type 2 Nonhomogeneity)

	Symmetric–Symmetric			Symmetric–Antisymmetric		
m	First	Second	Third	First	Second	Third
0.1	26.257	70.648	121.73	325.96	484.80	622.87
0.2	26.243	70.733	122.04	164.46	249.44	327.27
0.3	26.214	70.781	122.15	111.26	173.75	234.48
0.4	26.162	70.702	121.82	85.093	137.60	191.16
0.5	26.073	70.394	111.06	69.688	116.87	166.49
0.6	25.922	69.686	77.581	59.611	103.45	140.73
0.7	25.658	57.584	68.196	52.508	93.712	104.00
0.8	25.174	44.935	65.214	47.158	80.395	85.684
0.9	24.228	36.891	60.467	42.811	64.579	78.233
1.0	22.449	32.249	54.990	38.884	53.888	70.992

	Antisymmetric–Symmetric			Antisymmetric–Antisymmetric		
m	First	Second	Third	First	Second	Third
0.1	47.716	95.299	149.98	224.49	409.55	555.31
0.2	47.734	95.483	150.43	112.24	208.63	288.69
0.3	47.735	95.574	150.53	74.833	143.22	203.98
0.4	47.682	95.413	149.92	56.124	111.54	164.02
0.5	47.526	94.821	148.13	44.899	93.183	141.25
0.6	47.189	93.455	96.503	37.416	81.290	126.34
0.7	46.512	76.421	90.550	32.071	72.860	115.09
0.8	45.142	64.129	85.227	28.062	66.283	99.329
0.9	42.556	57.182	78.220	24.944	60.531	84.330
1.0	38.884	53.888	70.992	22.449	54.990	74.779

five frequencies, then the results can be obtained from the various combinations of symmetric–symmetric, symmetric–antisymmetric, antisymmetric–symmetric, and antisymmetric–antisymmetric modes about the two axes of the ellipse. Consequently, the first five natural frequencies can be chosen (as pointed out earlier) from them to ensure that none are left out. To fix N, i.e., the number of approximations needed, calculations were carried out for different values of N until the first five significant digits converged. The results converged in about 4–5 approximations for a nonhomogeneous circular plate (i.e. $m=1$), and in 10–12 approximations for a nonhomogeneous elliptic plate. However, to be on the safer side (similar to type 1 nonhomogeneity), all the results were obtained by carrying out up to 15 approximations. To get a feel of the convergence, i.e., how fast the results converge, the results for $\alpha=4$, $\beta=3$, and $m=0.5$ and 0.8 with increasing N from 2 to 15 for the case of symmetric–symmetric mode are given in Table 11.40.

Figures 11.14 through 11.16 give the effect of aspect ratio on the first three frequencies corresponding to the four separate modes, viz., symmetric–symmetric, symmetric–antisymmetric, antisymmetric–symmetric, and antisymmetric–antisymmetric modes, denoted by SS, SA, AS, and AA, respectively, for the different sets of degree of nonhomogeneity (α,β) as $(4,2)$, $(5,4)$, and $(10,8)$.

For the special case of homogeneous elliptic and circular plate $(\alpha=\beta=0)$, the obtained results were compared with those of Singh and Chakraverty (1991) and was found to be in exact agreement, as reported in Table 11.41.

TABLE 11.40

Convergence of First Three Modes Obtained from Symmetric–Symmetric Modes for $\alpha=4$, $\beta=3$, $m=b/a=0.5$, 0.8 (Type 2 Nonhomogeneity)

	$m=0.5$			$m=0.8$		
N	First	Second	Third	First	Second	Third
2	14.605			14.605		
3	13.766	59.498		13.136	24.357	
4	13.681	49.104	59.594	13.069	24.302	49.242
5	13.681	45.752	59.076	13.069	24.224	40.317
6	13.681	45.712	58.693	13.065	24.080	39.641
7	13.681	44.339	58.693	13.065	24.080	39.156
8	13.681	44.285	58.690	13.065	24.080	39.000
9	13.681	44.285	58.689	13.065	24.080	38.920
10	13.681	44.285	58.689	13.065	24.080	38.908
11	13.681	44.274	58.689	13.065	24.080	38.905
12	13.681	44.274	58.689	13.065	24.080	38.905
13	13.681	44.274	58.689	13.065	24.080	38.905
14	13.681	44.274	58.689	13.065	24.080	38.905
15	13.681	44.274	58.689	13.065	24.080	38.905

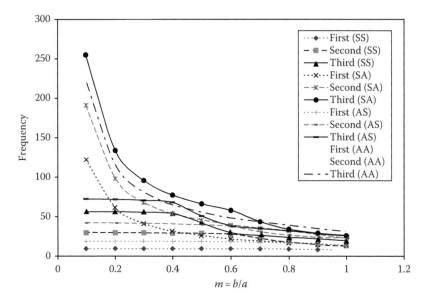

FIGURE 11.14
Variation of frequency versus aspect ratio (4,2) (type 2 nonhomogeneity).

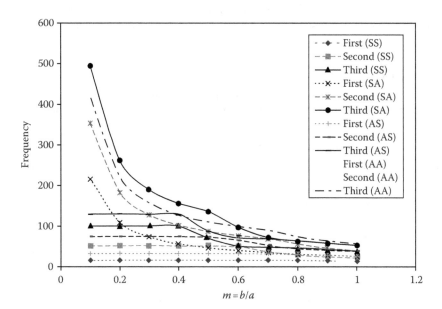

FIGURE 11.15
Variation of frequency versus aspect ratio (5,4) (type 2 nonhomogeneity).

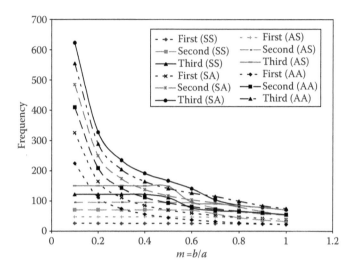

FIGURE 11.16
Variation of frequency versus aspect ratio (10,8) (type 2 nonhomogeneity).

From Tables 11.35 through 11.39 and 11.41, it can be observed that as *m* increases, the frequencies decrease. When α is fixed and β is increased, then for various values of *m*, the frequencies increase, as shown in Tables 11.35 and 11.36 ($\alpha = 4, \beta = 2, 3$) and Tables 11.37 and 11.38 ($\alpha = 5, \beta = 3, 4$). Furthermore, from Tables 11.36 and 11.37, it is also important to note that when β is fixed ($\beta = 3$) and α is increased from 4 to 5, the corresponding frequencies decrease

TABLE 11.41

First Three Frequency Parameters for Symmetric–Symmetric, Symmetric–Antisymmetric, Antisymmetric–Symmetric, and Antisymmetric–Antisymmetric Modes for $\alpha = 0, \beta = 0$ (Type 2 Nonhomogeneity)

	Symmetric–Symmetric			Symmetric–Antisymmetric		
m	First	Second	Third	First	Second	Third
0.1	6.7780	32.773	89.433	94.283	187.95	385.73
0.2	6.7777	32.853	89.171	48.582	102.93	212.81
0.3	6.7653	32.709	71.775	33.709	76.347	158.74
0.4	6.7321	32.287	41.937	26.392	63.386	107.60
0.5	6.6705	27.768	31.538	22.014	55.406	70.561
0.6	6.5711	19.921	30.355	19.060	49.637	50.200
0.7	6.4185	15.159	28.636	16.888	37.799	44.960
0.8	6.1860	12.128	26.454	15.174	29.743	40.816
0.9	5.8381	10.190	24.103	13.730	24.274	37.008
1.0	5.3583	9.0031	21.837	12.438	20.474	33.502

(continued)

TABLE 11.41 (continued)

First Three Frequency Parameters for Symmetric–Symmetric, Symmetric–Antisymmetric, Antisymmetric–Symmetric, and Antisymmetric–Antisymmetric Modes for $\alpha = 0$, $\beta = 0$ (Type 2 Nonhomogeneity)

m	Antisymmetric–Symmetric			Antisymmetric–Antisymmetric		
	First	Second	Third	First	Second	Third
0.1	17.364	53.128	131.64	51.196	138.31	242.52
0.2	17.388	53.277	130.77	25.747	73.648	136.59
0.3	17.343	52.920	95.383	17.301	53.017	103.95
0.4	17.195	51.997	60.264	13.084	42.929	87.968
0.5	16.921	42.991	50.410	10.547	36.828	77.858
0.6	16.483	33.239	47.937	8.8447	32.590	69.416
0.7	15.821	27.357	44.582	7.6161	29.319	55.089
0.8	14.879	23.789	40.775	6.6841	26.570	45.769
0.9	13.699	21.694	37.007	5.9508	24.107	39.524
1.0	12.438	20.474	33.502	5.3583	21.835	35.260

for various values of m. It is interesting to conclude, either from Tables 11.35 through 11.39 or from Figures 11.14 through 11.16, that the fundamental modes remain nearly a constant as the aspect ratio of the ellipse increases from 0.1 to 1.0 for any particular set of the degree of nonhomogeneity. However, the fundamental modes increase as the degree of the nonhomo-geneity changes from (0,0) to (10,8) for a particular value of aspect ratio of the ellipse. Furthermore, the higher modes (second to fifth) depend on the aspect ratio of the ellipse as shown in Figures 11.14 through 11.16 and Tables 11.35 through 11.39, and these also increase as the nonhomogeneity changes from set (0,0) to (10,8).

Thus, it can be confirmed that the natural frequencies of vibration depend on the present type of degree of the nonhomogeneity. A design engineer can directly see the presented plots of the figures to have the knowledge about a particular mode for any of the aspect ratio of the nonhomogeneous plate, to finalize the design of the structure. The methodology can be easily extended to various other types of nonhomogeneity in plates with complex geometry.

Bibliography

Chakraverty, S. and Petyt, M. 1997. Natural frequencies for free vibration of non-homogeneous elliptic and circular plates using two-dimensional orthogonal polynomials. *Applied Mathematical Modelling*, 21: 399–417.

Chakraverty, S. and Petyt, M. 1998. Vibration of non-homogeneous plates using two-dimensional orthogonal polynomials as shape functions in the Rayleigh-Ritz method. *Proceeding of Institution of Mechanical Engineers*; 213, Part C, 707–714.

Laura, P.A.A. and Gutierrez, R.H. 1984. Transverse vibrations of orthotropic, non-homogeneous rectangular plates. *Fibre Science and Technology*, 133: 125–133.

Leissa, A.W. 1978. Recent research in plate vibrations, 1973–76: Complicating effects. *The Shock and Vibration Digest*, 10(12): 21–35.

Leissa, A.W. 1981. Plate vibration research, 1976–80: Complicating effects. *The Shock Vibration and Digest*, 13(10): 19–36.

Leissa, A.W. 1987. Recent studies in plate vibrations: 1981–85 Part II. Complicating effects. *The Shock and Vibration Digest*, 19(3): 10–24.

Mishra, D.M. and Das, A.K. 1971. Free vibrations of an isotropic nonhomogeneous circular plate. *AIAA Journal*, 9(5): 963–964.

Pan, M. 1976. Note on the transverse vibration of an isotropic circular plate with density varying parabolically. *Indian Journal of Theoretical Physics*, 24(4): 179–182.

Rao, G.V., Rao, B.P., and Raju, I.S. 1974. Vibrations of inhomogeneous thin plates using a high precision triangular element. *Journal of Sound and Vibration*, 34(3): 444–445.

Rao, B.P., Venkateswara Rao, G., and Raju, I.S. 1976. A perturbation solution for the vibration of inhomogeneous rectangular plates. *Journal of Aeronautical Society of India*, 28(1): 121–125.

Singh, B. and Chakraverty, S. 1991. Transverse vibration of completely free elliptic and circular plates using orthogonal polynomials in Rayleigh-Ritz method. *International Journal of Mechanical Science*, 33(9): 741–751.

Singh, B. and Chakraverty, S. 1992a. On the use of orthogonal Polynomials in Rayleigh-Ritz method for the study of transverse vibration of elliptic plates. *Computers and Structures*, 43(3): 439–443.

Singh, B. and Chakraverty, S. 1992b. Transverse Vibration of Simply-Supported Elliptic and Circular Plates Using Orthogonal Polynomials in Two Variables. *Journal of Sound and Vibration*, 152(1): 149–155.

Tomar, J.S., Gupta, D.C., and Jain, N.C. 1982. Axisymmetric vibrations of an isotropic elastic non-homogeneous circular plate of linearly varying thickness. *Journal of Sound and Vibration*, 85(3): 365–370.

Tomar, J.S., Gupta, D.C., and Jain, N.C. 1982. Vibrations of non-homogeneous plates of variable thickness. *Journal of Acoustical Society of America*, 72(3): 851–855.

Tomar, J.S., Gupta, D.C., and Jain, N.C. 1983. Free vibrations of an isotropic non-homogeneous infinite plate of linearly varying thickness. *Meccanica*, 18(1): 30–33.

Tomar, J.S., Sharma, R.K., and Gupta, D.C. 1983. Transverse vibrations of nonuniform rectangular orthotropic plates. *American Institute of Aeronautics and Astronautics Journal*, 21(7): 1050–1053.

12

Plates with Variable Thickness

12.1 Introduction

The study of transverse vibration of plates of variable thickness is very important in a wide variety of applications in the industry and engineering. Lot of studies have been carried out on rectangular and circular plates with variable thickness. Again, in this chapter, the example of elliptic plate with variable thickness is taken into consideration to understand how the variable thickness plates are to be handled using the method of boundary characteristic orthogonal polynomials (BCOPs). Before going into the details of the methodology and the corresponding results, we will discuss some of the works specific to the elliptic plates with variable thickness.

Singh and Goel (1985b), Singh and Tyagi (1985a), and Singh and Chakraverty (1991a, 1992) considered the vibrations of elliptic and circular plates with clamped boundary and thickness varying according to different laws. In these studies, the deflections were assumed to be the same along the concentric ellipses. Previously, McNitt (1962) and Mazumdar (1971) obtained few frequencies for elliptic plates based on the above assumption. The approach, in general, in the works of Singh and Goel (1985b), Singh and Tyagi (1985a), and Singh and Chakraverty (1991a, 1992) was to take an N-term approximation and then work out various approximations by Ritz or Galerkin methods. The results obtained from these studies were doubtful, as the eccentricity of the domain was increased. However, when the ellipses were very close to a circular domain, the accuracy was better.

As such, orthogonal polynomials were used for elliptic and circular plates as mentioned in the previous chapters (Singh and Chakraverty (1991b, 1992a,b, 1993)), where we got excellent results. The basic aim of this chapter is to use the BCOPs in plates with variable thickness. No assumptions were made about the type of uniform displacement along the concentric ellipses. Hence, the procedure is valid for ellipses of all eccentricities. Again, we start from a set of linearly independent functions, and a set of orthogonal polynomials is generated by Gram–Schmidt process. These are used as the basis functions in the Rayleigh–Ritz method. The number of approximations is successively increased and the results are compared with the previous approximation. The process is truncated when the desired numbers of frequencies converged to the first three frequencies of each group, viz., symmetric–symmetric, symmetric–antisymmetric, antisymmetric–symmetric, and antisymmetric–antisymmetric modes, to four or five significant digits.

12.2 Generation of BCOPs for Variable Thickness Plates

The example domain was considered as ellipse, defined by

$$D = \left\{ (x,y), \ \frac{x^2}{a^2} + \frac{y^2}{b^2} \leq 1, \ x, y \in R \right\} \tag{12.1}$$

where a and b are semimajor and semiminor axes of the elliptic domain, respectively, as shown in Figure 12.1. We introduced variables u and v to obtain

$$\frac{x^2}{a^2} + \frac{y^2}{b^2} = 1 - u = v^2; \quad 0 \leq u \leq 1, \quad 0 \leq v \leq 1 \tag{12.2}$$

It may be noted that $u = 0$ or $v = 1$ designate the boundary ∂D of the domain. The center is given by $u = 1$ and $v = 0$, as shown in Figure 12.1. Hence, the curves $u = $ constant and $v = $ constant give the concentric ellipses.

As in the previous chapters, the orthogonal polynomials are generated from the following linearly independent functions:

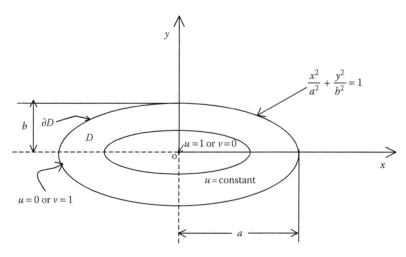

FIGURE 12.1
Domain of an elliptical plate.

$$F_i(x,y) = f(x,y)f_i(x,y), \quad i = 1, 2, \ldots, N \tag{12.3}$$

where the first term on the right-hand side of Equation 12.3, viz., $f(x,y)$, satisfies the essential boundary conditions on the boundary ∂D, and f_i are the suitably chosen linearly independent functions, involving the products of non-negative integral powers of x and y. As such, $f_i(x,y)$ will determine the symmetric–symmetric, symmetric–antisymmetric, antisymmetric–symmetric, and anti-symmetric–antisymmetric modes, corresponding to the integral powers of x and y as even–even, even–odd, odd–even, and odd–odd, respectively. This chapter introduces the method to handle any variable-thickness property of the plates. Accordingly, the inner product of the two functions $p(x,y)$ and $q(x,y)$ over the domain D can be defined as

$$\langle p,q \rangle = \iint\limits_{D} h(x,y)p(x,y)q(x,y)dxdy \tag{12.4}$$

where $h(x,y)$ in the above integral designates the thickness of the plate. For the particular simple functions, the above integral can be evaluated exactly, while in other cases one may use any suitable numerical methods. The norm of p is thus given by

$$\|p\| = \langle p,p \rangle^2 \tag{12.5}$$

The Gram–Schmidt process can be described, as in the previous chapters, by following the steps given below for generating the orthogonal polynomials ϕ_i:

$$\left. \begin{array}{l} \phi_1 = F_1 \\ \phi_i = F_i - \sum\limits_{j=1}^{i-1} \alpha_{ij}\phi_j \end{array} \right\} \tag{12.6}$$

where

$$\alpha_{ij} = \frac{\langle F_i,\phi_j \rangle}{\langle \phi_j,\phi_j \rangle}, \quad j = 1, 2, \ldots, i-1, \quad i = 2, 3, \ldots, N \tag{12.7}$$

The normalized polynomials can be generated by dividing each ϕ_i by its norm

$$\hat{\phi}_i = \frac{\phi_i}{\|\phi_i\|} \tag{12.8}$$

As mentioned earlier, all the integrals involved in the inner product can be evaluated in closed form, if the variable-thickness function $h(x,y)$ is chosen as polynomial in x, y, u, and v, where u and v are already defined. The result of the integrals can be written as given by Singh and Chakraverty (1994)

$$\iint_D x^i y^j u^k v^t \, dx \, dy = a^{i+1} b^{j+1} \frac{\left)\frac{i+j+t}{2}+1\right) \overline{k+1} \left)\frac{i+1}{2}\right) \left)\frac{j+1}{2}\right)}{\left)\frac{i+j+t}{2}+k+2\right) \left)\frac{i+j}{2}+1\right)} \qquad (12.9)$$

where i, j are the even integers and $i, j, k, t > 1$.

The following results are also very useful while evaluating the integrals:

$$\iint_D x^i y^j v^t \, dx \, dy = \frac{2a^{i+1} b^{j+1}}{(i+j+t+2)} \frac{\left)\frac{i+1}{2}\right) \left)\frac{j+1}{2}\right)}{\left)\frac{i+j+2}{2}\right)} \qquad (12.10)$$

and

$$\iint_D x^i y^j u^k \, dx \, dy = a^{i+1} b^{j+1} \frac{\left)\frac{i+1}{2}\right) \left)\frac{j+1}{2}\right) \overline{k+1}}{\left)\frac{i+j}{2}+k+1\right)} \qquad (12.11)$$

As mentioned in the previous chapters, the orthogonal polynomials can be written as

$$\phi_i = f \sum_{k=1}^{i} \beta_{ik} f_k \qquad (12.12)$$

Accordingly, from Equation 12.7 we can obtain

$$\alpha_{ij} = \frac{\langle ff_i, \phi_j \rangle}{\langle \phi_j, \phi_j \rangle}, \quad j = 1, 2, \ldots, i-1, \quad i = 2, 3, \ldots, N$$

$$= \frac{\langle ff_i, f \sum_{k=1}^{j} \beta_{jk} f_k \rangle}{\left\langle f \sum_{k=1}^{j} \beta_{jk} f_k, f \sum_{\mathit{l}=1}^{j} \beta_{j\mathit{l}} f_\mathit{l} \right\rangle} \qquad (12.13)$$

The function $f(x,y)$ defined by

$$f(x,y) = (1 - x^2 - y^2/m^2)^s = u^s \qquad (12.14)$$

designates the boundary condition cases of clamped, simply supported, and completely free edges, represented by the parameters as 2, 1, and 0, respectively. The functions f_i are suitably chosen linearly independent functions, involving the products of non-negative integral powers of x and y, i.e., they are of the form

$$f_i(x,y) = x^{m_i} y^{n_i} \tag{12.15}$$

By using Equations 12.14 and 12.15 in Equation 12.13, along with the definition of the inner product from Equation 12.4, we can obtain

$$\alpha_{ij} = \frac{\displaystyle\sum_{k=1}^{j} \beta_{jk} \iint_D h(x,y) u^{2s} x^{m_i+m_k} y^{n_i+n_k} \, dx \, dy}{\displaystyle\sum_{\xi=1}^{j} \sum_{\zeta=1}^{j} \beta_{j\xi}\beta_{j\zeta} \iint_D h(x,y) u^{2s} x^{m_\xi+m_\zeta} y^{n_\xi+n_\zeta} \, dx \, dy} \tag{12.16}$$

Thus, the constants in Equation 12.6 can be determined by using Equation 12.16 for various-thickness functions. If the thickness functions are polynomials, then the integrations can be computed exactly by utilizing the formulae given in Equations 12.9 through 12.11. If we have linear thickness variations as

$$h(x,y) = 1 + \eta x + \theta y \tag{12.17}$$

then we can write the numerator of Equation 12.16 as

$$\sum_{k=1}^{j} \beta_{jk} \left[\iint_D u^{2s} x^{m_i+m_k} y^{n_i+n_k} \, dx \, dy + \eta \iint_D u^{2s} x^{m_i+m_k+1} y^{n_i+n_k} \, dx \, dy \right. $$
$$\left. + \theta \iint_D u^{2s} x^{m_i+m_k} y^{n_i+n_k+1} \, dx \, dy \right] \tag{12.18}$$

while the denominator of Equation 12.16 can be written as

$$\sum_{\xi=1}^{j} \sum_{\zeta=1}^{j} \beta_{j\xi}\beta_{j\zeta} \left[\iint_D u^{2s} x^{m_\xi+m_\zeta} y^{n_\xi+n_\zeta} \, dx \, dy + \eta \iint_D u^{2s} x^{m_\xi+m_\zeta+1} y^{n_\xi+n_\zeta} \, dx \, dy \right. $$
$$\left. + \theta \iint_D u^{2s} x^{m_\xi+m_\zeta} y^{n_\xi+n_\zeta+1} \, dx \, dy \right] \tag{12.19}$$

Thus, the BCOPs for the thickness variation can be obtained.

12.3 Rayleigh–Ritz Method in the Variable Thickness Plates

Let us assume the displacement to be of the form

$$w(x,y,t) = W(x,y) \sin \omega t \tag{12.20}$$

where ω is the radian natural frequency and $W(x,y)$ is the maximum displacement. Equating the maximum strain energy and the maximum kinetic energy reduces the problem to the Rayleigh quotient

$$\omega^2 = \frac{E}{12\rho(1-\nu^2)} \frac{\iint_D h^3 \left[(\nabla^2 W)^2 + 2(1-\nu)(W_{xx}^2 - W_{xx}W_{yy}) \right] dxdy}{\iint_D hW^2 dxdy} \tag{12.21}$$

where E, ρ, and ν are Young's modulus, the mass per unit volume, and Poisson's ratio of the material of the plate, respectively. Now, substituting the N-term approximation

$$W(x,y) = \sum_{j=1}^{N} c_j \phi_j \tag{12.22}$$

in Equation 12.21 and minimizing ω^2 as a function of the coefficients $c_1, c_2, \ldots, c_N$ yields

$$\sum_{j=1}^{N} (a_{ij} - \lambda^2 b_{ij}) c_j = 0, \quad i = 1, 2, \ldots, N \tag{12.23}$$

where

$$a_{ij} = \iint_D H^3 \left[\phi_i^{XX} \phi_j^{XX} + \phi_i^{YY} \phi_j^{YY} + \nu \left(\phi_i^{XX} \phi_j^{YY} + \phi_i^{YY} \phi_j^{XX} \right) \right.$$
$$\left. + 2(1-\nu)\phi_i^{XY} \phi_j^{XY} \right] dXdY \tag{12.24}$$

$$b_{ij} = \iint_{D'} H \phi_i \phi_j dXdY \tag{12.25}$$

$$\lambda^2 = \frac{12\rho a^4 (1-\nu^2)\omega^2}{Eh_0^2} \tag{12.26}$$

$$X = \frac{x}{a}, \quad Y = \frac{y}{a}, \quad H = \frac{h}{h_0} \tag{12.27}$$

in which h_0 is the thickness at some standard point, taken here to be the origin, and the new domain D' is defined by

$$D' = \{(X,Y),\ X^2 + Y^2/m^2 \leq 1,\ X,Y \in R\},\quad m = b/a \tag{12.28}$$

In this chapter, the following two thickness variations have been considered for the analysis:

$$\text{(i)}\quad H = 1 + \alpha X + \beta X^2 \tag{12.29}$$

$$\text{(ii)}\quad H = 1 + \gamma v + \delta v^2 \tag{12.30}$$

where α, β, γ, and δ are the constants. The above-mentioned thickness variation gives both the linear as well as quadratic variation, i.e., if we take α and γ as zero, we obtain a pure quadratic variation of thickness in both the cases, whereas if β and δ are zero, then the problem is of linear thickness variation. In the former case, the lines of constant thickness are parallel to the minor axis and as X is varied, one has a parabolic variation parallel to the X-axis. In the latter case, the curves of constant thickness are concentric ellipses, where $v = $ constant. This is the analogous problem of axisymmetric vibrations of a circular plate. If $b = a$ or $m = 1$, one can observe from Equation 12.2 that the variable v is the nondimensional radial distance from the center, i.e.,

$$v^2 = \frac{x^2 + y^2}{a^2} = \left(\frac{r}{a}\right)^2 = R^2 \tag{12.31}$$

and hence, for a circular plate the thickness variation Equation 12.30 reduces to

$$H = 1 + \gamma R + \delta R^2 \tag{12.32}$$

Now, we can express Equation 12.24 in terms of orthogonal polynomials and show how to obtain the terms for the computation of the matrices involved in Equation 12.23 as discussed below:
We have

$$\phi_i = u^s \sum_{k=1}^{i} \beta_{ik} f_k \tag{12.33}$$

By substituting the variable thickness of Case (i), mentioned in Equation 12.29, and determining the double derivatives of ϕ_i with respect to x, we can write the first term of Equation 12.24 as

$$\iint\limits_{D} (1 + \alpha X + \beta X^2)^3 \phi_i^{XX} \phi_j^{XX} dXdY = PX1 + PX2 + PX3 + \cdots + PX9 \tag{12.34}$$

where

$$PX1 = \sum_{k=1}^{i} \beta_{ik} m_k (m_k - 1) \sum_{\ell=1}^{j} \beta_{j\ell} m_\ell (m_\ell - 1) \iint_D X^{m_k + m_\ell - 4} Y^{n_k + n_\ell} u^{2s} [\bar{A}] dX dY$$

(12.35)

$$\bar{A} = [1 + A_1 X + A_2 X^2 + A_3 X^3 + A_4 X^4 + A_5 X^5 + A_6 X^6]$$

(12.36)

$$\left.\begin{array}{l} A_1 = 3\alpha \\ A_2 = 3(\alpha^2 + \beta) \\ A_3 = \alpha(\alpha^2 + 6\beta) \\ A_4 = 3\beta(\alpha^2 + \beta) \\ A_5 = 3\beta^2 \alpha \\ A_6 = \beta^3 \end{array}\right\}$$

(12.37)

Similarly, the other terms can be written as

$$PX2 = -\sum_{k=1}^{i} \beta_{ik} m_k (m_k - 1) \sum_{\ell=1}^{j} \beta_{j\ell} s(4m_\ell + 2) \iint_D X^{m_k + m_\ell - 2} Y^{n_k + n_\ell} u^{2s-1} [\bar{A}] dX dY$$

(12.38)

$$PX3 = \sum_{k=1}^{i} \beta_{ik} m_k (m_k - 1) \sum_{\ell=1}^{j} \beta_{j\ell} 4s(s - 1) \iint_D X^{m_k + m_\ell} Y^{n_k + n_\ell} u^{2s-2} [\bar{A}] dX dY \quad (12.39)$$

$$PX4 = -\sum_{k=1}^{i} \beta_{ik} s(4m_k + 2) \sum_{\ell=1}^{j} \beta_{j\ell} m_\ell (m_\ell - 1) \iint_D X^{m_k + m_\ell - 2} Y^{n_k + n_\ell} u^{2s-1} [\bar{A}] dX dY$$

(12.40)

$$PX5 = \sum_{k=1}^{i} \beta_{ik} s(4m_k + 2) \sum_{\ell=1}^{j} \beta_{j\ell} s(4m_\ell + 2) \iint_D X^{m_k + m_\ell} Y^{n_k + n_\ell} u^{2s-2} [\bar{A}] dX dY$$

(12.41)

$$PX6 = -\sum_{k=1}^{i} \beta_{ik} s(4m_k + 2) \sum_{\ell=1}^{j} \beta_{j\ell} 4s(s - 1) \iint_D X^{m_k + m_\ell + 2} Y^{n_k + n_\ell} u^{2s-3} [\bar{A}] dX dY$$

(12.42)

$$PX7 = \sum_{k=1}^{i} \beta_{ik} 4s(s - 1) \sum_{\ell=1}^{j} \beta_{j\ell} m_\ell (m_\ell - 1) \iint_D X^{m_k + m_\ell} Y^{n_k + n_\ell} u^{2s-2} [\bar{A}] dX dY \quad (12.43)$$

$$PX8 = -\sum_{k=1}^{i}\beta_{ik}4s(s-1)\sum_{\ell=1}^{j}\beta_{j\ell}s(4m_\ell+2)\iint_D X^{m_k+m_\ell+2}Y^{n_k+n_\ell}u^{2s-3}[\bar{A}]dXdY$$

$$(12.44)$$

$$PX9 = \sum_{k=1}^{i}\beta_{ik}4s(s-1)\sum_{\ell=1}^{j}\beta_{j\ell}4s(s-1)\iint_D X^{m_k+m_\ell+4}Y^{n_k+n_\ell}u^{2s-4}[\bar{A}]dXdY \quad (12.45)$$

The second term of Equation 12.24 will be

$$\iint_D (1+\alpha X+\beta X^2)^3\phi_i^{YY}\phi_j^{YY}dXdY = PY1+PY2+PY3+\cdots+PY9 \quad (12.46)$$

where

$$PY1 = \sum_{k=1}^{i}\beta_{ik}n_k(n_k-1)\sum_{\ell=1}^{j}\beta_{j\ell}n_\ell(n_\ell-1)\iint_D X^{m_k+m_\ell}Y^{n_k+n_\ell-4}u^{2s}[\bar{A}]dXdY \quad (12.47)$$

$$PY2 = -\sum_{k=1}^{i}\beta_{ik}n_k(n_k-1)\sum_{\ell=1}^{j}\beta_{j\ell}\frac{s}{m^2}(4n_\ell+2)\iint_D X^{m_k+m_\ell}Y^{n_k+n_\ell-2}u^{2s-1}[\bar{A}]dXdY$$

$$(12.48)$$

$$PY3 = \sum_{k=1}^{i}\beta_{ik}n_k(n_k-1)\sum_{\ell=1}^{j}\beta_{j\ell}\frac{4s}{m^4}(s-1)\iint_D X^{m_k+m_\ell}Y^{n_k+n_\ell}u^{2s-2}[\bar{A}]dXdY \quad (12.49)$$

$$PY4 = -\sum_{k=1}^{i}\beta_{ik}\frac{s}{m^2}(4n_k+2)\sum_{\ell=1}^{j}\beta_{j\ell}n_\ell(n_\ell-1)\iint_D X^{m_k+m_\ell}Y^{n_k+n_\ell-2}u^{2s-1}[\bar{A}]dXdY$$

$$(12.50)$$

$$PY5 = \sum_{k=1}^{i}\beta_{ik}\frac{s}{m^2}(4n_k+2)\sum_{\ell=1}^{j}\beta_{j\ell}\frac{s}{m^2}(4n_\ell+2)\iint_D X^{m_k+m_\ell}Y^{n_k+n_\ell}u^{2s-2}[\bar{A}]dXdY$$

$$(12.51)$$

$$PY6 = -\sum_{k=1}^{i}\beta_{ik}\frac{s}{m^2}(4n_k+2)\sum_{\ell=1}^{j}\beta_{j\ell}\frac{4s}{m^4}(s-1)\iint_D X^{m_k+m_\ell}Y^{n_k+n_\ell+2}u^{2s-3}[\bar{A}]dXdY$$

$$(12.52)$$

$$PY7 = \sum_{k=1}^{i}\beta_{ik}\frac{4s}{m^4}(s-1)\sum_{\ell=1}^{j}\beta_{j\ell}n_\ell(n_\ell-1)\iint_D X^{m_k+m_\ell}Y^{n_k+n_\ell}u^{2s-2}[\bar{A}]dXdY \quad (12.53)$$

$$PY8 = -\sum_{k=1}^{i} \beta_{ik} \frac{4s}{m^4}(s-1) \sum_{\ell=1}^{j} \beta_{j\ell} \frac{s}{m^2}(4n_\ell + 2) \iint_D X^{m_k + m_\ell} Y^{n_k + n_\ell + 2} u^{2s-3}[\bar{A}] dX dY$$

$$(12.54)$$

$$PY9 = \sum_{k=1}^{i} \beta_{ik} \frac{4s}{m^4}(s-1) \sum_{\ell=1}^{j} \beta_{j\ell} \frac{4s}{m^4}(s-1) \iint_D X^{m_k + m_\ell} Y^{n_k + n_\ell + 4} u^{2s-4}[\bar{A}] dX dY$$

$$(12.55)$$

Similarly, the third term of Equation 12.24 can be written as

$$\iint_D (1 + \alpha X + \beta X^2)^3 \phi_i^{XX} \phi_j^{YY} dX dY = PXI1 + PXI2 + PXI3 + \cdots + PXI9$$

$$(12.56)$$

where

$$PXI1 = \sum_{k=1}^{i} \beta_{ik} m_k (m_k - 1) \sum_{\ell=1}^{j} \beta_{j\ell} n_\ell (n_\ell - 1) \iint_D X^{m_k + m_\ell - 2} Y^{n_k + n_\ell - 2} u^{2s}[\bar{A}] dX dY$$

$$(12.57)$$

$$PXI2 = -\sum_{k=1}^{i} \beta_{ik} m_k (m_k - 1) \sum_{\ell=1}^{j} \beta_{j\ell} \frac{s}{m^2}(4n_\ell + 2) \iint_D X^{m_k + m_\ell - 2} Y^{n_k + n_\ell} u^{2s-1}[\bar{A}] dX dY$$

$$(12.58)$$

$$PXI3 = \sum_{k=1}^{i} \beta_{ik} m_k (m_k - 1) \sum_{\ell=1}^{j} \beta_{j\ell} \frac{4s}{m^4}(s-1) \iint_D X^{m_k + m_\ell - 2} Y^{n_k + n_\ell + 2} u^{2s-2}[\bar{A}] dX dY$$

$$(12.59)$$

$$PXI4 = -\sum_{k=1}^{i} \beta_{ik} s(4m_k + 2) \sum_{\ell=1}^{j} \beta_{j\ell} n_\ell (n_\ell - 1) \iint_D X^{m_k + m_\ell} Y^{n_k + n_\ell - 2} u^{2s-1}[\bar{A}] dX dY$$

$$(12.60)$$

$$PXI5 = \sum_{k=1}^{i} \beta_{ik} s(4m_k + 2) \sum_{\ell=1}^{j} \beta_{j\ell} \frac{s}{m^2}(4n_\ell + 2) \iint_D X^{m_k + m_\ell} Y^{n_k + n_\ell} u^{2s-2}[\bar{A}] dX dY$$

$$(12.61)$$

$$PXI6 = -\sum_{k=1}^{i} \beta_{ik} s(4m_k + 2) \sum_{\ell=1}^{j} \beta_{j\ell} \frac{4s}{m^4}(s-1) \iint_D X^{m_k + m_\ell} Y^{n_k + n_\ell + 2} u^{2s-3}[\bar{A}] dX dY$$

$$(12.62)$$

$$PXI7 = \sum_{k=1}^{i} \beta_{ik} 4s(s-1) \sum_{\ell=1}^{j} \beta_{j\ell} n_\ell(n_\ell - 1) \iint_D X^{m_k+m_\ell+2} Y^{n_k+n_\ell-2} u^{2s-2} [\bar{A}] dX dY$$

(12.63)

$$PXI8 = -\sum_{k=1}^{i} \beta_{ik} 4s(s-1) \sum_{\ell=1}^{j} \beta_{j\ell} \frac{s}{m^2} (4n_\ell + 2) \iint_D X^{m_k+m_\ell+2} Y^{n_k+n_\ell} u^{2s-3} [\bar{A}] dX dY$$

(12.64)

$$PXI9 = \sum_{k=1}^{i} \beta_{ik} 4s(s-1) \sum_{\ell=1}^{j} \beta_{j\ell} \frac{4s}{m^4} (s-1) \iint_D X^{m_k+m_\ell+2} Y^{n_k+n_\ell+2} u^{2s-4} [\bar{A}] dX dY$$

(12.65)

We can obtain the fourth term of Equation 12.24 as

$$\iint_D (1 + \alpha X + \beta X^2)^3 \phi_i^{YY} \phi_j^{XX} dX dY = PYI1 + PYI2 + PYI3 + \cdots + PYI9$$

(12.66)

where

$$PYI1 = \sum_{k=1}^{i} \beta_{ik} n_k(n_k - 1) \sum_{\ell=1}^{j} \beta_{j\ell} m_\ell(m_\ell - 1) \iint_D X^{m_k+m_\ell-2} Y^{n_k+n_\ell-2} u^{2s} [\bar{A}] dX dY$$

(12.67)

$$PYI2 = -\sum_{k=1}^{i} \beta_{ik} n_k(n_k - 1) \sum_{\ell=1}^{j} \beta_{j\ell} s(4m_\ell + 2) \iint_D X^{m_k+m_\ell} Y^{n_k+n_\ell-2} u^{2s-1} [\bar{A}] dX dY$$

(12.68)

$$PYI3 = \sum_{k=1}^{i} \beta_{ik} n_k(n_k - 1) \sum_{\ell=1}^{j} \beta_{j\ell} 4s(s-1) \iint_D X^{m_k+m_\ell+2} Y^{n_k+n_\ell-2} u^{2s-2} [\bar{A}] dX dY$$

(12.69)

$$PYI4 = -\sum_{k=1}^{i} \beta_{ik} \frac{s}{m^2} (4n_k + 2) \sum_{\ell=1}^{j} \beta_{j\ell} m_\ell(m_\ell - 1) \iint_D X^{m_k+m_\ell-2} Y^{n_k+n_\ell} u^{2s-1} [\bar{A}] dX dY$$

(12.70)

$$PYI5 = \sum_{k=1}^{i} \beta_{ik} \frac{s}{m^2} (4n_k + 2) \sum_{\ell=1}^{j} \beta_{j\ell} s(4m_\ell + 2) \iint_D X^{m_k+m_\ell} Y^{n_k+n_\ell} u^{2s-2} [\bar{A}] dX dY$$

(12.71)

$$PYI6 = -\sum_{k=1}^{i} \beta_{ik} \frac{s}{m^2}(4n_k+2) \sum_{\ell=1}^{j} \beta_{j\ell} 4s(s-1) \iint_D X^{m_k+m_\ell+2} Y^{n_k+n_\ell} u^{2s-3}[\bar{A}]dXdY$$

(12.72)

$$PYI7 = \sum_{k=1}^{i} \beta_{ik} \frac{4s}{m^4}(s-1) \sum_{\ell=1}^{j} \beta_{j\ell} m_\ell(m_\ell-1) \iint_D X^{m_k+m_\ell-2} Y^{n_k+n_\ell+2} u^{2s-2}[\bar{A}]dXdY$$

(12.73)

$$PYI8 = -\sum_{k=1}^{i} \beta_{ik} \frac{4s}{m^4}(s-1) \sum_{\ell=1}^{j} \beta_{j\ell} s(4m_\ell+2) \iint_D X^{m_k+m_\ell} Y^{n_k+n_\ell+2} u^{2s-3}[\bar{A}]dXdY$$

(12.74)

$$PYI9 = \sum_{k=1}^{i} \beta_{ik} \frac{4s}{m^4}(s-1) \sum_{\ell=1}^{j} \beta_{j\ell} 4s(s-1) \iint_D X^{m_k+m_\ell+2} Y^{n_k+n_\ell+2} u^{2s-4}[\bar{A}]dXdY$$

(12.75)

Finally, the last term of Equation 12.24 would be

$$\iint_D (1+\alpha X+\beta X^2)^3 \phi_i^{XY} \phi_j^{XY} dXdY = PXY1 + PXY2 + PXY3 + \cdots + PXY16$$

(12.76)

where

$$PXY1 = \sum_{k=1}^{i} \beta_{ik} m_k n_k \sum_{\ell=1}^{j} \beta_{j\ell} m_\ell n_\ell \iint_D X^{m_k+m_\ell-2} Y^{n_k+n_\ell-2} u^{2s}[\bar{A}]dXdY \quad (12.77)$$

$$PXY2 = -\sum_{k=1}^{i} \beta_{ik} m_k n_k \sum_{\ell=1}^{j} \beta_{j\ell} \frac{2sm_\ell}{m^2} \iint_D X^{m_k+m_\ell-2} Y^{n_k+n_\ell} u^{2s-1}[\bar{A}]dXdY \quad (12.78)$$

$$PXY3 = -\sum_{k=1}^{i} \beta_{ik} m_k n_k \sum_{\ell=1}^{j} \beta_{j\ell} 2sn_\ell \iint_D X^{m_k+m_\ell} Y^{n_k+n_\ell-2} u^{2s-2}[\bar{A}]dXdY \quad (12.79)$$

$$PXY4 = \sum_{k=1}^{i} \beta_{ik} m_k n_k \sum_{\ell=1}^{j} \beta_{j\ell} \frac{4s(s-1)}{m^2} \iint_D X^{m_k+m_\ell} Y^{n_k+n_\ell} u^{2s-2}[\bar{A}]dXdY \quad (12.80)$$

$$PXY5 = -\sum_{k=1}^{i} \beta_{ik} \frac{2sm_k}{m^2} \sum_{\ell=1}^{j} \beta_{j\ell} m_\ell n_\ell \iint_D X^{m_k+m_\ell-2} Y^{n_k+n_\ell} u^{2s-1}[\bar{A}]dXdY \quad (12.81)$$

$$PXY6 = \sum_{k=1}^{i} \beta_{ik} \frac{2sm_k}{m^2} \sum_{\ell=1}^{j} \beta_{j\ell} \frac{2sm_\ell}{m^2} \iint_D X^{m_k+m_\ell-2} Y^{n_k+n_\ell+2} u^{2s-2} [\bar{A}] dXdY \quad (12.82)$$

$$PXY7 = \sum_{k=1}^{i} \beta_{ik} \frac{2sm_k}{m^2} \sum_{\ell=1}^{j} \beta_{j\ell} 2sn_\ell \iint_D X^{m_k+m_\ell} Y^{n_k+n_\ell} u^{2s-2} [\bar{A}] dXdY \quad (12.83)$$

$$PXY8 = -\sum_{k=1}^{i} \beta_{ik} \frac{2sm_k}{m^2} (s-1) \sum_{\ell=1}^{j} \beta_{j\ell} \frac{4s(s-1)}{m^2} \iint_D X^{m_k+m_\ell} Y^{n_k+n_\ell+2} u^{2s-3} [\bar{A}] dXdY$$

$$(12.84)$$

$$PXY9 = \sum_{k=1}^{i} \beta_{ik} 2sn_k \sum_{\ell=1}^{j} \beta_{j\ell} m_\ell n_\ell \iint_D X^{m_k+m_\ell} Y^{n_k+n_\ell-2} u^{2s-1} [\bar{A}] dXdY \quad (12.85)$$

$$PXY10 = \sum_{k=1}^{i} \beta_{ik} 2sn_k \sum_{\ell=1}^{j} \beta_{j\ell} \frac{2sm_\ell}{m^2} \iint_D X^{m_k+m_\ell} Y^{n_k+n_\ell} u^{2s-2} [\bar{A}] dXdY \quad (12.86)$$

$$PXY11 = \sum_{k=1}^{i} \beta_{ik} 2sn_k \sum_{\ell=1}^{j} \beta_{j\ell} 2sn_\ell \iint_D X^{m_k+m_\ell+2} Y^{n_k+n_\ell-2} u^{2s-2} [\bar{A}] dXdY \quad (12.87)$$

$$PXY12 = -\sum_{k=1}^{i} \beta_{ik} 2sn_k \sum_{\ell=1}^{j} \beta_{j\ell} \frac{4s(s-1)}{m^2} \iint_D X^{m_k+m_\ell+2} Y^{n_k+n_\ell} u^{2s-3} [\bar{A}] dXdY$$

$$(12.88)$$

$$PXY13 = \sum_{k=1}^{i} \beta_{ik} \frac{4s(s-1)}{m^2} \sum_{\ell=1}^{j} \beta_{j\ell} m_\ell n_\ell \iint_D X^{m_k+m_\ell} Y^{n_k+n_\ell} u^{2s-2} [\bar{A}] dXdY \quad (12.89)$$

$$PXY14 = -\sum_{k=1}^{i} \beta_{ik} \frac{4s(s-1)}{m^2} \sum_{\ell=1}^{j} \beta_{j\ell} \frac{2sm_\ell}{m^2} \iint_D X^{m_k+m_\ell} Y^{n_k+n_\ell+2} u^{2s-3} [\bar{A}] dXdY$$

$$(12.90)$$

$$PXY15 = -\sum_{k=1}^{i} \beta_{ik} \frac{4s(s-1)}{m^2} \sum_{\ell=1}^{j} \beta_{j\ell} 2sn_\ell \iint_D X^{m_k+m_\ell+2} Y^{n_k+n_\ell} u^{2s-3} [\bar{A}] dXdY$$

$$(12.91)$$

$$PXY16 = \sum_{k=1}^{i} \beta_{ik} \frac{4s(s-1)}{m^2} \sum_{\ell=1}^{j} \beta_{j\ell} \frac{4s(s-1)}{m^2} \iint_D X^{m_k+m_\ell+2} Y^{n_k+n_\ell+2} u^{2s-4} [\bar{A}] dXdY$$

$$(12.92)$$

Similarly, the terms for the thickness variation of Case (ii), mentioned in Equation 12.30, can be written. The double integrals involved in the above-mentioned terms can be accurately found out using the integral formulae given earlier. Accordingly, all the matrix elements of a_{ij} and b_{ij} could be determined and thus the standard eigenvalue problem can be solved for the vibration characteristics.

12.4 Numerical Results for Variable Thickness Plates

The two types of thickness variations mentioned earlier are considered as Case 1 and Case 2, such that

$$\text{Case 1:} \quad H = 1 + \alpha X + \beta X^2 \tag{12.93}$$

$$\text{Case 2:} \quad H = 1 + \gamma v + \delta v^2 \tag{12.94}$$

All the boundary conditions, such as clamped, simply supported, and free are considered as per the value of the parameter s that may have the values of 2, 1, and 0. On considering the parameters of the variable thickness in Equations 12.93 and 12.94, the boundary condition, aspect ratio of the domain, and Poisson's ratio, we will have a total of seven parameters, α, β, γ, δ, s, m, and v. However, results for only few selected parameters are incorporated.

12.4.1 Variable Thickness (Case 1)

In this case, the results were computed with all the mode groups, viz., symmetric–symmetric, symmetric–antisymmetric, antisymmetric–symmetric, and antisymmetric–antisymmetric, and in each mode, the first three frequencies were obtained with $\alpha = 0.1$, $\beta = 0.1$; $\alpha = 0.0$, $\beta = 0.1$; $\alpha = 0.1$, $\beta = 0.0$. The aspect ratio is varied from 0.1 to 1.0 with an interval of 0.1. In all the computations, Poisson's ratio v was taken as 0.3. Tables 12.1 through 12.3 give the results for the clamped boundary with $\alpha = 0.1$, $\beta = 0.1$; $\alpha = 0.0$, $\beta = 0.1$; $\alpha = 0.1$, $\beta = 0.0$, respectively. Table 12.4 shows the result for uniform thickness plate, i.e., when $\alpha = \beta = 0.0$ for the clamped boundary.

The results for frequency parameters of the simply supported boundary with the thickness parameters as mentioned are shown in Tables 12.5 through 12.7. The results for uniform thickness for the simply supported boundary are given in Table 12.8. The frequency parameters of a free boundary plate for all the mode groups with the variable-thickness parameters are given in Tables 12.9 through 12.11. Table 12.12 depicts the result for uniform thickness plate with free boundary.

TABLE 12.1

First Three Frequency Parameters for Symmetric–Symmetric, Symmetric–
Antisymmetric, Antisymmetric–Symmetric, and Antisymmetric–Antisymmetric
Modes for Variable Thickness Plates (Clamped Boundary, $\nu=0.3$), $H=1+\alpha x+\beta x^2$
($\alpha=0.1$, $\beta=0.1$)

	Symmetric–Symmetric			Symmetric–Antisymmetric		
m	First	Second	Third	First	Second	Third
0.1	581.04	719.35	1083.5	1586.9	1898.8	2719.6
0.2	150.34	203.08	312.12	404.88	505.39	728.93
0.3	69.662	107.49	175.01	184.56	247.09	363.30
0.4	41.089	73.540	129.92	106.75	155.74	237.32
0.5	27.784	57.850	110.40	70.429	112.72	179.97
0.6	20.584	49.540	94.643	50.573	88.981	148.98
0.7	16.311	44.680	71.378	38.562	74.500	114.29
0.8	13.613	41.528	56.362	30.771	64.958	89.562
0.9	11.830	39.049	46.444	25.454	58.127	72.887
1.0	10.608	35.845	40.822	21.683	52.511	61.695

	Antisymmetric–Symmetric			Antisymmetric–Antisymmetric		
m	First	Second	Third	First	Second	Third
0.1	628.73	808.41	1256.0	1688.0	2100.0	3097.6
0.2	173.09	236.97	372.80	445.86	568.08	841.47
0.3	86.407	132.85	217.70	212.60	285.68	428.08
0.4	55.174	96.432	167.59	129.07	186.53	286.23
0.5	40.561	79.862	145.98	89.561	140.03	221.81
0.6	32.726	71.112	117.69	67.763	114.37	181.51
0.7	28.151	65.824	92.571	54.518	98.543	140.69
0.8	25.304	61.937	76.712	45.920	87.736	114.49
0.9	23.438	57.796	67.317	40.058	79.329	97.412
1.0	22.156	52.515	62.925	35.903	71.754	86.692

TABLE 12.2

First Three Frequency Parameters for Symmetric–Symmetric, Symmetric–
Antisymmetric, Antisymmetric–Symmetric, and Antisymmetric–Antisymmetric
Modes for Variable Thickness Plates (Clamped Boundary, $\nu=0.3$), $H=1+\alpha x+\beta x^2$
($\alpha=0.0$, $\beta=0.1$)

	Symmetric–Symmetric			Symmetric–Antisymmetric		
m	First	Second	Third	First	Second	Third
0.1	580.83	717.44	1076.8	1586.4	1894.4	2704.4
0.2	150.26	202.53	310.21	404.73	504.20	724.85
0.3	69.604	107.17	173.98	184.47	246.48	361.30
0.4	41.040	73.289	129.19	106.68	155.33	236.05

(*continued*)

TABLE 12.2 (continued)

First Three Frequency Parameters for Symmetric–Symmetric, Symmetric–Antisymmetric, Antisymmetric–Symmetric, and Antisymmetric–Antisymmetric Modes for Variable Thickness Plates (Clamped Boundary, $\nu = 0.3$), $H = 1 + \alpha x + \beta x^2$ ($\alpha = 0.0$, $\beta = 0.1$)

	Symmetric–Symmetric			Symmetric–Antisymmetric		
m	First	Second	Third	First	Second	Third
0.5	27.741	57.634	109.79	70.365	112.40	179.04
0.6	20.544	49.345	94.568	50.517	88.706	148.24
0.7	16.273	44.497	71.311	38.511	74.256	114.20
0.8	13.575	41.355	56.298	30.724	64.736	89.484
0.9	11.792	38.894	46.372	25.409	57.927	72.808
1.0	10.571	35.745	40.702	21.640	52.346	61.596

	Antisymmetric–Symmetric			Antisymmetric–Antisymmetric		
m	First	Second	Third	First	Second	Third
0.1	628.05	805.40	1247.1	1686.4	2092.9	3077.6
0.2	172.85	236.11	370.21	445.39	566.20	836.06
0.3	86.237	132.35	216.25	212.31	284.74	425.38
0.4	55.036	96.049	166.53	128.86	185.90	284.49
0.5	40.441	79.531	145.10	89.382	139.54	220.52
0.6	32.617	70.810	117.48	67.607	113.95	181.23
0.7	28.048	65.542	92.382	54.379	98.176	140.45
0.8	25.206	61.676	76.530	45.792	87.406	114.27
0.9	23.342	57.580	67.112	39.939	79.040	97.190
1.0	22.062	52.349	62.685	35.790	71.515	86.450

TABLE 12.3

First Three Frequency Parameters for Symmetric–Symmetric, Symmetric–Antisymmetric, Antisymmetric–Symmetric, and Antisymmetric–Antisymmetric Modes for Variable Thickness Plates (Clamped Boundary, $\nu = 0.3$), $H = 1 + \alpha x + \beta x^2$ ($\alpha = 0.1$, $\beta = 0.0$)

	Symmetric–Symmetric			Symmetric–Antisymmetric		
m	First	Second	Third	First	Second	Third
0.1	579.56	705.41	1033.0	1583.5	1866.3	2605.8
0.2	149.74	199.11	297.69	403.78	496.84	698.50
0.3	69.208	105.14	167.22	183.84	242.80	348.40
0.4	40.697	71.679	124.36	106.17	152.81	227.89
0.5	27.422	56.215	105.79	69.925	110.37	173.05
0.6	20.237	48.030	94.059	50.119	86.954	143.44
0.7	15.968	43.248	70.846	38.141	72.676	113.64

TABLE 12.3 (continued)

First Three Frequency Parameters for Symmetric–Symmetric, Symmetric–Antisymmetric, Antisymmetric–Symmetric, and Antisymmetric–Antisymmetric Modes for Variable Thickness Plates (Clamped Boundary, $\nu = 0.3$), $H = 1 + \alpha x + \beta x^2$ ($\alpha = 0.1, \beta = 0.0$)

	Symmetric–Symmetric			Symmetric–Antisymmetric		
m	First	Second	Third	First	Second	Third
0.8	13.268	40.160	55.850	30.371	63.282	88.958
0.9	11.481	37.800	45.872	25.068	56.597	72.271
1.0	10.255	34.997	39.890	21.305	51.217	60.941

	Antisymmetric–Symmetric			Antisymmetric–Antisymmetric		
m	First	Second	Third	First	Second	Third
0.1	623.91	786.32	1189.3	1676.7	2048.6	2948.2
0.2	171.35	230.69	353.19	442.54	554.47	800.99
0.3	85.158	129.22	206.76	210.57	278.96	407.87
0.4	54.128	93.608	159.58	127.50	182.05	273.22
0.5	39.625	77.391	139.28	88.237	136.50	212.18
0.6	31.853	68.831	116.13	66.594	111.34	178.80
0.7	27.314	63.672	91.170	53.454	95.833	138.95
0.8	24.488	59.934	75.355	44.928	85.281	112.87
0.9	22.634	56.115	65.774	39.117	77.157	95.776
1.0	21.361	51.217	61.097	34.998	69.938	84.895

TABLE 12.4

First Three Frequency Parameters for Symmetric–Symmetric, Symmetric–Antisymmetric, Antisymmetric–Symmetric, and Antisymmetric–Antisymmetric Modes for Variable Thickness Plates (Clamped Boundary, $\nu = 0.3$), $H = 1 + \alpha x + \beta x^2$ ($\alpha = 0.0, \beta = 0.0$) – Uniform Thickness

	Symmetric–Symmetric			Symmetric–Antisymmetric		
m	First	Second	Third	First	Second	Third
0.1	579.36	703.49	1026.1	1583.1	1861.8	2590.1
0.2	149.66	198.55	295.71	403.63	495.64	694.30
0.3	69.147	104.80	166.16	183.73	242.19	346.34
0.4	40.646	71.414	123.61	106.09	152.38	226.58
0.5	27.377	55.985	105.17	69.858	110.03	172.09
0.6	20.195	47.820	93.980	50.060	86.663	142.67
0.7	15.928	43.050	70.776	38.087	72.415	113.55
0.8	13.229	39.972	55.784	30.322	63.045	88.878
0.9	11.442	37.628	45.800	25.021	56.382	72.191
1.0	10.216	34.878	39.773	21.260	51.033	60.844

(continued)

TABLE 12.4 (continued)

First Three Frequency Parameters for Symmetric–Symmetric, Symmetric–Antisymmetric, Antisymmetric–Symmetric, and Antisymmetric–Antisymmetric Modes for Variable Thickness Plates (Clamped Boundary, $\nu = 0.3$), $H = 1 + \alpha x + \beta x^2$ ($\alpha = 0.0$, $\beta = 0.0$) – Uniform Thickness

	Antisymmetric–Symmetric			Antisymmetric–Antisymmetric		
m	First	Second	Third	First	Second	Third
0.1	623.23	783.26	1180.1	1675.1	2041.4	2927.4
0.2	171.10	229.81	350.50	442.06	552.56	795.37
0.3	84.979	128.70	205.26	210.28	278.01	405.08
0.4	53.982	93.202	158.49	127.27	181.40	271.42
0.5	39.497	77.037	138.36	88.048	135.99	210.85
0.6	31.736	68.506	115.91	66.430	110.90	177.84
0.7	27.204	63.367	90.972	53.306	95.441	138.70
0.8	24.383	59.650	75.165	44.792	84.926	112.64
0.9	22.532	55.875	65.562	38.990	76.843	95.546
1.0	21.260	51.033	60.844	34.877	69.675	84.644

TABLE 12.5

First Three Frequency Parameters for Symmetric–Symmetric, Symmetric–Antisymmetric, Antisymmetric–Symmetric, and Antisymmetric–Antisymmetric Modes for Variable Thickness Plates (Simply Supported Boundary, $\nu = 0.3$), $H = 1 + \alpha x + \beta x^2$ ($\alpha = 0.1$, $\beta = 0.1$)

	Symmetric–Symmetric			Symmetric–Antisymmetric		
m	First	Second	Third	First	Second	Third
0.1	264.27	369.98	678.08	1027.7	1324.00	2221.2
0.2	70.317	116.46	222.74	264.04	360.59	612.29
0.3	33.279	66.831	136.68	121.18	180.70	314.04
0.4	19.877	48.327	107.03	70.474	116.37	209.20
0.5	13.497	39.567	93.675	46.671	85.705	160.20
0.6	9.9844	34.836	70.904	33.573	68.536	122.18
0.7	7.8763	31.991	53.477	25.589	57.879	91.563
0.8	6.5357	30.067	42.196	20.365	50.715	71.720
0.9	5.6461	28.460	34.712	16.765	45.479	58.306
1.0	5.0354	26.278	30.360	14.185	41.120	49.235

	Antisymmetric–Symmetric			Antisymmetric–Antisymmetric		
m	First	Second	Third	First	Second	Third
0.1	303.11	437.92	828.75	1119.4	1511.9	2616.6
0.2	90.420	146.88	285.67	302.02	421.52	736.24
0.3	47.942	90.181	183.25	147.08	218.48	387.31
0.4	31.968	69.371	148.44	90.944	146.11	264.31
0.5	24.274	59.671	124.47	64.044	111.66	204.35

TABLE 12.5 (continued)

First Three Frequency Parameters for Symmetric–Symmetric, Symmetric–Antisymmetric, Antisymmetric–Symmetric, and Antisymmetric–Antisymmetric Modes for Variable Thickness Plates (Simply Supported Boundary, $\nu=0.3$), $H=1+\alpha x+\beta x^2$ ($\alpha=0.1$, $\beta=0.1$)

	Antisymmetric–Symmetric			Antisymmetric–Antisymmetric		
m	First	Second	Third	First	Second	Third
0.6	20.028	54.378	92.484	48.994	92.324	150.24
0.7	17.602	51.005	73.073	39.711	80.145	116.72
0.8	16.037	48.311	60.788	33.588	71.614	95.182
0.9	14.992	45.194	53.505	29.344	64.822	81.096
1.0	14.260	41.121	50.046	26.284	58.652	72.145

TABLE 12.6

First Three Frequency Parameters for Symmetric–Symmetric, Symmetric–Antisymmetric, Antisymmetric–Symmetric, and Antisymmetric–Antisymmetric Modes for Variable Thickness Plates (Simply Supported Boundary, $\nu=0.3$), $H=1+\alpha x+\beta x^2$ ($\alpha=0.0$, $\beta=0.1$)

	Symmetric–Symmetric			Symmetric–Antisymmetric		
m	First	Second	Third	First	Second	Third
0.1	264.12	368.64	672.76	1027.3	1319.8	2205.5
0.2	70.245	116.03	221.03	263.90	359.45	607.98
0.3	33.228	66.576	135.69	121.09	180.12	311.86
0.4	19.838	48.141	106.29	70.404	115.99	207.78
0.5	13.466	39.415	93.053	46.613	85.413	159.16
0.6	9.9605	34.703	70.832	33.523	68.296	122.08
0.7	7.8572	31.869	53.415	25.546	57.672	91.481
0.8	6.5201	29.952	42.140	20.327	50.532	71.647
0.9	5.6331	28.354	34.656	16.732	45.316	58.237
1.0	5.0242	26.201	30.283	14.155	40.983	49.158

	Antisymmetric–Symmetric			Antisymmetric–Antisymmetric		
m	First	Second	Third	First	Second	Third
0.1	302.63	435.80	821.63	1118.0	1505.3	2596.1
0.2	90.223	146.20	283.28	301.59	419.72	730.49
0.3	47.809	89.777	181.79	146.82	217.57	384.34
0.4	31.869	69.070	147.32	90.753	145.51	262.35
0.5	24.196	59.417	124.21	63.891	111.21	203.50
0.6	20.017	54.148	92.281	48.865	91.947	149.96
0.7	17.546	50.789	72.901	39.599	79.818	116.48
0.8	15.985	48.110	60.632	33.489	71.326	94.973
0.9	14.944	45.022	53.342	29.255	64.570	80.896
1.0	14.214	40.984	49.865	26.203	58.442	71.938

TABLE 12.7

First Three Frequency Parameters for Symmetric–Symmetric, Symmetric–Antisymmetric, Antisymmetric–Symmetric, and Antisymmetric–Antisymmetric Modes for Variable Thickness Plates (Simply Supported Boundary, $\nu = 0.3$), $H = 1 + \alpha x + \beta x^2$ ($\alpha = 0.1$, $\beta = 0.0$)

m	Symmetric–Symmetric			Symmetric–Antisymmetric		
	First	Second	Third	First	Second	Third
0.1	263.14	360.03	636.93	1024.6	1292.9	2101.4
0.2	69.759	113.36	209.54	262.99	352.23	579.33
0.3	32.866	65.011	129.10	120.49	176.52	297.38
0.4	19.555	47.018	101.44	69.928	113.62	198.43
0.5	13.245	38.514	88.960	46.211	83.616	152.30
0.6	9.7875	33.918	70.345	33.174	66.823	121.44
0.7	7.7203	31.148	52.990	25.237	56.407	90.933
0.8	6.4096	29.271	41.757	20.051	49.413	71.158
0.9	5.5417	27.720	34.282	16.483	44.318	57.774
1.0	4.9468	25.704	29.814	13.930	40.129	48.659

m	Antisymmetric–Symmetric			Antisymmetric–Antisymmetric		
	First	Second	Third	First	Second	Third
0.1	299.72	422.12	773.73	1109.2	1463.1	2460.1
0.2	88.999	141.93	267.17	298.97	408.24	692.27
0.3	46.971	87.300	172.09	145.25	211.84	364.56
0.4	31.252	67.252	139.83	89.568	141.78	249.33
0.5	23.724	57.893	122.58	62.928	108.40	195.07
0.6	19.634	52.771	91.025	48.054	89.640	148.21
0.7	17.217	49.498	71.833	38.899	77.827	115.02
0.8	15.689	46.901	59.665	32.873	69.565	93.682
0.9	14.666	43.968	52.349	28.702	63.028	79.669
1.0	13.948	40.129	48.771	25.699	70.678	57.138

TABLE 12.8

First Three Frequency Parameters for Symmetric–Symmetric, Symmetric–Antisymmetric, Antisymmetric–Symmetric, and Antisymmetric–Antisymmetric Modes for Variable Thickness Plates (Simply Supported Boundary, $\nu = 0.3$), $H = 1 + \alpha x + \beta x^2$ ($\alpha = 0.0$, $\beta = 0.0$) – Uniform Thickness

m	Symmetric–Symmetric			Symmetric–Antisymmetric		
	First	Second	Third	First	Second	Third
0.1	262.98	358.67	631.36	1024.1	1288.6	2085.0
0.2	69.684	112.92	207.75	262.85	351.07	574.82
0.3	32.813	64.746	128.07	120.39	175.93	295.11
0.4	19.514	46.823	100.67	69.855	113.23	196.96
0.5	13.213	38.354	88.314	46.151	83.313	151.21

TABLE 12.8 (continued)

First Three Frequency Parameters for Symmetric–Symmetric, Symmetric–
Antisymmetric, Antisymmetric–Symmetric, and Antisymmetric–Antisymmetric
Modes for Variable Thickness Plates (Simply Supported Boundary, $\nu = 0.3$),
$H = 1 + \alpha x + \beta x^2$ ($\alpha = 0.0$, $\beta = 0.0$) – Uniform Thickness

	Symmetric–Symmetric			Symmetric–Antisymmetric		
m	First	Second	Third	First	Second	Third
0.6	9.7629	33.777	70.271	33.122	66.572	121.34
0.7	7.7007	31.017	52.927	25.192	56.190	90.849
0.8	6.3935	29.148	41.699	20.012	49.220	71.083
0.9	5.5282	27.607	34.225	16.448	44.145	57.703
1.0	4.9351	25.619	29.736	13.898	39.981	48.582

	Antisymmetric–Symmetric			Antisymmetric–Antisymmetric		
m	First	Second	Third	First	Second	Third
0.1	299.24	419.95	766.22	1107.7	1456.4	2438.5
0.2	88.792	141.23	264.65	298.53	406.40	686.21
0.3	46.830	86.882	170.57	144.98	210.91	361.44
0.4	31.146	66.937	138.65	89.367	141.16	247.27
0.5	23.641	57.625	122.27	62.766	107.92	193.57
0.6	19.566	52.527	90.816	47.917	89.247	147.92
0.7	17.157	49.269	71.655	38.781	77.484	114.78
0.8	15.634	46.687	59.503	32.769	69.260	93.467
0.9	14.615	43.783	52.181	28.609	62.759	79.464
1.0	13.898	39.981	48.582	25.613	56.911	70.467

TABLE 12.9

First Three Frequency Parameters for Symmetric–Symmetric, Symmetric–
Antisymmetric, Antisymmetric–Symmetric, and Antisymmetric–Antisymmetric
Modes for Variable Thickness Plates (Free Boundary, $\nu = 0.3$), $H = 1 + \alpha x + \beta x^2$
($\alpha = 0.1$, $\beta = 0.1$)

	Symmetric–Symmetric			Symmetric–Antisymmetric		
m	First	Second	Third	First	Second	Third
0.1	6.7169	35.595	188.22	95.889	224.95	875.83
0.2	6.7189	35.619	156.74	49.449	120.08	414.96
0.3	6.7103	35.417	72.806	34.359	86.978	188.80
0.4	6.6824	34.896	42.677	26.945	70.895	108.88
0.5	6.6283	28.340	33.983	22.512	61.083	71.464
0.6	6.5397	20.380	32.529	19.517	50.873	54.124
0.7	6.4031	15.523	30.478	17.312	38.374	48.527
0.8	6.1955	12.404	27.993	15.570	30.228	43.721
0.9	5.8838	10.382	25.419	14.102	24.687	39.443
1.0	5.4447	9.1143	22.991	12.787	20.828	35.613

(continued)

TABLE 12.9 (continued)

First Three Frequency Parameters for Symmetric–Symmetric, Symmetric–Antisymmetric, Antisymmetric–Symmetric, and Antisymmetric–Antisymmetric Modes for Variable Thickness Plates (Free Boundary, $\nu = 0.3$), $H = 1 + \alpha x + \beta x^2$ ($\alpha = 0.1$, $\beta = 0.1$)

	Antisymmetric–Symmetric			Antisymmetric–Antisymmetric		
m	First	Second	Third	First	Second	Third
0.1	17.591	60.639	306.40	51.039	142.74	310.24
0.2	17.624	60.596	195.37	25.701	75.871	168.34
0.3	17.590	59.955	98.063	17.303	54.598	123.83
0.4	17.456	58.448	62.004	13.116	44.228	101.95
0.5	17.199	44.129	56.174	10.600	37.969	88.160
0.6	16.784	34.091	52.618	8.9117	33.622	71.094
0.7	16.149	28.002	48.236	7.6931	30.268	56.416
0.8	15.233	24.283	43.693	6.7677	27.445	46.862
0.9	14.061	22.089	39.442	6.0384	24.912	40.456
1.0	12.787	20.815	35.613	5.4476	22.570	36.084

TABLE 12.10

First Three Frequency Parameters for Symmetric–Symmetric, Symmetric–Antisymmetric, Antisymmetric–Symmetric, and Antisymmetric–Antisymmetric Modes for Variable Thickness Plates (Free Boundary, $\nu = 0.3$), $H = 1 + \alpha x + \beta x^2$ ($\alpha = 0.0$, $\beta = 0.1$)

	Symmetric–Symmetric			Symmetric–Antisymmetric		
m	First	Second	Third	First	Second	Third
0.1	6.7093	35.416	186.30	95.644	223.51	866.53
0.2	6.7111	35.439	156.57	49.315	119.32	414.60
0.3	6.7021	35.238	72.702	34.259	86.446	188.63
0.4	6.6739	34.719	42.599	26.861	70.472	108.76
0.5	6.6192	28.280	33.810	22.437	60.729	71.373
0.6	6.5299	20.332	32.365	19.449	50.801	53.818
0.7	6.3924	15.484	30.326	17.249	38.314	48.262
0.8	6.1834	12.373	27.857	15.511	30.176	43.490
0.9	5.8699	10.357	25.298	14.048	24.642	39.240
1.0	5.4291	9.0960	22.884	12.737	20.788	35.436

	Antisymmetric–Symmetric			Antisymmetric–Antisymmetric		
m	First	Second	Third	First	Second	Third
0.1	17.542	60.266	303.13	50.985	142.26	307.94
0.2	17.574	60.224	194.89	25.671	75.604	167.13
0.3	17.539	59.589	97.797	17.279	54.396	122.96
0.4	17.404	58.104	61.810	13.095	44.058	101.26
0.5	17.146	43.994	55.838	10.580	37.817	87.602

TABLE 12.10 (continued)

First Three Frequency Parameters for Symmetric–Symmetric, Symmetric–Antisymmetric, Antisymmetric–Symmetric, and Antisymmetric–Antisymmetric Modes for Variable Thickness Plates (Free Boundary, $v = 0.3$), $H = 1 + \alpha x + \beta x^2$ ($\alpha = 0.0$, $\beta = 0.1$)

	Antisymmetric–Symmetric			Antisymmetric–Antisymmetric		
m	First	Second	Third	First	Second	Third
0.6	16.730	33.985	52.314	8.8930	33.484	70.913
0.7	16.094	27.918	47.969	7.6751	30.139	56.267
0.8	15.177	24.214	43.461	6.7503	27.326	46.735
0.9	14.007	22.028	39.240	6.0216	24.803	40.345
1.0	12.737	35.436	20.759	5.4315	22.471	35.983

TABLE 12.11

First Three Frequency Parameters for Symmetric–Symmetric, Symmetric–Antisymmetric, Antisymmetric–Symmetric, and Antisymmetric–Antisymmetric Modes for Variable Thickness Plates (Free Boundary, $v = 0.3$), $H = 1 + \alpha x + \beta x^2$ ($\alpha = 0.1$, $\beta = 0.0$)

	Symmetric–Symmetric			Symmetric–Antisymmetric		
m	First	Second	Third	First	Second	Third
0.1	6.7860	34.455	172.76	94.684	214.58	801.92
0.2	6.7863	34.477	155.40	48.769	114.67	405.24
0.3	6.7742	34.278	71.966	33.833	83.183	187.50
0.4	6.7414	33.768	42.037	26.490	67.907	107.97
0.5	6.6804	27.835	32.875	22.098	58.607	70.751
0.6	6.5819	19.973	31.472	19.135	50.294	52.009
0.7	6.4303	15.201	29.499	16.956	37.886	46.711
0.8	6.1994	12.160	27.114	15.237	29.812	42.155
0.9	5.8534	10.215	24.640	13.789	24.331	38.089
1.0	5.3753	9.0217	22.306	12.492	20.524	34.440

	Antisymmetric–Symmetric			Antisymmetric–Antisymmetric		
m	First	Second	Third	First	Second	Third
0.1	17.419	28.010	58.030	51.254	139.80	293.16
0.2	17.444	57.999	191.87	25.780	74.236	159.35
0.3	17.400	57.399	96.126	17.326	53.354	117.48
0.4	17.253	56.035	60.623	13.107	43.168	96.977
0.5	16.980	43.175	53.841	10.569	37.020	84.134
0.6	16.544	33.371	50.524	8.8646	32.755	69.804
0.7	15.882	27.458	46.415	7.6352	29.466	55.365
0.8	14.940	23.873	42.125	6.7024	26.702	45.986
0.9	13.757	21.767	38.089	5.9686	24.227	39.707
1.0	12.492	20.543	34.440	5.3753	21.943	35.425

TABLE 12.12

First Three Frequency Parameters for Symmetric–Symmetric, Symmetric–Antisymmetric, Antisymmetric–Symmetric, and Antisymmetric–Antisymmetric Modes for Variable Thickness Plates (Free Boundary, $v = 0.3$), $H = 1 + \alpha x + \beta x^2$ ($\alpha = 0.0$, $\beta = 0.0$) – Uniform Thickness

	Symmetric–Symmetric			Symmetric–Antisymmetric		
m	First	Second	Third	First	Second	Third
0.1	6.7781	34.270	170.71	94.427	213.10	791.94
0.2	6.7778	34.292	155.22	48.628	113.89	400.64
0.3	6.7654	34.093	71.859	33.728	82.631	187.33
0.4	6.7322	33.585	41.958	26.401	67.468	107.85
0.5	6.6706	27.773	32.696	22.019	58.238	70.659
0.6	6.5712	19.924	31.301	19.063	50.219	51.690
0.7	6.4185	15.161	29.340	16.889	37.824	46.432
0.8	6.1861	12.129	26.970	15.175	29.759	41.911
0.9	5.8382	10.191	24.513	13.731	24.285	37.875
1.0	5.3583	9.0035	22.193	12.439	20.484	34.253

	Antisymmetric–Symmetric			Antisymmetric–Antisymmetric		
m	First	Second	Third	First	Second	Third
0.1	17.367	57.644	276.60	51.196	139.30	290.76
0.2	17.392	57.614	191.39	25.747	73.957	158.08
0.3	17.346	57.019	95.853	17.301	53.142	116.58
0.4	17.198	55.671	60.429	13.084	42.988	96.261
0.5	16.923	43.035	53.491	10.548	36.858	83.543
0.6	16.485	33.261	50.206	8.8447	32.607	69.617
0.7	15.823	27.371	46.134	7.6161	29.329	55.210
0.8	14.880	23.800	41.880	6.6841	26.575	45.854
0.9	13.700	21.704	37.875	5.9509	24.110	39.592
1.0	12.439	20.484	34.253	5.3583	21.837	35.319

The frequency results in the last row of each of the above-mentioned tables correspond to circular plate, i.e., $m = 1.0$. It can also be noted that as the aspect ratio increased, the frequency decreased in each of the mode groups for all the boundary conditions. As expected, the clamped boundary results are the maximum, while the free boundary results are the minimum for each parameter. The tables also show the result of the linear and the quadratic variation results, when $\alpha = 0.1$, $\beta = 0.0$ and $\alpha = 0.0$, $\beta = 0.1$, respectively. In addition, the results are also given when the thickness vary in such a way that it contains linear as well as the quadratic terms. The uniform thickness plate results given in Tables 12.4, 12.8, and 12.12 agree very well with the earlier results obtained in previous chapters.

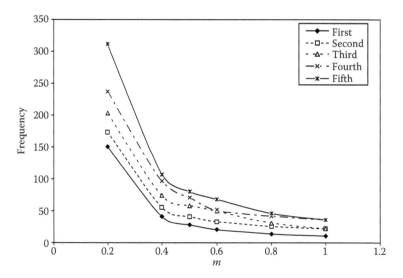

FIGURE 12.2
First five frequencies for $(\alpha, \beta) = (0.1, 0.1)$, clamped boundary (variable thickness: Case 1).

The first five frequencies can be chosen from the mode groups of symmetric–symmetric, symmetric–antisymmetric, antisymmetric–symmetric, and antisymmetric–antisymmetric of each boundary condition. The variations of first five frequencies with $\alpha = 0.1$, $\beta = 0.1$ for clamped, simply supported,

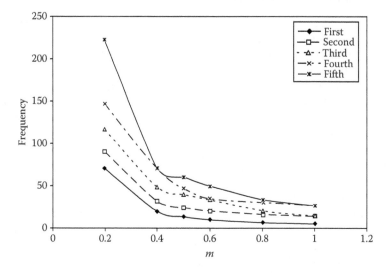

FIGURE 12.3
First five frequencies for $(\alpha, \beta) = (0.1, 0.1)$, simply supported boundary (variable thickness: Case 1).

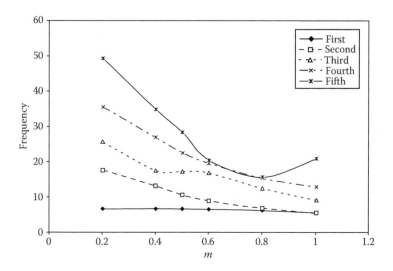

FIGURE 12.4
First five frequencies for $(\alpha, \beta) = (0.1, 0.1)$, free boundary (variable thickness: Case 1).

and free boundary are shown in Figures 12.2 through 12.4, demonstrating that the frequencies decrease as we increase the aspect ratio of the domain for all the boundary conditions.

12.4.2 Variable Thickness (Case 2)

The results obtained were computed with all the mode groups, viz., symmetric–symmetric, symmetric–antisymmetric, antisymmetric–symmetric, and antisymmetric–antisymmetric, and in each case, the first three frequencies were obtained with $\gamma = 0.1$, $\delta = 0.1$; $\gamma = 0.0$, $\delta = 0.1$; $\gamma = 0.1$, $\delta = 0.0$. The aspect ratio varied again from 0.1 to 1.0 with an interval of 0.1. In all the computations, Poisson's ratio ν was taken as 0.3. Tables 12.13 through 12.15 give the results for clamped boundary with $\gamma = 0.1$, $\delta = 0.1$; $\gamma = 0.0$, $\delta = 0.1$; $\gamma = 0.1$, $\delta = 0.0$, respectively.

The results for the frequency parameters of the simply supported boundary with the thickness parameters as mentioned are shown in Tables 12.16 through 12.18. Also, the frequency parameters of a free boundary plate for all the mode groups with the variable-thickness parameters of Case 2 are given in Tables 12.19 through 12.21. Frequency results in the last row of each of the above-mentioned tables correspond to the circular plate, i.e., $m = 1.0$. Table 12.22 shows few frequency parameters for uniform thickness circular plate when zero values of the parameters γ and δ are introduced in the model, illustrating the results of all the three boundary conditions.

TABLE 12.13

First Three Frequency Parameters for Symmetric–Symmetric, Symmetric–Antisymmetric, Antisymmetric–Symmetric, and Antisymmetric–Antisymmetric Modes for Variable Thickness Plates (Clamped Boundary, $\nu = 0.3$), $H = 1 + \gamma v + \delta v^2$ ($\gamma = 0.1, \delta = 0.1$)

	Symmetric–Symmetric			Symmetric–Antisymmetric		
m	First	Second	Third	First	Second	Third
0.2	171.37	228.00	346.30	449.70	559.54	804.14
0.4	46.598	80.881	141.46	118.38	170.96	260.06
0.5	31.409	63.107	119.45	78.047	123.30	196.60
0.6	23.183	53.695	103.78	55.996	97.049	162.43
0.8	15.197	44.679	61.675	34.003	70.574	97.741
1.0	11.738	39.028	43.887	23.904	57.176	66.918

	Antisymmetric–Symmetric			Antisymmetric–Antisymmetric		
m	First	Second	Third	First	Second	Third
0.2	195.65	264.31	411.62	493.88	627.66	925.64
0.4	61.484	105.09	181.90	142.20	204.08	312.98
0.5	44.881	86.383	157.70	98.429	152.64	241.97
0.6	35.969	76.535	128.57	74.291	124.33	198.06
0.8	27.505	66.525	83.193	50.109	95.222	124.30
1.0	23.905	57.176	66.918	39.018	78.275	93.210

It can again be noted that as the aspect ratio is increased, the frequency is decreased in each of the mode groups for all the boundary conditions. It is also worth mentioning that clamped boundary results are the maximum,

TABLE 12.14

First Three Frequency Parameters for Symmetric–Symmetric, Symmetric–Antisymmetric, Antisymmetric–Symmetric, and Antisymmetric–Antisymmetric Modes for Variable Thickness Plates (Clamped Boundary, $\nu = 0.3$), $H = 1 + \gamma v + \delta v^2$ ($\gamma = 0.0, \delta = 0.1$)

	Symmetric–Symmetric			Symmetric–Antisymmetric		
m	First	Second	Third	First	Second	Third
0.2	159.77	212.08	319.66	424.00	523.99	745.28
0.4	43.412	75.677	131.68	111.52	160.52	241.82
0.5	29.251	59.167	111.52	73.480	115.83	183.12
0.6	21.583	50.427	98.207	52.689	91.200	151.49
0.8	14.144	42.043	58.323	31.958	66.329	92.601
1.0	10.923	36.707	41.537	22.440	53.714	63.394

(*continued*)

TABLE 12.14 (continued)

First Three Frequency Parameters for Symmetric–Symmetric, Symmetric–Antisymmetric, Antisymmetric–Symmetric, and Antisymmetric–Antisymmetric Modes for Variable Thickness Plates (Clamped Boundary, $\nu = 0.3$), $H = 1 + \gamma v + \delta v^2$ ($\gamma = 0.0, \delta = 0.1$)

m	Antisymmetric–Symmetric			Antisymmetric–Antisymmetric		
	First	Second	Third	First	Second	Third
0.2	182.40	245.66	379.60	464.88	586.16	856.42
0.4	57.409	98.484	169.12	133.84	191.35	290.51
0.5	41.947	81.141	147.01	92.622	143.25	224.98
0.6	33.653	72.004	121.31	69.896	116.74	187.09
0.8	25.785	62.632	78.576	47.138	89.401	117.52
1.0	22.440	53.714	63.394	36.704	73.419	88.216

while the free boundary results are the minimum for each parameter. The tables also show the result of linear and quadratic variation results, when $\gamma = 0.1, \delta = 0.0$ and $\gamma = 0.0, \delta = 0.1$, respectively. In this case also, the results are presented when the thickness vary in such a way that it contains linear as

TABLE 12.15

First Three Frequency Parameters for Symmetric–Symmetric, Symmetric–Antisymmetric, Antisymmetric–Symmetric, and Antisymmetric–Antisymmetric Modes for Variable Thickness Plates (Clamped Boundary, $\nu = 0.3$), $H = 1 + \gamma v + \delta v^2$ ($\gamma = 0.1, \delta = 0.0$)

m	Symmetric–Symmetric			Symmetric–Antisymmetric		
	First	Second	Third	First	Second	Third
0.2	161.24	214.38	322.12	429.67	531.20	752.87
0.4	43.831	76.633	133.25	113.04	162.85	244.68
0.5	29.535	59.945	112.99	74.489	117.53	185.45
0.6	21.794	51.110	99.617	53.412	92.552	153.53
0.8	14.282	42.632	59.169	32.392	67.332	94.985
1.0	11.030	37.219	42.148	22.740	54.612	64.964

m	Antisymmetric–Symmetric			Antisymmetric–Antisymmetric		
	First	Second	Third	First	Second	Third
0.2	185.63	270.85	541.90	471.34	593.97	864.30
0.4	58.076	99.824	171.09	135.72	194.13	293.72
0.5	42.449	82.302	148.90	93.922	145.39	227.70
0.6	34.068	73.067	123.21	70.874	118.51	190.17
0.8	26.118	63.580	79.811	47.794	90.788	119.45
1.0	22.739	54.523	64.398	37.215	74.563	89.664

TABLE 12.16

First Three Frequency Parameters for Symmetric–Symmetric, Symmetric–Antisymmetric, Antisymmetric–Symmetric, and Antisymmetric–Antisymmetric Modes for Variable Thickness Plates (Simply Supported Boundary, $\nu = 0.3$), $H = 1 + \gamma v + \delta v^2$ ($\gamma = 0.1$, $\delta = 0.1$)

	Symmetric–Symmetric			Symmetric–Antisymmetric		
m	First	Second	Third	First	Second	Third
0.2	74.238	125.95	241.88	284.56	391.41	666.76
0.4	21.104	52.334	116.17	76.019	126.35	228.14
0.5	14.349	42.764	101.53	50.355	93.029	174.84
0.6	10.628	37.564	76.799	36.227	74.368	132.83
0.8	6.9747	32.311	45.639	21.978	55.020	77.874
1.0	5.3861	28.453	32.469	15.313	44.743	53.226

	Antisymmetric–Symmetric			Antisymmetric–Antisymmetric		
m	First	Second	Third	First	Second	Third
0.2	97.073	159.31	310.18	326.49	458.00	800.85
0.4	34.531	75.040	161.00	98.419	158.79	288.01
0.5	26.233	64.357	135.22	69.318	121.30	222.93
0.6	21.693	58.508	100.36	53.027	100.26	163.55
0.8	17.279	51.982	65.616	36.340	77.854	103.23
1.0	15.313	44.743	53.226	28.419	64.088	77.684

well as quadratic terms, i.e., when $\gamma = 0.1$, $\delta = 0.1$. The uniform thickness plate results given in Table 12.22 agree very well with the results obtained in the previous chapters.

TABLE 12.17

First Three Frequency Parameters for Symmetric–Symmetric, Symmetric–Antisymmetric, Antisymmetric–Symmetric, and Antisymmetric–Antisymmetric Modes for Variable Thickness Plates (Simply Supported Boundary, $\nu = 0.3$), $H = 1 + \gamma v + \delta v^2$ ($\gamma = 0.0$, $\delta = 0.1$)

	Symmetric–Symmetric			Symmetric–Antisymmetric		
m	First	Second	Third	First	Second	Third
0.2	71.298	118.34	223.63	270.98	367.82	617.39
0.4	20.123	49.159	107.75	72.223	118.75	211.38
0.5	13.656	40.215	94.307	47.785	87.421	162.10
0.6	10.103	35.366	72.858	34.342	69.879	125.89
0.8	6.6238	30.466	43.265	20.797	51.688	73.781
1.0	5.1141	26.809	30.807	14.470	42.013	50.425

(*continued*)

TABLE 12.17 (continued)

First Three Frequency Parameters for Symmetric–Symmetric, Symmetric–Antisymmetric, Antisymmetric–Symmetric, and Antisymmetric–Antisymmetric Modes for Variable Thickness Plates (Simply Supported Boundary, $\nu = 0.3$), $H = 1 + \gamma v + \delta v^2$ ($\gamma = 0.0$, $\delta = 0.1$)

	Antisymmetric–Symmetric			Antisymmetric–Antisymmetric		
m	First	Second	Third	First	Second	Third
0.2	92.031	149.01	286.16	309.46	428.56	740.08
0.4	32.545	70.401	149.02	93.045	148.75	266.35
0.5	24.716	60.477	127.60	65.464	113.68	207.82
0.6	20.445	55.044	94.711	50.041	93.991	154.30
0.8	16.307	48.918	61.981	34.268	72.965	97.448
1.0	14.470	42.013	50.425	26.794	60.014	73.393

As mentioned earlier, from the mode groups of symmetric–symmetric, symmetric–antisymmetric, antisymmetric–symmetric, and antisymmetric–antisymmetric of each boundary conditions, the first few frequencies can be chosen. The variations of the first five frequencies with $\gamma = 0.1$, $\delta = 0.1$ for

TABLE 12.18

First Three Frequency Parameters for Symmetric–Symmetric, Symmetric–Antisymmetric, Antisymmetric–Symmetric, and Antisymmetric–Antisymmetric Modes for Variable Thickness Plates (Simply Supported Boundary, $\nu = 0.3$), $H = 1 + \gamma v + \delta v^2$ ($\gamma = 0.1$, $\delta = 0.0$)

	Symmetric–Symmetric			Symmetric–Antisymmetric		
m	First	Second	Third	First	Second	Third
0.2	72.661	120.46	225.86	276.72	374.64	624.00
0.4	20.497	49.983	108.98	73.722	120.81	213.61
0.5	13.906	40.896	95.440	48.764	88.909	163.85
0.6	10.286	35.972	74.214	35.036	71.054	128.24
0.8	6.7431	30.995	44.073	21.207	52.550	75.159
1.0	5.2061	27.264	31.398	14.748	42.710	51.372

	Antisymmetric–Symmetric			Antisymmetric–Antisymmetric		
m	First	Second	Third	First	Second	Third
0.2	93.839	151.43	288.50	315.75	435.76	746.80
0.4	33.133	71.548	150.50	94.790	151.14	268.80
0.5	25.161	61.489	129.95	66.650	115.50	210.04
0.6	20.817	55.982	96.452	50.923	95.500	157.11
0.8	16.612	49.754	63.120	34.852	74.144	99.215
1.0	14.748	42.710	51.372	27.247	60.982	74.727

TABLE 12.19

First Three Frequency Parameters for Symmetric–Symmetric, Symmetric–Antisymmetric, Antisymmetric–Symmetric, and Antisymmetric–Antisymmetric Modes for Variable Thickness Plates (Free Boundary, $\nu = 0.3$), $H = 1 + \gamma v + \delta v^2$ ($\gamma = 0.1$, $\delta = 0.1$)

	Symmetric–Symmetric			Symmetric–Antisymmetric		
m	First	Second	Third	First	Second	Third
0.2	7.3114	38.797	159.82	53.235	129.76	438.32
0.4	7.2657	38.037	43.899	29.130	77.287	115.25
0.5	7.1987	29.227	37.062	24.376	66.834	75.708
0.6	7.0903	21.055	35.533	21.160	53.939	59.389
0.8	6.6743	12.874	30.721	16.910	32.065	48.242
1.0	5.7831	9.5640	25.306	13.884	22.130	39.446

	Antisymmetric–Symmetric			Antisymmetric–Antisymmetric		
m	First	Second	Third	First	Second	Third
0.2	19.216	65.949	204.17	27.554	81.899	182.05
0.4	19.015	63.719	65.720	14.052	48.102	111.29
0.5	18.721	46.929	61.347	11.346	41.402	96.646
0.6	18.258	36.300	57.665	9.5265	36.739	76.359
0.8	16.564	25.847	48.206	7.2102	30.080	50.311
1.0	13.884	22.130	39.446	5.7825	24.763	38.731

clamped, simply supported, and free boundary are shown in Figures 12.5 through 12.7. These figures again show that the frequencies decrease as we increase the aspect ratio of the elliptic domain for all the boundary conditions except the free boundary condition as is clear from Figure 12.7.

TABLE 12.20

First Three Frequency Parameters for Symmetric–Symmetric, Symmetric–Antisymmetric, Antisymmetric–Symmetric, and Antisymmetric–Antisymmetric Modes for Variable Thickness Plates (Free Boundary, $\nu = 0.3$), $H = 1 + \gamma v + \delta v^2$ ($\gamma = 0.0$, $\delta = 0.1$)

	Symmetric–Symmetric			Symmetric–Antisymmetric		
m	First	Second	Third	First	Second	Third
0.2	6.9670	36.241	155.66	50.382	120.92	419.89
0.4	6.9224	35.520	42.425	27.484	71.886	110.20
0.5	6.8590	28.170	34.599	22.969	62.123	72.311
0.6	6.7565	20.254	33.155	19.918	51.468	55.177
0.8	6.3604	12.359	28.629	15.892	30.553	44.791
1.0	5.5103	9.1785	23.573	13.040	21.062	36.618

(continued)

TABLE 12.20 (continued)

First Three Frequency Parameters for Symmetric–Symmetric, Symmetric–Antisymmetric, Antisymmetric–Symmetric, and Antisymmetric–Antisymmetric Modes for Variable Thickness Plates (Free Boundary, $\nu = 0.3$), $H = 1 + \gamma v + \delta v^2$ ($\gamma = 0.0$, $\delta = 0.1$)

	Antisymmetric–Symmetric			Antisymmetric–Antisymmetric		
m	First	Second	Third	First	Second	Third
0.2	18.117	61.344	195.31	26.353	77.174	169.01
0.4	17.925	59.297	62.345	13.418	45.145	103.18
0.5	17.645	44.485	57.031	10.826	38.799	89.594
0.6	17.202	34.400	53.582	9.0854	34.387	72.223
0.8	15.574	24.545	44.758	6.8719	28.102	47.587
1.0	13.040	21.062	36.618	5.5102	23.116	36.645

In conclusion, it can be mentioned that the procedure of analysis discussed in this chapter can be carried out practically for any thickness variation, provided the integrals involved are evaluated accurately. For polynomial variation of thickness, the integrals can be obtained accurately using the

TABLE 12.21

First Three Frequency Parameters for Symmetric–Symmetric, Symmetric–Antisymmetric, Antisymmetric–Symmetric, and Antisymmetric–Antisymmetric Modes for Variable Thickness Plates (Free Boundary, $\nu = 0.3$), $H = 1 + \gamma v + \delta v^2$ ($\gamma = 0.1$, $\delta = 0.0$)

	Symmetric–Symmetric			Symmetric–Antisymmetric		
m	First	Second	Third	First	Second	Third
0.2	7.1085	36.807	159.18	51.418	122.62	429.86
0.4	7.0623	36.065	43.368	28.018	72.818	112.84
0.5	6.9974	28.788	35.125	23.403	62.908	74.021
0.6	6.8926	20.694	33.649	20.286	52.672	55.862
0.8	6.4881	12.625	29.040	16.176	31.256	45.336
1.0	5.6205	9.3761	23.908	13.270	21.540	37.061

	Antisymmetric–Symmetric			Antisymmetric–Antisymmetric		
m	First	Second	Third	First	Second	Third
0.2	18.470	62.159	200.00	26.895	78.600	171.00
0.4	18.270	60.076	63.702	13.691	45.910	104.31
0.5	17.983	45.434	57.763	11.046	39.435	90.566
0.6	17.527	35.130	54.253	9.2684	34.937	73.715
0.8	15.855	25.084	45.303	7.0094	28.536	48.550
1.0	13.270	21.540	37.061	5.6203	23.469	37.385

TABLE 12.22

First Three Frequency Parameters for Symmetric–Symmetric (S–S), Symmetric–Antisymmetric (S–A), Antisymmetric–Symmetric (A–S), and Antisymmetric–Antisymmetric (A–A) Modes for Constant Thickness Plates (Clamped, Simply Supported, and Free Boundary, $v=0.3$), $H=1+\gamma v+\delta v^2$ ($\gamma=0.0$, $\delta=0.0$) for a Circular Plate ($m=1.0$)

B.C.	Mode	First	Second	Third
Clamped	S–S	10.216	34.877	39.773
	S–A	21.260	51.033	60.844
	A–S	21.260	51.033	60.844
	A–A	34.877	69.675	84.644
Simply supported	S–S	4.9351	25.618	29.736
	S–A	13.898	39.981	48.582
	A–S	13.898	39.981	48.582
	A–A	25.613	56.910	70.466
Free	S–S	5.3583	9.0034	22.193
	S–A	12.439	20.483	34.252
	A–S	12.439	20.483	34.252
	A–A	5.3583	21.837	35.319

formulae given; however, for other variations one can use other efficient numerical methods to compute the integrals involved. This method can be easily utilized for other domain plates by following the same procedure as explained.

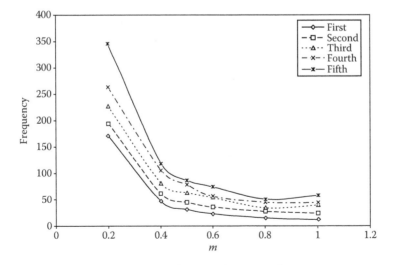

FIGURE 12.5
First five frequencies for $(\gamma, \delta) = (0.1, 0.1)$, clamped boundary (variable thickness: Case 2).

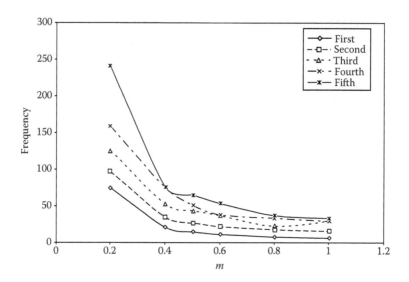

FIGURE 12.6
First five frequencies for $(\gamma, \delta) = (0.1, 0.1)$, simply supported boundary (variable thickness: Case 2).

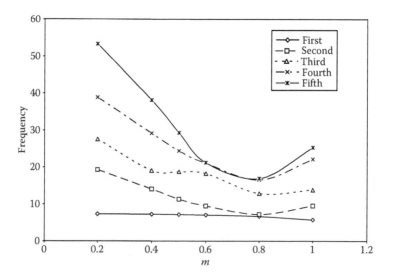

FIGURE 12.7
First five frequencies for $(\gamma, \delta) = (0.1, 0.1)$, free boundary (variable thickness: Case 2).

Bibliography

Bhat, R.B. 1985. Natural frequencies of rectangular plates using characteristic orthogonal polynomials in Rayleigh-Ritz method. *Journal of Sound and Vibration*, 102: 493–499.

Bhat, R.B., Laura, P.A.A., Gutierrez, R.C., Cortinez, V.H., and Sanzi, H.C. 1990. Numerical experiments on the determination of natural frequencies of transverse vibrations of rectangular plates of non-uniform thickness. *Journal of Sound and Vibration*, 138: 205–219.

Leissa, A.W. 1969. *Vibration of Plates*, NASA SP-160, U.S. Government Printing Office, Washington DC.

Mazumdar, J. 1971. Transverse vibrations of elastic plates by the method of constant deflection lines. *Journal of Sound and Vibration*, 18: 147.

McNitt, R.P. 1962. Free vibrations of a clamped elliptic plate. *Journal of Aerospace Science*, 29(9): 1124–1125.

Prasad, C., Jain, R.K., and Soni, S.R. 1972. Axisymmetric vibrations of circular plates of linearly varying thickness. *Zeitschrift fur Angewandte Mathematik und Physik*, 23: 941.

Singh, B. and Chakraverty, S. 1991a. Transverse vibration of circular and elliptic plates with variable thickness. *Indian Journal of Pure and Applied Mathematics*, 22: 787–803.

Singh, B. and Chakraverty, S. 1991b. Transverse vibration of completely free elliptic and circular plates using orthogonal polynomials in Rayleigh-Ritz method. *International Journal of Mechanical Sciences*, 33: 741–751.

Singh, B. and Chakraverty, S. 1992.Transverse vibration of circular and elliptic plates with quadratically varying thickness. *Applied Mathematical Modeling*, 16: 269–274.

Singh, B. and Chakraverty, S. 1992a. Transverse vibration of simply-supported elliptic and circular plates using boundary characteristic orthogonal polynomials in two dimensions. *Journal of Sound and Vibration*, 152: 149–155.

Singh, B. and Chakraverty, S. 1992b. On the use of orthogonal polynomials in Rayleigh-Ritz method for the study of transverse vibration of elliptic plates. *International Journal of Computers and Structures*, 43: 439–443.

Singh, B. and Chakraverty, S. 1993. Transverse vibration of annular circular and elliptic plates using characteristic orthogonal polynomials in two dimensions. *Journal of Sound and Vibration*, 162: 537–546.

Singh, B. and Chakraverty, S. 1994. Use of characteristic orthogonal polynomials in two dimensions for transverse vibration of elliptic and circular plates with variable thickness. *Journal of Sound and Vibration*, 173: 289–299.

Singh, B. and Goel, R. 1985b. Transverse vibrations of an elliptic plate with variable thickness. *Proceedings of the Workshop on Solid Mechanics*, University of Roorkee, March 13–16, pp. 19–27.

Singh, B. and Tyagi, D.K. 1985a. Transverse vibration of an elliptic plate with variable thickness. *Journal of Sound and Vibration*, 99: 379–391.

13

Plates with Orthotropic Material Properties

13.1 Introduction

In recent years, lightweight structures have been widely used in many engineering fields, and hence vibration analysis of differently shaped plates has been studied extensively owing to its practical applications. The applications of composite materials in engineering structures require information about the vibration characteristics of anisotropic materials. The free vibration of orthotropic plates is an important area of such behavior. Orthotropic materials have extensive application in the modern technology, such as in modern missiles, space crafts, nuclear reactors, and printed circuit boards. Their high strength along with small specific mass makes the composite materials ideal for applications in space crafts, vehicle systems, nuclear reactors, etc. Most of the applications subject the composite materials to dynamic loading. It is known that the orthotropic materials exhibit a different dynamic response when compared with that of similar isotropic structures. A vast amount of work has been done for theoretical and experimental results for vibration of orthotropic skew, triangular, circular, annular, and polygonal plates as mentioned by Leissa (1969, 1978, 1981, 1987) and Bert (1976, 1979, 1980, 1982, 1985, 1991). The survey of literature reveals that elliptic orthotropic geometry studies are very few. That is why, in this chapter, an example of the elliptic plates is given. Many studies for other shapes with orthotropy are already available in the literature. The investigation presented in this chapter gives extensive and a wide variety of results to study the free vibration of specially rectilinear orthotropic (i.e., whose symmetrical axes coincide with the principal elastic axes of the plate material): (1) elliptic and circular plates and (2) annular elliptic plates.

For a circular plate with rectangular orthotropy, only a few results are available in the existing literature, namely Rajappa (1963), Leissa (1969), Sakata (1976), Narita (1983), Dong and Lopez (1985) and also some of the references mentioned therein. Rajappa (1963) had used Galerkin's method and reported only the fundamental frequency of circular orthotropic plates with clamped and simply supported boundaries. Reduction methods have been used by Sakata (1976), who had given only the fundamental frequency for a clamped orthotropic circular plate. Narita (1983) gave some higher modes for a circular plate with clamped boundary by a series type method. Dong and Lopez (1985) had analyzed a clamped circular plate with rectilinear orthotropy by the modified application of the interior collocation method.

Compared with circular plates, even less work has been carried out for elliptical plates with rectangular orthotropy. The main difficulty in studying the elliptic plates is the choice of coordinates. Elliptic coordinates may be used with the exact mode shape in the form of Mathieu functions. In this respect, Sakata (1976) dealt with only the fundamental frequency for a clamped elliptic plate. He obtained his results by using a simple coordinate transformation of a clamped orthotropic circular plate to give a reduction formula for the fundamental frequency of an elliptic orthotropic plate with the same boundary. The Ritz method analysis was carried out by Narita (1985) to obtain the first few natural frequencies for an orthotropic elliptical plate with a free boundary by taking a complete power series as a trial function. Numerical results are illustrated there by two figures only, for two types of orthotropic material properties.

Sakata (1979a,b), in a two-part article, describes three exact reduction methods in Part I and a generalized reduction method in Part II. The reduction method is used to derive an approximate formula for estimating the natural frequencies of orthotropic plates. Vibration of an orthotropic elliptical plate with a similar hole was analyzed by Irie and Yamada (1979). In another paper, Irie et al. (1983) had dealt with the free vibration of circular-segment-shaped membranes and clamped plates of rectangular orthotropy. An interesting paper is that of Narita (1986), who had analyzed the free vibration of orthotropic elliptical plates with point supports of arbitrary location. Only those papers that deal with rectangular orthotropic circular or elliptical geometries are mentioned in this chapter. Recently Chakraverty (1996) and Chakraverty and Petyt (1999) had used two-dimensional boundary characteristic orthogonal polynomials (BCOPs) in the Rayleigh–Ritz method to study the free vibration of full elliptic and circular plates having rectangular orthotropy. They have reported a variety of results for seven types of orthotropic material properties with different boundary conditions, viz., clamped, simply supported, and free at the edges.

Owing to the difficulty in studying elliptic plates, annular elliptic plates have not been studied in detail. Recently, Singh and Chakraverty (1993) and Chakraverty et al. (2001) studied isotropic annular elliptic plates. Singh and Chakraverty (1993) supplied only the fundamental frequencies for various boundary conditions at the inner and outer edges for different shapes of elliptic plates with various hole sizes. Chakraverty et al. (2001) gave 12 higher modes of vibration for all the boundary conditions. To the best of the author's knowledge, the problem of transverse vibration of annular elliptic plate of rectangular orthotropy has not been reported so far except in only one paper by Chakraverty et al. (2000).

13.2 Domain Definitions

13.2.1 Elliptic Orthotropic Plates

The domain occupied by the elliptic plate may be written as

$$S = \{(x,y),\ x^2/a^2 + y^2/b^2 \leq 1,\quad x,y \in R\} \tag{13.1}$$

where a and b are the semi-major and semi-minor axes of the ellipse, respectively.

13.2.2 Annular Elliptic Orthotropic Plates

Let the outer boundary of the elliptic plate be defined as

$$R = \left\{ (x,y),\ \frac{x^2}{a^2} + \frac{y^2}{b^2} \leq 1,\quad x,y \in \mathbf{R} \right\} \tag{13.2}$$

where a and b are the semi-major and semi-minor axes, respectively. A family of concentric ellipses is defined by introducing a variable C,

$$x^2 + \frac{y^2}{m^2} = 1 - C,\quad 0 \leq C \leq C_0 \tag{13.3}$$

where $m = b/a$ and C_0 define the inner boundary of the ellipse. The eccentricity of the inner boundary is defined by k, where

$$k = \sqrt{1 - C_0}$$

13.3 Basic Equations and Method of Solutions

The maximum strain energy, $V_{\max}$, of the deformed orthotropic plate is given (Timoshenko and Woinowsky (1953)) as

$$V_{\max} = (1/2) \iint_R \left[D_x W_{xx}^2 + 2\nu_x D_y W_{xx} W_{yy} + D_y W_{yy}^2 + 4 D_{xy} W_{xy}^2 \right] dy\,dx \tag{13.4}$$

where $W(x,y)$ is the deflection of the plate, W_{xx} is the second derivative of W with respect to x. The D coefficients are bending rigidities defined by

$$
\left.\begin{array}{l}
D_x = E_x h^3/(12(1 - \nu_x \nu_y)) \\
D_y = E_y h^3/(12(1 - \nu_y \nu_x)) \\
D_y \nu_x = D_x \nu_y \\
\text{and} \\
D_{xy} = G_{xy} h^3/12
\end{array}\right\}
\tag{13.5}
$$

where
  E_x, E_y are Young's moduli
  ν_x, ν_y are Poisson's ratios in the x and y directions
  G_{xy} is the shear modulus
  h is the uniform thickness

The maximum kinetic energy is given by

$$
T_{max} = (1/2)\rho h \omega^2 \iint\limits_{R} W^2 dy dx
\tag{13.6}
$$

where
  ρ is the mass density per unit volume
  ω is the radian natural frequency of the plate

Equating the maximum strain and kinetic energies, we have the Rayleigh quotient as

$$
\omega^2 = \frac{\iint\limits_{R} \left[D_x W_{xx}^2 + 2\nu_x D_y W_{xx} W_{yy} + D_y W_{yy}^2 + 4 D_{xy} W_{xy}^2 \right] dy dx}{h\rho \iint\limits_{R} W^2 dy dx}
\tag{13.7}
$$

Substituting the N-term approximation

$$
W(x,y) = \sum_{j=1}^{N} c_j \phi_j(x,y)
\tag{13.8}
$$

and minimizing ω^2 as a function of the coefficients c_j's, we obtain

$$
\sum_{j=1}^{N} (a_{ij} - \lambda^2 b_{ij}) c_j = 0, \quad i = 1, 2, \ldots, N
\tag{13.9}
$$

where

$$a_{ij} = \iint\limits_{R'} \left[(D_x/H)\phi_i^{XX}\phi_j^{XX} + (D_y/H)\phi_i^{YY}\phi_j^{YY} + \nu_x(D_y/H)\left(\phi_i^{XX}\phi_j^{YY} + \phi_i^{YY}\phi_j^{XX}\right) \right.$$

$$\left. + 2(1 - \nu_x(D_y/H))\phi_i^{XY}\phi_j^{XY} \right] dY dX \tag{13.10}$$

$$b_{ij} = \iint\limits_{R'} \phi_i \phi_j dY dX \tag{13.11}$$

$$\lambda^2 = \frac{a^4 \rho h \omega^2}{H} \tag{13.12}$$

and

$$H = D_y \nu_x + 2D_{xy} \tag{13.13}$$

The ϕ_i's are orthogonal polynomials and are described in the next section. ϕ_i^{XX} is the second derivative of ϕ_i with respect to X and the new domain R' is defined by

$$R' = \{(X,Y), X^2 + Y^2/m^2 \leq 1, \quad X,Y \in R\}$$

where
$X = x/a$
$Y = y/a$
$m = b/a$

If the ϕ_i's are orthogonal, Equation 13.9 reduces to

$$\sum_{j=1}^{N} (a_{ij} - \lambda^2 \delta_{ij})c_j = 0, \quad i = 1, 2, \ldots, N \tag{13.14}$$

where

$$\delta_{ij} = 0, \quad \text{if} \quad i \neq j$$

$$\delta_{ij} = 1, \quad \text{if} \quad i = j$$

The three parameters D_x/H, D_y/H, and ν_x define the orthotropic property of the material under consideration. It is interesting to note here that for an isotropic plate, these parameters reduce to $D_x/H = D_y/H = 1$ and $\nu_x = \nu_y = \nu$. Equation 13.14 is a standard eigenvalue problem and can be solved for the vibration characteristics.

13.4   Generation of BCOPs

13.4.1   Orthogonal Polynomials for Elliptic Orthotropic Plates

The present polynomials have been generated in exactly the same way as described in the earlier chapters (Chakraverty (1992), Singh and Chakraverty (1991, 1992a,b)), and the method is described below for the sake of completeness.

We start with a linearly independent set

$$F_i(x,y) = f(x,y)\{f_i(x,y)\}, \quad i = 1, 2, \ldots, N \tag{13.15}$$

where $f(x,y)$ satisfies the essential boundary conditions and $f_i(x,y)$ are taken as the combinations of terms of the form $x^{\ell_i} y^{n_i}$, where ℓ_i and n_i are non-negative integers. The function f is defined by

$$f(x,y) = (1 - x^2 - y^2/m^2)^p \tag{13.16}$$

If we take the right-hand side of Equation 13.16 as u^p, where $u = 1 - x^2 - y^2/m^2$, then it is clear that the boundary of the ellipse ∂S is given by $u = 0$ and at the center $u = 1$. The curves $u = $ constant will be concentric ellipses. From Equation 13.16, it is to be noted that

1. If $p = 0, f = 1$ on ∂S
2. If $p = 1, f = 0$ and $\partial f/\partial n = 0$ on ∂S
3. If $p = 2, f = 0$ on ∂S

Hence, the functions ff_i also satisfy the same conditions on ∂S. When $p = 0$, $f = 1$ on ∂S and so ff_i are free since their values on ∂S depend on ℓ_i and n_i. Therefore, it is clear that p takes the value of 0, 1, or 2 depending on whether the boundary of the elliptic (or circular) plate is free, simply supported, or clamped.

From $F_i(x,y)$, we generate the orthogonal set. For this, we define the inner product of two functions f and g (as done earlier) by

$$<f,g> = \iint\limits_{R} f(x,y)g(x,y)dxdy \tag{13.17}$$

The norm of f is, therefore, given by

$$\| f \| = \langle f, f \rangle^{1/2} \tag{13.18}$$

Following the procedure as in Chakraverty (1992) and Singh and Chakraverty (1991, 1992a,b), the Gram–Schmidt orthogonalization process can be written as

$$\phi_1 = F_1$$

$$\left.\begin{array}{l} \phi_i = F_i - \displaystyle\sum_{j=1}^{i-1} \alpha_{ij}\phi_j \\[2mm] \alpha_{ij} = \langle F_i,\phi_j\rangle/\langle\phi_j,\phi_j\rangle, \quad j = 1, 2, \ldots, (i-1) \end{array}\right\}, \quad i = 2, \ldots, N \qquad (13.19)$$

where ϕ_i's are orthogonal polynomials. The normalized polynomials are generated by

$$\hat{\phi}_i = \phi_i/\|\phi_i\| \qquad (13.20)$$

All the integrals involved in the inner product are evaluated in closed form by the formulae given before (Chakraverty (1992), Singh and Chakraverty (1991, 1992a,b)).

13.4.2 Orthogonal Polynomials for Annular Elliptic Orthotropic Plates

For the generation of the two-dimensional orthogonal polynomials in this case, the following linearly independent set of functions is employed:

$$F_i(X,Y) = f(X,Y)\{f_i(X,Y)\}, \quad i = 1, 2, \ldots, N \qquad (13.21)$$

where $f(X,Y)$ satisfies the essential boundary conditions and $f_i(X,Y)$ are taken as the combinations of terms of the form $x^{\ell_i} y^{n_i}$, where ℓ_i and n_i are non-negative positive integers. The function $f(X,Y)$ in this case is defined by

$$f(X,Y) = C^s(C_0 - C)^t \qquad (13.22)$$

where s takes the value of 0,1, or 2 to define free, simply supported, or clamped conditions, respectively, at the outer boundary of the annular elliptic plate. Similarly, $t = 0$, 1, or 2 will define the corresponding boundary conditions at the inner edge of the annular elliptic plate. From $F_i(X,Y)$, an orthogonal set can be generated by the well-known Gram–Schmidt process as discussed in Section 13.4.1.

13.5 Numerical Results and Discussions

In all, there are five parameters, viz., D_x/H, D_y/H, ν_x, p, and m. Seven different types of materials have been selected here, the properties of which are given in Table 13.1. The first three (M1, M2, and M3) have been taken from a paper

TABLE 13.1

Material Properties

Material	D_x/H	D_y/H	ν_x
M1: Graphite–epoxy	13.90	0.79	0.28
M2: Glass–epoxy	3.75	0.80	0.26
M3: Boron–epoxy	13.34	1.21	0.23
M4: Carbon–epoxy	15.64	0.91	0.32
M5: Kevlar	2.60	2.60	0.14
M6	2.0	0.5	0.3
M7	0.5	2.0	0.075

by Lam et al. (1990), M4 from Kim and Dickinson (1990), and M5 from Dong and Lopez (1985). Lastly, M6 and M7 are taken from Narita (1985).

In the following sections, the results for (1) elliptic and circular plates and (2) annular elliptic plates with different boundary conditions are discussed.

13.5.1 Results for Elliptic and Circular Plates with Rectangular Orthotropy

In this section, the results for all the seven types of materials have been given for different boundary conditions.

Numerical investigations are discussed in terms of the various boundary conditions in the following headings.

13.5.1.1 Clamped Boundary

Depending on the values of ℓ_i and n_i, the modes are computed in terms of symmetric–symmetric, symmetric–antisymmetric, antisymmetric–symmetric, and antisymmetric–antisymmetric groups. Accordingly, Tables 13.2 through 13.8 incorporate the results for clamped boundary conditions

TABLE 13.2

First Three Frequency Parameters for Symmetric–Symmetric, Symmetric–Antisymmetric, Antisymmetric–Symmetric, and Antisymmetric–Antisymmetric Modes for M1: Graphite–Epoxy (Elliptic Orthotropic Plate, Clamped Boundary)

m	Symmetric–Symmetric			Symmetric–Antisymmetric		
	First	Second	Third	First	Second	Third
0.2	121.58	204.06	346.85	319.22	431.17	601.63
0.4	38.515	119.43	164.07	88.500	177.02	263.31
0.5	29.820	105.79	116.23	61.192	147.99	175.54
0.6	25.727	79.329	109.92	46.958	120.66	140.51
0.8	22.362	51.829	105.04	33.992	75.747	124.46
1.0	21.053	40.024	74.173	28.712	55.124	97.124

TABLE 13.2 (continued)

First Three Frequency Parameters for Symmetric–Symmetric, Symmetric–Antisymmetric, Antisymmetric–Symmetric, and Antisymmetric–Antisymmetric Modes for M1: Graphite–Epoxy (Elliptic Orthotropic Plate, Clamped Boundary)

	Antisymmetric–Symmetric			Antisymmetric–Antisymmetric		
m	First	Second	Third	First	Second	Third
0.2	155.69	267.26	442.90	368.56	506.82	709.09
0.4	70.581	181.60	210.35	124.81	243.40	316.14
0.5	62.876	146.28	181.74	97.197	203.06	235.20
0.6	59.260	117.95	176.46	83.345	162.00	211.51
0.8	55.961	90.777	147.61	70.674	116.33	184.06
1.0	54.409	78.743	116.69	64.959	95.781	141.08

with respect to the six material properties, respectively. The first three natural frequencies for various values of $m = 0.2$, 0.4, 0.5, 0.6, 0.8, and 1.0 for the materials M1 to M7 are given in Tables 13.2 through 13.8 for the four groups of vibration of modes. As done previously, the first few natural frequency results may be obtained from various combinations of symmetric–symmetric, symmetric–antisymmetric, antisymmetric–symmetric, and antisymmetric–antisymmetric modes about the two axes of the ellipse.

TABLE 13.3

First Three Frequency Parameters for Symmetric–Symmetric, Symmetric–Antisymmetric, Antisymmetric–Symmetric, and Antisymmetric–Antisymmetric Modes for M2: Glass–Epoxy (Elliptic Orthotropic Plate, Clamped Boundary)

	Symmetric–Symmetric			Symmetric–Antisymmetric		
m	First	Second	Third	First	Second	Third
0.2	136.51	196.95	291.55	364.55	456.68	596.59
0.4	38.849	88.032	175.06	97.536	158.66	253.75
0.5	27.376	76.667	121.46	65.146	122.38	195.41
0.6	21.476	70.761	87.569	47.635	103.04	139.32
0.8	16.242	51.940	67.479	30.676	78.191	90.103
1.0	14.225	36.986	64.880	23.344	54.917	79.607

	Antisymmetric–Symmetric			Antisymmetric–Antisymmetric		
m	First	Second	Third	First	Second	Third
0.2	162.86	238.88	352.62	406.59	514.20	679.50
0.4	58.675	126.87	218.64	124.00	201.72	314.22
0.5	47.134	115.14	152.39	89.380	163.80	232.66
0.6	41.499	105.58	119.96	70.937	142.15	175.12
0.8	36.617	76.914	107.70	53.560	105.69	133.56
1.0	34.608	61.595	103.98	46.113	81.019	122.55

TABLE 13.4

First Three Frequency Parameters for Symmetric–Symmetric, Symmetric–Antisymmetric, Antisymmetric–Symmetric, and Antisymmetric–Antisymmetric Modes for M3: Boron–Epoxy (Elliptic Orthotropic Plate, Clamped Boundary)

	Symmetric–Symmetric			Symmetric–Antisymmetric		
m	First	Second	Third	First	Second	Third
0.2	168.53	259.89	417.53	446.53	570.14	766.41
0.4	50.152	138.17	226.22	120.56	215.37	365.09
0.5	37.181	127.35	150.14	81.506	176.07	239.73
0.6	30.966	106.38	124.77	60.849	154.46	173.50
0.8	25.907	66.946	119.38	41.777	100.80	141.73
1.0	24.057	49.722	97.930	34.098	70.999	129.77

	Antisymmetric–Symmetric			Antisymmetric–Antisymmetric		
m	First	Second	Third	First	Second	Third
0.2	206.27	329.49	523.81	500.33	654.56	888.81
0.4	84.244	210.08	272.81	158.93	289.41	419.82
0.5	72.884	185.47	207.78	119.23	247.49	292.42
0.6	67.751	146.85	199.14	99.312	206.47	243.29
0.8	63.371	108.29	185.81	81.689	143.13	219.32
1.0	61.434	91.881	142.42	74.210	114.43	175.55

TABLE 13.5

First Three Frequency Parameters for Symmetric–Symmetric, Symmetric–Antisymmetric, Antisymmetric–Symmetric, and Antisymmetric–Antisymmetric Modes for M4: Carbon–Epoxy (Elliptic Orthotropic Plate, Clamped Boundary)

	Symmetric–Symmetric			Symmetric–Antisymmetric		
m	First	Second	Third	First	Second	Third
0.2	148.67	244.77	415.10	391.04	517.88	717.60
0.4	46.501	143.48	199.71	107.40	210.22	321.29
0.5	35.876	128.84	139.51	73.839	176.76	212.63
0.6	30.937	96.030	132.47	56.392	146.84	166.96
0.8	26.959	62.217	127.14	40.652	91.465	148.67
1.0	25.442	47.887	89.431	34.364	66.178	117.56

	Antisymmetric–Symmetric			Antisymmetric–Antisymmetric		
m	First	Second	Third	First	Second	Third
0.2	187.90	319.84	530.72	445.98	605.84	844.50
0.4	84.407	219.51	250.96	148.68	290.14	378.82
0.5	75.401	174.91	218.56	115.37	244.40	278.54
0.6	71.279	140.23	213.05	98.952	193.79	251.59
0.8	67.571	107.78	175.94	84.248	138.23	220.09
1.0	65.832	93.714	138.69	77.704	113.79	167.95

TABLE 13.6

First Three Frequency Parameters for Symmetric–Symmetric, Symmetric–Antisymmetric, Antisymmetric–Symmetric, and Antisymmetric–Antisymmetric Modes for M5: Kevlar (Elliptic Orthotropic Plate, Clamped Boundary)

	Symmetric–Symmetric			Symmetric–Antisymmetric		
m	First	Second	Third	First	Second	Third
0.2	236.43	295.66	395.79	640.34	741.34	936.60
0.4	63.077	104.17	175.08	165.91	221.90	307.34
0.5	42.043	81.320	152.46	108.39	158.20	236.51
0.6	30.666	69.813	141.28	77.014	123.40	199.17
0.8	19.726	59.825	85.039	45.765	89.649	136.66
1.0	15.142	53.990	58.822	31.502	74.705	90.725

	Antisymmetric–Symmetric			Antisymmetric–Antisymmetric		
m	First	Second	Third	First	Second	Third
0.2	262.28	335.29	456.00	682.84	805.73	1040.75
0.4	80.103	135.56	222.77	190.59	259.68	361.09
0.5	57.827	112.77	200.25	129.90	193.35	286.96
0.6	46.109	101.65	170.61	96.627	157.44	247.98
0.8	35.566	91.085	107.66	63.754	122.76	162.78
1.0	31.502	74.705	90.725	49.291	103.25	119.14

TABLE 13.7

First Three Frequency Parameters for Symmetric–Symmetric, Symmetric–Antisymmetric, Antisymmetric–Symmetric, and Antisymmetric–Antisymmetric Modes for M6 (Elliptic Orthotropic Plate, Clamped Boundary)

	Symmetric–Symmetric			Symmetric–Antisymmetric		
m	First	Second	Third	First	Second	Third
0.2	109.24	161.12	237.51	291.05	374.93	493.11
0.4	31.435	70.860	136.290	78.762	131.93	207.87
0.5	22.195	60.510	98.269	52.925	101.27	157.55
0.6	17.358	54.865	71.309	38.906	84.234	113.09
0.8	12.910	41.976	52.078	25.172	62.475	73.718
1.0	11.097	30.255	49.134	19.062	44.590	63.338

	Antisymmetric–Symmetric			Antisymmetric–Antisymmetric		
m	First	Second	Third	First	Second	Third
0.2	132.54	195.22	285.74	330.37	424.70	562.52
0.4	47.970	100.24	178.96	102.66	166.78	254.91
0.5	38.125	89.181	126.67	74.357	133.67	192.79
0.6	33.082	81.906	98.700	58.993	114.26	146.17
0.8	28.453	63.340	82.260	43.962	86.222	108.01
1.0	26.462	50.635	78.776	37.178	66.838	96.398

TABLE 13.8

First Three Frequency Parameters for Symmetric–Symmetric, Symmetric–Antisymmetric, Antisymmetric–Symmetric, and Antisymmetric–Antisymmetric Modes for M7 (Elliptic Orthotropic Plate, Clamped Boundary)

	Symmetric–Symmetric			Symmetric–Antisymmetric		
m	First	Second	Third	First	Second	Third
0.2	207.23	252.19	330.19	562.96	650.62	819.83
0.4	54.622	80.560	118.75	145.52	187.46	246.55
0.5	36.017	58.970	94.062	94.835	130.19	178.43
0.6	25.847	47.096	80.878	67.127	98.380	141.32
0.8	15.717	35.430	68.145	39.381	65.965	103.93
1.0	11.097	30.255	49.134	26.462	50.635	78.776

	Antisymmetric–Symmetric			Antisymmetric–Antisymmetric		
m	First	Second	Third	First	Second	Third
0.2	227.47	279.71	373.65	600.40	704.35	908.03
0.4	66.273	97.613	142.87	165.18	212.35	281.26
0.5	46.109	74.768	116.67	111.26	151.71	207.48
0.6	35.008	62.283	102.88	81.484	117.94	167.56
0.8	23.985	50.122	89.484	51.331	83.394	127.45
1.0	19.062	44.590	63.338	37.178	66.838	96.398

13.5.1.2 *Simply Supported Boundary*

Again, the first three natural frequencies are presented as per the four groups of modes for various values of m for the mentioned material properties and the corresponding results are cited in Tables 13.9 through 13.15.

13.5.1.3 *Free Boundary*

The results for this boundary condition are given in Tables 13.16 through 13.22. These tables show the numerical results for the first three frequencies for each of the modes of symmetric–symmetric, symmetric–antisymmetric, antisymmetric–symmetric, and antisymmetric–antisymmetric groups for all the seven material properties.

As mentioned in earlier sections, the present problem reduces to that of an isotropic plate if $D_x/H = D_y/H = 1$ and $\nu_x = \nu_y = \nu$, for which results are already reported by Chakraverty (1992) and Singh and Chakraverty (1991, 1992a,b), where they have already made comparison with all the existing results for the isotropic case. Using the present computer program taking $D_x/H = D_y/H = 1$ and $\nu_x = \nu_y = 0.3$, again the results have been computed for various values of m and are found to be exactly the same as reported in the earlier chapters (Chakraverty (1992) and Singh and Chakraverty (1991, 1992a,b)).

TABLE 13.9

First Three Frequency Parameters for Symmetric–Symmetric, Symmetric–
Antisymmetric, Antisymmetric–Symmetric, and Antisymmetric–Antisymmetric
Modes for M1: Graphite–Epoxy (Elliptic Orthotropic Plate, Simply Supported
Boundary)

	Symmetric–Symmetric			Symmetric–Antisymmetric		
m	First	Second	Third	First	Second	Third
0.2	57.833	130.17	265.24	209.02	312.42	480.77
0.4	18.365	87.547	122.47	58.266	136.99	210.02
0.5	14.198	78.259	86.709	39.739	115.40	139.78
0.6	12.268	58.119	82.053	29.821	94.851	110.25
0.8	10.704	36.411	78.871	20.481	58.270	96.290
1.0	10.068	26.703	58.813	16.555	40.692	80.881

	Antisymmetric–Symmetric			Antisymmetric–Antisymmetric		
m	First	Second	Third	First	Second	Third
0.2	86.990	188.75	359.05	254.22	383.84	589.06
0.4	44.864	142.74	165.14	89.945	197.79	259.52
0.5	40.999	113.96	144.84	70.259	165.44	193.81
0.6	39.022	91.252	140.75	60.050	131.61	173.60
0.8	36.915	68.665	122.69	50.188	92.531	156.11
1.0	35.754	58.180	95.042	45.393	74.225	119.22

TABLE 13.10

First Three Frequency Parameters for Symmetric–Symmetric, Symmetric–
Antisymmetric, Antisymmetric–Symmetric, and Antisymmetric–Antisymmetric
Modes for M2: Glass–Epoxy (Elliptic Orthotropic Plate, Simply Supported Boundary)

	Symmetric–Symmetric			Symmetric–Antisymmetric		
m	First	Second	Third	First	Second	Third
0.2	64.107	117.60	206.17	237.80	323.19	463.43
0.4	18.605	61.498	137.37	64.360	120.21	210.69
0.5	13.072	55.477	90.674	42.980	94.408	155.89
0.6	10.229	52.034	65.331	31.252	80.222	111.11
0.8	7.7392	38.107	50.355	19.633	61.317	71.100
1.0	6.8010	26.488	48.426	14.441	42.678	62.094

	Antisymmetric–Symmetric			Antisymmetric–Antisymmetric		
m	First	Second	Third	First	Second	Third
0.2	87.157	156.01	265.68	276.48	377.44	547.70
0.4	35.348	96.930	169.57	88.311	160.37	268.84
0.5	29.539	90.005	119.25	64.463	132.54	190.96
0.6	26.678	82.696	94.951	51.470	115.78	144.36
0.8	24.063	59.798	85.914	38.830	86.032	110.01
1.0	22.848	47.253	83.453	33.131	65.229	100.40

TABLE 13.11

First Three Frequency Parameters for Symmetric–Symmetric, Symmetric–Antisymmetric, Antisymmetric–Symmetric, and Antisymmetric–Antisymmetric Modes for M3: Boron–Epoxy (Elliptic Orthotropic Plate, Simply Supported Boundary)

	Symmetric–Symmetric			Symmetric–Antisymmetric		
m	First	Second	Third	First	Second	Third
0.2	79.662	159.90	309.73	291.26	405.06	601.48
0.4	24.296	100.48	168.95	79.785	165.11	291.35
0.5	18.022	94.475	112.24	53.656	136.83	191.37
0.6	15.009	78.933	93.283	39.459	120.62	138.06
0.8	12.545	48.243	89.098	25.840	78.970	109.90
1.0	11.606	34.186	79.308	20.093	53.895	101.87

	Antisymmetric–Symmetric			Antisymmetric–Antisymmetric		
m	First	Second	Third	First	Second	Third
0.2	112.08	224.45	414.71	340.30	484.74	726.09
0.4	52.741	165.33	212.19	113.73	233.72	342.80
0.5	47.243	144.87	165.30	85.975	201.78	240.33
0.6	44.593	114.14	158.97	71.615	168.34	199.97
0.8	41.923	82.586	153.06	58.260	115.01	178.83
1.0	40.465	68.452	117.18	52.088	89.690	149.65

TABLE 13.12

First Three Frequency Parameters for Symmetric–Symmetric, Symmetric–Antisymmetric, Antisymmetric–Symmetric, and Antisymmetric–Antisymmetric Modes for M4: Carbon–Epoxy (Elliptic Orthotropic Plate, Simply Supported Boundary)

	Symmetric–Symmetric			Symmetric–Antisymmetric		
m	First	Second	Third	First	Second	Third
0.2	70.180	155.31	317.48	255.74	372.74	570.93
0.4	22.549	105.64	149.30	71.006	162.52	256.41
0.5	17.401	95.743	104.32	48.245	137.82	169.62
0.6	15.002	70.739	99.033	36.044	115.78	130.99
0.8	13.043	43.931	95.385	24.582	70.748	114.73
1.0	12.234	32.004	71.262	19.805	49.051	98.323

	Antisymmetric–Symmetric			Antisymmetric–Antisymmetric		
m	First	Second	Third	First	Second	Third
0.2	104.36	225.31	431.03	305.93	455.93	699.28
0.4	53.901	173.12	196.78	106.98	235.90	310.45
0.5	49.407	136.28	174.43	83.310	199.30	229.33
0.6	47.116	108.45	170.03	71.212	157.60	206.21
0.8	44.632	81.361	146.30	59.687	109.87	186.26
1.0	43.250	68.968	112.79	54.093	87.934	141.88

TABLE 13.13

First Three Frequency Parameters for Symmetric–Symmetric, Symmetric–Antisymmetric, Antisymmetric–Symmetric, and Antisymmetric–Antisymmetric Modes for M5: Kevlar (Elliptic Orthotropic Plate, Simply Supported Boundary)

	Symmetric–Symmetric			Symmetric–Antisymmetric		
m	First	Second	Third	First	Second	Third
0.2	108.59	159.70	248.31	414.80	507.25	698.95
0.4	30.141	66.453	133.80	108.70	160.46	245.18
0.5	20.323	54.991	121.91	71.395	117.34	193.37
0.6	14.915	49.238	108.72	50.936	93.385	165.63
0.8	9.6424	43.998	63.624	30.316	69.490	109.14
1.0	7.4094	40.097	44.034	20.671	58.381	72.425

	Antisymmetric–Symmetric			Antisymmetric–Antisymmetric		
m	First	Second	Third	First	Second	Third
0.2	130.90	194.97	304.89	453.52	566.52	801.38
0.4	44.829	95.651	180.55	131.33	196.16	299.54
0.5	33.810	84.169	167.69	91.060	150.48	243.37
0.6	28.002	78.398	132.02	68.696	125.24	211.76
0.8	22.771	71.829	84.531	46.093	99.875	133.67
1.0	20.671	58.381	72.425	35.796	84.274	98.325

TABLE 13.14

First Three Frequency Parameters for Symmetric–Symmetric, Symmetric–Antisymmetric, Antisymmetric–Symmetric and Antisymmetric–Antisymmetric Modes for M6 (Elliptic Orthotropic Plate, Simply Supported Boundary)

	Symmetric–Symmetric			Symmetric–Antisymmetric		
m	First	Second	Third	First	Second	Third
0.2	51.430	97.665	168.77	190.28	268.42	386.53
0.4	14.789	48.970	109.09	51.835	100.35	172.41
0.5	10.325	43.171	73.196	34.730	78.206	125.55
0.6	8.0119	39.805	53.012	25.356	65.539	90.018
0.8	5.9520	30.525	38.698	16.039	48.691	58.311
1.0	5.1628	21.585	36.598	11.813	34.423	49.684

	Antisymmetric–Symmetric			Antisymmetric–Antisymmetric		
m	First	Second	Third	First	Second	Third
0.2	72.006	128.84	215.27	226.67	315.54	456.93
0.4	28.765	75.558	140.48	73.472	132.72	218.04
0.5	23.627	68.846	99.369	53.812	107.96	158.62
0.6	20.985	63.801	77.935	42.941	92.847	120.68
0.8	18.501	49.328	65.448	32.025	70.103	89.236
1.0	17.365	39.103	62.775	26.906	53.982	79.312

TABLE 13.15

First Three Frequency Parameters for Symmetric–Symmetric, Symmetric–Antisymmetric, Antisymmetric–Symmetric, and Antisymmetric–Antisymmetric Modes for M7 (Elliptic Orthotropic Plate, Simply Supported Boundary)

	Symmetric–Symmetric			Symmetric–Antisymmetric		
m	First	Second	Third	First	Second	Third
0.2	95.060	134.10	198.58	364.80	445.42	610.10
0.4	25.715	48.832	84.387	95.140	134.21	193.26
0.5	17.060	37.324	70.030	62.195	95.016	142.69
0.6	12.256	30.877	62.271	44.120	72.995	114.79
0.8	7.3947	24.485	54.548	25.917	50.176	86.207
1.0	5.1628	21.585	36.598	17.365	39.103	62.775

	Antisymmetric–Symmetric			Antisymmetric–Antisymmetric		
m	First	Second	Third	First	Second	Third
0.2	113.07	157.92	236.76	399.27	494.69	695.50
0.4	36.003	64.422	107.63	113.33	157.77	228.46
0.5	25.847	51.751	91.864	77.352	115.32	171.98
0.6	20.128	44.731	83.409	57.297	91.367	140.92
0.8	14.382	37.779	70.241	36.736	66.364	109.02
1.0	11.813	34.423	49.684	26.906	53.982	79.312

TABLE 13.16

First Three Frequency Parameters for Symmetric–Symmetric, Symmetric–Antisymmetric, Antisymmetric–Symmetric, and Antisymmetric–Antisymmetric Modes for M1: Graphite–Epoxy (Elliptic Orthotropic Plate, Completely Free Boundary)

	Symmetric–Symmetric			Symmetric–Antisymmetric		
m	First	Second	Third	First	Second	Third
0.2	22.847	106.95	125.20	64.334	173.41	323.15
0.4	21.656	34.778	79.096	41.528	86.597	122.21
0.5	18.945	25.569	63.385	36.285	58.102	98.970
0.6	14.373	23.283	52.718	31.342	43.621	80.683
0.8	8.3939	21.926	36.934	20.637	33.432	56.533
1.0	5.4281	21.101	25.402	13.653	29.671	40.044

	Antisymmetric–Symmetric			Antisymmetric–Antisymmetric		
m	First	Second	Third	First	Second	Third
0.2	57.912	160.63	177.74	29.174	112.06	248.40
0.4	47.015	65.575	121.13	14.989	77.940	122.74
0.5	35.683	60.149	99.198	12.013	64.864	94.197
0.6	27.616	58.018	81.375	9.9919	51.847	82.022
0.8	18.419	52.963	92.615	7.4302	34.049	70.686
1.0	13.597	38.206	55.054	5.8856	24.357	54.888

TABLE 13.17

First Three Frequency Parameters for Symmetric–Symmetric, Symmetric–
Antisymmetric, Antisymmetric–Symmetric, and Antisymmetric–Antisymmetric
Modes for M2: Glass–Epoxy (Elliptic Orthotropic Plate, Completely Free Boundary)

	Symmetric–Symmetric			Symmetric–Antisymmetric		
m	First	Second	Third	First	Second	Third
0.2	13.630	65.255	140.95	55.710	132.43	281.28
0.4	13.390	38.493	60.220	33.001	91.089	98.916
0.5	13.073	25.723	53.370	28.399	64.504	80.961
0.6	12.414	18.980	45.614	25.147	46.342	70.328
0.8	9.1133	14.525	34.126	19.938	28.951	52.304
1.0	6.1159	13.667	26.022	14.750	22.825	39.816

	Antisymmetric–Symmetric			Antisymmetric–Antisymmetric		
m	First	Second	Third	First	Second	Third
0.2	34.772	105.06	178.77	27.290	90.026	183.43
0.4	33.162	59.800	93.637	14.292	58.831	125.60
0.5	30.769	45.188	81.390	11.519	51.703	90.463
0.6	26.583	39.211	70.075	9.6328	45.609	70.521
0.8	18.638	35.777	52.475	7.2211	33.912	53.418
1.0	13.835	34.391	39.401	5.7466	25.057	46.966

TABLE 13.18

First Three Frequency Parameters for Symmetric–Symmetric, Symmetric–
Antisymmetric, Antisymmetric–Symmetric, and Antisymmetric–Antisymmetric
Modes for M3: Boron–Epoxy (Elliptic Orthotropic Plate, Completely Free Boundary)

	Symmetric–Symmetric			Symmetric–Antisymmetric		
m	First	Second	Third	First	Second	Third
0.2	25.786	121.82	171.30	64.448	184.99	409.16
0.4	24.892	46.821	94.403	44.449	117.41	143.84
0.5	23.421	32.225	75.029	40.186	78.216	119.93
0.6	19.551	26.810	62.517	36.604	56.673	98.476
0.8	11.801	24.569	47.020	27.803	38.399	70.647
1.0	7.6796	23.623	34.689	19.073	33.135	53.310

	Antisymmetric–Symmetric			Antisymmetric–Antisymmetric		
m	First	Second	Third	First	Second	Third
0.2	65.513	190.89	219.02	27.580	116.64	269.58
0.4	56.785	76.730	142.68	14.225	87.836	153.72
0.5	43.991	67.836	118.11	11.420	77.257	112.52
0.6	33.554	64.987	99.948	9.5176	64.437	93.694
0.8	21.502	61.419	70.563	7.1015	42.433	79.320
1.0	15.338	48.004	62.033	5.6353	29.649	68.283

TABLE 13.19

First Three Frequency Parameters for Symmetric–Symmetric, Symmetric–
Antisymmetric, Antisymmetric–Symmetric, and Antisymmetric–Antisymmetric
Modes for M4: Carbon–Epoxy (Elliptic Orthotropic Plate, Completely Free Boundary)

	Symmetric–Symmetric			Symmetric–Antisymmetric		
m	First	Second	Third	First	Second	Third
0.2	27.860	129.62	151.97	66.096	193.76	394.04
0.4	26.366	41.955	91.637	46.107	104.31	143.89
0.5	23.185	30.711	72.916	41.403	69.757	115.85
0.6	17.656	27.854	61.078	36.797	51.652	94.153
0.8	10.341	26.145	44.451	25.217	38.539	67.101
1.0	6.6946	25.142	31.189	16.792	34.167	48.804

	Antisymmetric–Symmetric			Antisymmetric–Antisymmetric		
m	First	Second	Third	First	Second	Third
0.2	70.552	187.56	215.63	27.536	121.34	282.92
0.4	55.251	77.259	141.09	14.150	90.253	142.46
0.5	40.715	71.644	116.20	11.339	75.870	108.75
0.6	30.838	69.198	96.040	9.4344	60.324	94.747
0.8	19.865	61.510	69.180	7.0217	38.717	81.647
1.0	14.295	43.710	66.367	5.5628	27.067	63.616

TABLE 13.20

First Three Frequency Parameters for Symmetric–Symmetric, Symmetric–
Antisymmetric, Antisymmetric–Symmetric, and Antisymmetric–Antisymmetric
Modes for M5: Kevlar (Elliptic Orthotropic Plate, Completely Free Boundary)

	Symmetric–Symmetric			Symmetric–Antisymmetric		
m	First	Second	Third	First	Second	Third
0.2	11.327	54.504	147.44	49.587	116.34	247.76
0.4	11.268	52.720	64.491	29.458	82.408	167.44
0.5	11.195	41.805	51.783	25.555	75.182	109.39
0.6	11.077	30.031	48.282	22.968	68.848	78.245
0.8	10.611	18.074	38.472	19.656	44.723	59.631
1.0	9.3658	13.194	30.784	17.372	30.048	48.478

	Antisymmetric–Symmetric			Antisymmetric–Antisymmetric		
m	First	Second	Third	First	Second	Third
0.2	28.958	88.055	215.40	24.927	79.523	160.55
0.4	28.536	80.065	88.619	12.894	52.512	119.11
0.5	28.005	58.386	81.079	10.441	47.201	108.68
0.6	27.078	44.908	73.747	8.7754	43.456	94.139
0.8	22.910	33.063	59.001	6.6412	37.594	62.149
1.0	17.372	30.048	48.478	5.3248	31.609	47.338

TABLE 13.21

First Three Frequency Parameters for Symmetric–Symmetric, Symmetric–Antisymmetric, Antisymmetric–Symmetric, and Antisymmetric–Antisymmetric Modes for M6 (Elliptic Orthotropic Plate, Completely Free Boundary)

	Symmetric–Symmetric			Symmetric–Antisymmetric		
m	First	Second	Third	First	Second	Third
0.2	9.9168	47.656	114.02	54.402	119.69	296.23
0.4	9.7549	31.231	45.287	30.158	76.424	79.266
0.5	9.5450	20.771	42.340	25.263	52.365	66.385
0.6	9.1259	15.208	37.888	21.801	37.716	57.983
0.8	7.0021	11.161	28.646	16.372	24.034	43.763
1.0	4.7725	10.385	21.170	11.725	19.154	32.867

	Antisymmetric–Symmetric			Antisymmetric–Antisymmetric		
m	First	Second	Third	First	Second	Third
0.2	25.337	76.951	151.29	28.376	84.154	161.37
0.4	24.502	51.543	71.822	14.373	50.683	107.37
0.5	23.450	38.308	65.735	11.547	43.512	77.301
0.6	21.564	31.744	57.965	9.6444	37.952	60.248
0.8	16.451	27.450	43.767	7.2355	28.659	44.758
1.0	12.631	26.182	33.192	5.7735	21.756	38.651

TABLE 13.22

First Three Frequency Parameters for Symmetric–Symmetric, Symmetric–Antisymmetric, Antisymmetric–Symmetric, and Antisymmetric–Antisymmetric Modes for M7 (Elliptic Orthotropic Plate, Completely Free Boundary)

	Symmetric–Symmetric			Symmetric–Antisymmetric		
m	First	Second	Third	First	Second	Third
0.2	4.9659	23.959	65.277	52.205	105.53	571.19
0.4	4.9583	23.828	57.010	27.201	59.847	124.71
0.5	4.9483	23.665	37.499	22.316	51.235	96.514
0.6	4.9325	23.414	26.687	19.092	45.551	68.189
0.8	4.8774	15.615	22.643	15.079	38.212	39.633
1.0	4.7725	10.385	21.170	12.631	26.182	33.192

	Antisymmetric–Symmetric			Antisymmetric–Antisymmetric		
m	First	Second	Third	First	Second	Third
0.2	12.709	38.781	95.973	28.211	77.147	137.15
0.4	12.668	38.475	75.645	14.188	42.077	80.682
0.5	12.614	38.117	52.707	11.390	35.353	70.134
0.6	12.531	37.525	39.679	9.5247	30.924	63.109
0.8	12.251	25.771	35.911	7.1867	25.341	53.684
1.0	11.725	19.154	32.867	5.7735	21.756	38.651

To fix the number of approximations N needed, calculations were carried out for different values of N until the first five significant digits had converged. It was found that the results converged in about 8–10 approximations for clamped and simply supported boundary, and in 12–15 approximations for free boundary. Table 13.23 gives results for the convergence of the first three natural frequencies for clamped, simply supported, and free (i.e., $p=2$, 1, and 0, respectively) boundaries with N increasing from 2 to 15. These results were obtained for the material M6 and taking the aspect ratio of the ellipse to be 0.5. The first three frequencies are chosen from the four groups of modes as discussed earlier. This table also shows from which mode groups the frequencies are obtained. These are denoted by S–S, A–S, S–A, and A–A, i.e. those are obtained from symmetric–symmetric, antisymmetric–symmetric, symmetric–antisymmetric, and antisymmetric–antisymmetric groups, respectively.

It is seen from Tables 13.2 through 13.22 (for materials M1 to M7) that for any boundary condition, i.e. for clamped, simply supported, or free, the frequencies decrease as m is increased. For a clamped boundary, the frequencies are a maximum, and for a free boundary, these are a minimum for each m for all the materials considered in this section.

13.5.2 Results for Annular Elliptic Plates with Rectangular Orthotropy

Various results in detail may be obtained in Chakraverty et al. (2000). In this chapter, only few results are incorporated. Table 13.24 demonstrates the results for first three frequency parameters of symmetric–symmetric, symmetric–antisymmetric, antisymmetric–symmetric, and antisymmetric–antisymmetric groups. The first three natural frequencies have been computed taking the aspect ratio of the outer boundary as $b/a = m = 0.5$ and the inner boundary k is also 0.5. The orthotropic material property of carbon–epoxy (i.e., $\frac{D_x}{H} = 15.64$, $\frac{D_y}{H} = 0.91$ and $v_x = 0.32$) has been considered in the computations. Results are provided for boundary conditions of C–C, F–F, C–F, and F–C. Here C, S, and F designate clamped, simply supported, and free boundary, and first and second letters denote the conditions at the outer and inner edges, respectively. Table 13.25 shows the first 12 natural frequencies arranged in ascending order by choosing the frequencies from each of the above mode groups taking $m = k = 0.5$ for carbon–epoxy materials.

It is interesting to note the effect of hole sizes on the first eight natural frequencies from Figures 13.1 through 13.9 for the nine possible types of boundary conditions on the inner and outer edges of the annular elliptic plate. The case of $m = 0.5$ (outer boundary) is considered in all the figures. It may be seen that as k increases, the frequencies increase; for all the boundary conditions except for the exceptional case of FF boundary in Figure 13.1. For FF boundary condition, frequencies decrease as k increases. These figures give an idea as to how the hole sizes affect the frequency parameters for annular orthotropic elliptic plates.

TABLE 13.23

Convergence of First Three Frequency Parameters Chosen from the Sets of Symmetric–Symmetric (S–S), Symmetric–Antisymmetric (S–A), Antisymmetric–Symmetric (A–S), and Antisymmetric–Antisymmetric (A–A) Modes for M6 (Elliptic Orthotropic Plate, Simply Supported Boundary)

N	Clamped Boundary			Simply Supported Boundary			Completely Free Boundary		
	First (From S-S Mode)	Second (From A-A Mode)	Third (From S-A Mode)	First (From S-S Mode)	Second (From A-A Mode)	Third (From S-A Mode)	First (From S-S Mode)	Second (From A-A Mode)	Third (From S-S Mode)
2	22.267	38.268	53.920	11.037	25.130	42.168	11.313	11.933	—
3	22.199	38.161	52.977	10.339	23.786	35.004	10.650	11.607	25.206
4	22.196	38.131	52.958	10.331	23.675	34.956	9.6378	11.587	24.828
5	22.196	38.126	52.944	10.329	23.641	34.868	9.6195	11.569	23.693
6	22.195	38.125	52.926	10.325	23.628	34.732	9.5982	11.547	21.029
7	22.195	38.125	52.925	10.325	23.627	34.732	9.5572	11.547	21.020
8	22.195	38.125	52.925	10.325	23.627	34.731	9.5483	11.547	20.993
9	22.195	38.125	52.925	10.325	23.627	34.730	9.5462	11.547	20.941
10	22.195	38.125	52.925	10.325	23.627	34.730	9.5451	11.547	20.772
11	22.195	38.125	52.925	10.325	23.627	34.730	9.5450	11.547	20.772
12	22.195	38.125	52.925	10.325	23.627	34.730	9.5450	11.547	20.771
13	22.195	38.125	52.925	10.325	23.627	34.730	9.5450	11.547	20.771
14	22.195	38.125	52.925	10.325	23.627	34.730	9.5450	11.547	20.771
15	22.195	38.125	52.925	10.325	23.627	34.730	9.5450	11.547	20.771

TABLE 13.24

First Three Frequency Parameters of Symmetric–Symmetric (S–S), Symmetric–Antisymmetric (S–A), Antisymmetric–Symmetric (A–S), and Antisymmetric–Antisymmetric (A–A) Modes for Carbon–Epoxy Material for $m = k = 0.5$ (Annular Elliptic Orthotropic Plate)

Boundary Condition	Mode Numbers	S–S Mode	S–A Mode	A–S Mode	A–A Mode
C–C (2,2)	First	125.95	193.37	124.91	195.37
	Second	234.92	282.13	233.21	282.74
	Third	308.04	342.74	304.74	348.03
F–F (0,0)	First	5.7328	18.581	15.862	8.4687
	Second	24.637	47.490	41.095	34.585
	Third	33.508	53.929	52.196	75.815
C–F (2,0)	First	46.075	64.448	47.927	79.468
	Second	72.249	103.48	91.824	129.65
	Third	115.24	160.18	140.17	194.12
F–C (0,2)	First	17.034	30.614	16.754	30.223
	Second	45.397	54.255	46.313	60.776
	Third	62.234	77.561	75.740	97.489

TABLE 13.25

First Twelve Frequency Parameters of Annular Elliptic Orthotropic Plate Chosen from Symmetric–Symmetric (S–S), Symmetric–Antisymmetric (S–A), Antisymmetric–Symmetric (A–S), and Antisymmetric–Antisymmetric (A–A) Modes for Carbon–Epoxy Material for $m = k = 0.5$

Frequency Parameter	C–C (2,2)	F–F (0,0)	C–F (2,0)	F–C (0,2)
λ_1	124.91	5.7328	46.075	16.754
λ_2	125.95	8.4687	47.927	17.034
λ_3	193.37	15.862	64.448	30.223
λ_4	195.37	18.581	72.249	30.614
λ_5	233.21	24.637	79.468	45.397
λ_6	234.92	33.508	91.824	46.313
λ_7	282.13	34.585	103.48	54.255
λ_8	282.74	41.095	115.24	60.776
λ_9	304.74	47.490	129.65	62.234
λ_{10}	308.04	52.196	140.17	75.740
λ_{11}	342.74	53.929	160.18	77.561
λ_{12}	348.03	75.815	194.12	97.489

Note: Chosen from Table 13.24.

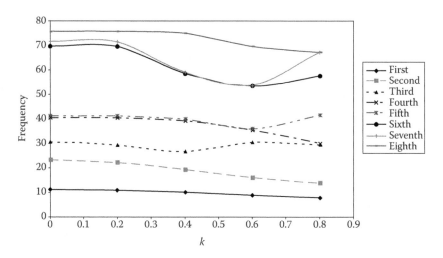

FIGURE 13.1
Effect of hole sizes on the natural frequencies for annular elliptic orthotropic plate ($m = 0.5$, FF boundary).

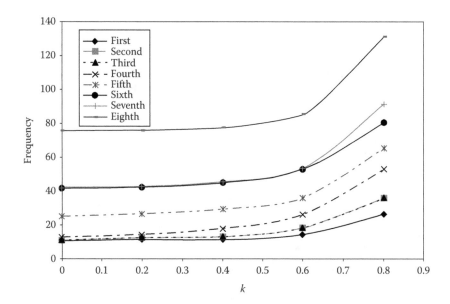

FIGURE 13.2
Effect of hole sizes on the natural frequencies for annular elliptic orthotropic plate ($m = 0.5$, FS boundary).

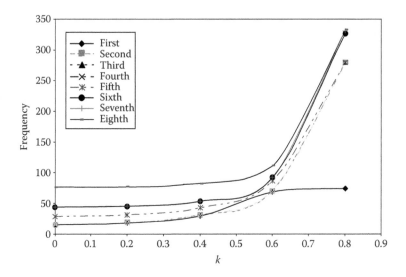

FIGURE 13.3
Effect of hole sizes on the natural frequencies for annular elliptic orthotropic plate ($m = 0.5$, FC boundary).

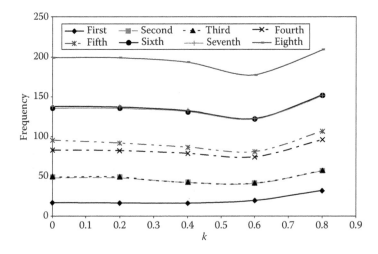

FIGURE 13.4
Effect of hole sizes on the natural frequencies for annular elliptic orthotropic plate ($m = 0.5$, SF boundary).

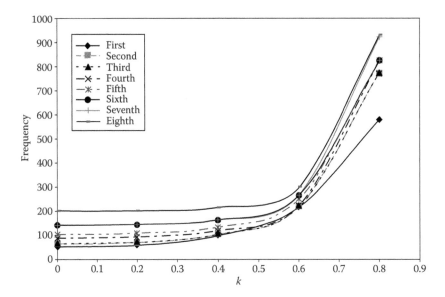

FIGURE 13.5
Effect of hole sizes on the natural frequencies for annular elliptic orthotropic plate ($m = 0.5$, SS boundary).

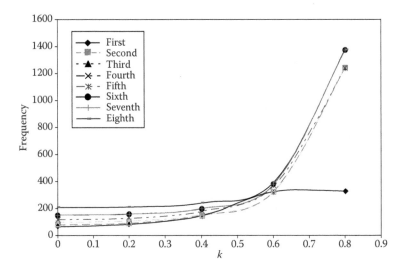

FIGURE 13.6
Effect of hole sizes on the natural frequencies for annular elliptic orthotropic plate ($m = 0.5$, SC boundary).

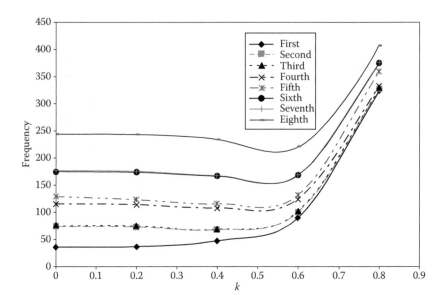

FIGURE 13.7

Effect of hole sizes on the natural frequencies for annular elliptic orthotropic plate ($m = 0.5$, CF boundary).

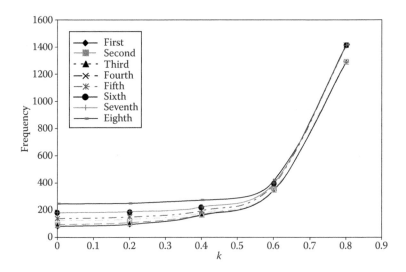

FIGURE 13.8

Effect of hole sizes on the natural frequencies for annular elliptic orthotropic plate ($m = 0.5$, CS boundary).

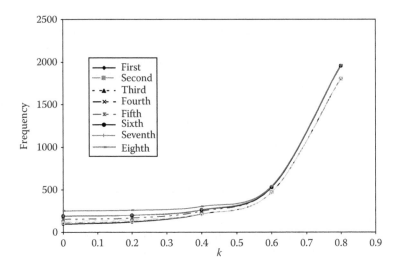

FIGURE 13.9

Effect of hole sizes on the natural frequencies for annular elliptic orthotropic plate ($m = 0.5$, CC boundary).

Bibliography

Bert, C.W. 1976. Dynamics of composite and sandwich panels-parts I and II. *The Shock Vibration Digest*, 8: 37–48, 15–24.

Bert, C.W. 1979. Recent research in composite and sandwich plate dynamics. *The Shock Vibration Digest*, 11: 13–23.

Bert, C.W. 1980. Vibration of composite structures. In *Recent Advances in Structural Dynamics*, Vol. 2, M. Petyt (Ed.), pp. 693–712, Institute of Sound and Vibration Research Southampton, United Kingdom.

Bert, C.W. 1982. Research on dynamics of composite and sandwich plates, 1979–1981. *The Shock Vibration Digest*, 14: 17–34.

Bert, C.W. 1985. Research on dynamic behavior of composite and sandwich plates-IV. *The Shock Vibration Digest*, 17: 3–15.

Bert, C.W. 1991. Research on dynamic behavior of composite and sandwich plates-V: Part I. *The Shock Vibration Digest*, 23: 3–14.

Bhat, R.B. 1985. Natural frequencies of rectangular plates using characteristic orthogonal polynomials in Rayleigh–Ritz method. *Journal of Sound and Vibration*, 102: 493–499.

Bhat, R.B. 1987. Flexural vibration of polygonal plates using characteristic orthogonal polynomials in two variables. *Journal of Sound and Vibration*, 114: 65–71.

Chakraverty, S. 1992. Numerical solution of vibration of plates, PhD Thesis, Department of Applied Mathematics, University of Roorkee, Roorkee, India.

Chakraverty, S. 1996. Effect of orthotropy on natural frequencies of vibration of elliptic plates, *Proceedings of Conference on Mathematics and Its Applications in Engineering and Industry*, Department of Mathematics, University of Roorkee, Roorkee, India.

Chakraverty, S. and Chakrabarti, S.C. 1993. Deflection of circular plate using orthogonal polynomials. *ISIAM Conference*, University of Roorkee, Roorkee, India.

Chakraverty, S. and Petyt, M. 1997. Natural frequencies for free vibration of non-homogeneous elliptic and circular plates using two dimensional orthogonal polynomials. *Applied Mathematical Modeling*, 21: 399–417.

Chakraverty, S. and Petyt, M. 1999. Free vibration analysis of elliptic and circular plates having rectangular orthotropy. *International Journal of Structural Engineering and Mechanics*, 7(1): 53–67.

Chakraverty, S., Bhat, R.B., and Stiharu, I. 2000. Vibration of annular elliptic orthotropic plates using two dimensional orthogonal polynomials. *Applied Mechanics and Engineering*, 5(4): 843–866.

Chakraverty, S., Bhat, R.B., and Stiharu, I. 2001. Free vibration of annular elliptic plates using boundary characteristic orthogonal polynomials as shape functions in Rayleigh–Ritz method. *Journal of Sound and Vibration*, 241(3): 524–539.

Dickinson, S.M. and Blasio, A.Di. 1986. On the use of orthogonal polynomials in Rayleigh–Ritz method for the study of the flexural vibration and buckling of isotropic and orthotropic rectangular plates. *Journal of Sound and Vibration*, 108: 51–62.

Dong, S.B. and Lopez, A.E. 1985. Natural vibrations of a clamped circular plate with rectilinear orthotropy by least-squares collocation. *International Journal of Solids Structure*, 21: 515–526.

Irie, T. and Yamada, G. 1979. Free vibration of an orthotropic elliptical plate with a similar hole. *Bulletin of the Japan Society of Mechanical Engineers*, 22: 1456–1462.

Irie, T., Yamada, G., and Kobayashi, Y. 1983. Free vibration of circular-segment-shaped membranes and plates of rectangular orthotropy. *Journal of Acoustical Society America*, 73: 2034–2040.

Kim, C.S. and Dickinson, S.M. 1987. The flexural vibration of rectangular plates with point supports. *Journal of Sound and Vibration*, 117: 249–261.

Kim, C.S. and Dickinson, S.M. 1989. On the lateral vibration of thin annular and circular composite plates subject to certain complicating effects. *Journal of Sound and Vibration*, 130: 363–377.

Kim, C.S. and Dickinson, S.M. 1990. The free flexural vibration of right triangular isotropic and orthotropic plates. *Journal of Sound and Vibration*, 141: 291–311.

Lam, K.Y., Liew, K.M., and Chow, S.T. 1990. Free vibration analysis of isotropic and orthotropic triangular plates. *International Journal of Mechanical Science*, 32: 455–464.

Laura, P.A.A., Gutierrez, R.H., and Bhat, R.B. 1989. Transverse vibration of a trapezoidal cantilever plate of variable thickness. *American Institute of Aeronautics and Astronautics Journal*, 27: 921–922.

Leissa, A.W. 1969. *Vibration of Plates*, NASA SP 160, U.S. Government Printing Office, Washington DC.

Leissa, A.W. 1978. Recent research in plate vibrations, 1973–76: Complicating effects. *The Shock Vibration Digest*, 10: 21–35.

Leissa, A.W. 1981. Plate vibration research, 1976–80: Complicating effects. *The Shock Vibration Digest*, 13: 19–36.

Leissa, A.W. 1987. Recent studies in plate vibrations: 1981–85 Part II. Complicating effects. *The Shock Vibration Digest*, 19: 10–24.

Liew, K.M. and Lam, K.Y. 1990. Application of two dimensional plate functions to flexural vibrations of skew plates. *Journal of Sound Vibration*, 139: 241–252.

Liew, K.M., Lam, K.Y., and Chow, S.T. 1990. Free vibration analysis of rectangular plates using orthogonal plate functions. *Computers and Structure*, 34: 79–85.

McNitt, R.P. 1962. Free vibration of a clamped elliptical plate. *Journal of Aeronautical Science*, 29: 1124–1125.

Narita, Y. 1983. Flexural vibrations of clamped polygonal and circular plates having rectangular orthotropy. *Journal of Applied Mechanics, Transactions on ASME*, 50: 691–692.

Narita, Y. 1985. Natural frequencies of free, orthotropic elliptical plates. *Journal of Sound and Vibration*, 100: 83–89.

Narita, Y. 1986. Free vibration analysis of orthotropic elliptical plates resting on arbitrary distributed point supports. *Journal of Sound and Vibration*, 108: 1–10.

Rajappa, N.R. 1963. Free vibration of rectangular and circular orthotropic plates. *American Institute of Aeronautics and Astronautics Journal*, 1: 1194–1195.

Sakata, T. 1976. A reduction method for problems of vibration of orthotropic plates. *Journal of Sound and Vibration*, 48: 405–412.

Sakata, T. 1979a. Reduction methods for problems of vibration of orthotropic plates. Part I: Exact methods. *The Shock Vibration Digest*, 5: 19–26.

Sakata, T. 1979b. Reduction methods for problems of vibration of orthotropic plates. Part II: Generalized reduction method for generally orthotropic plates with arbitrary shape. *The Shock Vibration Digest*, 6: 19–22.

Singh, B. and Chakraverty, S. 1991. Transverse vibration of completely free elliptic and circular plates using orthogonal polynomials in Rayleigh–Ritz method. *International Journal of Mechanical Science*, 33: 741–751.

Singh, B. and Chakraverty, S. 1992a. On the use of orthogonal polynomials in Rayleigh–Ritz method for the study of transverse vibration of elliptic plates. *Computers and Structure*, 43: 439–443.

Singh, B. and Chakraverty, S. 1992b. Transverse vibration of simply-supported elliptic and circular plates using orthogonal polynomials in two variables. *Journal of Sound and Vibration*, 152: 149–155.

Singh, B. and Chakraverty, S. 1992c. Transverse vibration of triangular plates using characteristic orthogonal polynomials in two variables. *International Journal of Mechanical Science*, 34: 947–955.

Singh, B. and Chakraverty, S. 1993. Transverse vibration of annular circular and elliptic plates using characteristic orthogonal polynomials in two dimensions. *Journal of Sound and Vibration*, 162: 537–546.

Singh, B. and Chakraverty, S. 1994a. Flexural vibration of skew plates using orthogonal polynomials in two variables. *Journal of Sound and Vibration*, 173: 157–178.

Singh, B. and Chakraverty, S. 1994b. Use of characteristic orthogonal polynomials in two dimensions for transverse vibrations of elliptic and circular plates with variable thickness. *Journal of Sound and Vibration*, 173: 289–299.

Singh, B. and Chakraverty, S. 1994c. Boundary characteristic orthogonal polynomials in numerical approximation. *Communications in Numerical Methods in Engineering*, 10: 1027–1043.

Timoshenko, S. and Woinowsky-Krieger, S. 1953. *Theory of Plates and Shells*, McGraw-Hill Book Co., New York.

14

Plates with Hybrid Complicating Effects

14.1 Introduction

Practical applications of orthotropic plates in civil, marine, and aerospace engineering are numerous. Plates with variable thickness are often used in machine design, nuclear reactor technology, naval structures, and acoustical components. The development of fiber-reinforced materials and plates fabricated out of modern composites, such as boron–epoxy, glass–epoxy, and Kevlar, and their increasing usage in various technological situations have necessitated the study of nonhomogeneity in anisotropic plates. The hybrid complicating effects include the simultaneous consideration of two or more complicating effects on the vibration behavior. For example, if variable thickness and nonhomogeneity are considered simultaneously, then we will name the behavior as hybrid complicating effect. Naturally, consideration of two or more complicating effects makes the corresponding governing equation complex owing to the complexities in the effects considered. As such, this chapter illustrates the analysis of vibration of plates with hybrid effects, viz., simultaneous effects of variable thickness, nonhomogeneity, and orthotropic material.

As mentioned in all the previous chapters, Leissa (1969, 1977, 1978, 1981a, 1981b) had done extensive review on the vibration of plates. As a brief review, notable contributions dealing with various types of nonhomogeneity considerations and variable thickness are mentioned there. Tomar, Gupta, and Jain (1982) had studied vibration of isotropic plates with variable thickness using Frobenius method. Singh and Tyagi (1985) had used Galerkin's method, while Singh and Chakraverty (1994) had used boundary characteristic orthogonal polynomials (BCOPs) in two dimensions to analyze the vibration of isotropic elliptic plates with variable thickness. BCOPs method was further used by Chakraverty and Petyt (1997) to study vibration of nonhomogeneous isotropic elliptic plates. Rayleigh–Ritz method has been used by Hassan and Makray (2003) to analyze the elliptic plates with linear thickness variation and mixed boundary conditions. Kim and Dickinson (1989) had also used Rayleigh–Ritz method to study vibrations of polar orthotropic plates with thickness varying along radius and concentric ring support, whereas Lal and Sharma (2004) had used Chebyshev collocation technique to study the polar orthotropic plates possessing nonhomogeneity and variable thickness. Vibrations of orthotropic elliptic plates resting on point supports by using Ritz–Lagrange multiplier method were studied by Narita (1986). Laura and Gutierrez (2002) had investigated vibration of rectangular plates with generalized anisotropy having discontinuously varying thickness using Rayleigh–Ritz method. Very recently, Gupta and Bhardwaj (2004) have used BCOPs method to study the vibration of orthotropic plates with variable thickness resting on Winkler foundation. Other studies on homogeneous elliptic plates with different boundary conditions have also been carried out by Sakata (1976), Rajappa (1963), Mcnitt (1962), and Kim (2003).

During the review of literature, very few works dealing with vibration of nonhomogeneous (in Young's modulus and density) orthotropic plates of varying thickness were found. It will not be possible to consider the problem in its full generality because of the vast number of parameters involved. Hence, in this chapter only the quadratic variation in thickness (h), density (ρ), and Young's modulus (E_x) along x-axis; quadratic variation in Young's modulus (E_y) along y-axis, and linear variation in shear modulus (G_{xy}) along both x and y axes have been considered. Two-dimensional boundary characteristic orthogonal polynomials (2D BCOPs) in the Rayleigh–Ritz method have been used for the analysis. The use of these polynomials makes the computation of the vibration characteristics simpler by reducing the problem to standard eigenvalue problem and by giving faster rate of convergence, in this case too.

14.2   Basic Equations for the Hybrid Complicating Effects

Let us consider a nonhomogeneous, elliptic plate of variable thickness made up of rectangular orthotropic material lying in x–y plane.

The domain occupied by the plate element is given as

$$S = \{(x,y),\ x^2/a^2 + y^2/b^2 \leq 1;\quad x,y \in R\} \tag{14.1}$$

where a and b are the semimajor and semiminor axes of the ellipse. On equating the maximum strain and kinetic energies, one can get (as done previously) the Rayleigh quotient,

$$\omega^2 = \frac{\iint\limits_{S} \left[D_x W_{xx}^2 + D_y W_{yy}^2 + 2\nu_x D_y W_{xx} W_{yy} + 4 D_{xy} W_{xy}^2 \right] dy dx}{h \iint\limits_{S} \rho W^2 dy dx} \tag{14.2}$$

where
  $W(x,y)$ is the deflection of the plate
  W_{xx} is the second derivative of W with respect to x

The D coefficients are bending rigidities defined by

$$\left.\begin{array}{l} D_x = E_x h^3/(12(1 - \nu_x \nu_y)) \\ D_y = E_y h^3/(12(1 - \nu_x \nu_y)) \\ D_y \nu_x = D_x \nu_y \\ D_{xy} = G_{xy} h^3/12 \end{array}\right\} \qquad (14.3)$$

If the variations of the parameters are taken as mentioned, then the non-homogeneity and thickness of the orthotropic structural element can be characterized by taking

$$E_x = (1 + \alpha_1 x + \alpha_2 x^2)E_{x0} \qquad (14.4)$$

$$E_y = (1 + \beta_1 y + \beta_2 y^2)E_{y0} \qquad (14.5)$$

$$G_{xy} = (1 + \gamma_1 x + \gamma_2 y)G_{xy0} \qquad (14.6)$$

$$\rho = (1 + \delta_1 x + \delta_2 x^2)\rho_0 \qquad (14.7)$$

$$h = (1 + \psi_1 x + \psi_2 x^2)h_0 \qquad (14.8)$$

Here, E_{x0}, E_{y0}, G_{xy0}, ρ_0, and h_0 are constants and α_1, α_2, β_1, β_2, γ_1, γ_2, δ_1, δ_2, ψ_1, ψ_2 are the parameters designating the nonhomogeneity and variable thickness of the orthotropic plate. The flexural rigidities may be obtained from Equation 14.3 by substituting the parameters defined in Equations 14.4 through 14.8 as

$$D_x = p t^3 D_{x0} \qquad (14.9)$$

$$D_y = q t^3 D_{y0} \qquad (14.10)$$

$$D_{xy} = r t^3 D_{xy0} \qquad (14.11)$$

where p, q, r, and t are the coefficients of E_{x0}, E_{y0}, G_{xy0}, and h_0 of Equations 14.4 through 14.6, and 14.8, respectively. The terms D_{x0}, D_{y0}, and D_{xy0} are the expressions similar to D_x, D_y, and D_{xy} when the constants E_x, E_y, G_{xy}, and h are, respectively, replaced by E_{x0}, E_{y0}, G_{xy0}, and h_0. Substituting Equations 14.9 through 14.11 in Equation 14.2, assuming the N-term approximation

$$W(x,y) = \sum_{j=1}^{N} c_j \varphi_j \qquad (14.12)$$

and minimizing ω^2 as a function of the coefficients c_j's following matrix equation may be obtained:

$$\sum_{j=1}^{N} (a_{ij} - \lambda^2 b_{ij})c_j = 0, \quad i = 1, 2, \ldots, N \qquad (14.13)$$

where

$$a_{ij} = H_0 \iint\limits_{S'} t^3 [p(D_{x0}/H_0)\phi_i^{xx}\phi_j^{xx} + q(D_{y0}/H_0)\phi_i^{yy}\phi_j^{yy} + q\nu_x(D_{y0}/H_0)$$

$$(\phi_i^{xx}\phi_j^{yy} + \phi_j^{xx}\phi_i^{yy}) + 2r(1 - \nu_x(D_{y0}/H_0))\phi_j^{xy}\phi_i^{xy}]dydx \qquad (14.14)$$

$$b_{ij} = \iint\limits_{S'} st\phi_i\phi_j dydx \qquad (14.15)$$

and

$$\lambda^2 = a^4 \rho_0 h_0 \omega^2 / H_0 \qquad (14.16)$$

where s is the coefficient of ρ_0 in Equation 14.7, $H_0 = D_{y0}\nu_x + 2D_{xy0}$, ϕ_i^{XX} are the second derivative of ϕ_i with respect to X and the new domain S' is defined by

$$S' = \{(X,Y), \quad X^2 + Y^2/m^2 \le 1, \quad X,Y \in R\} \qquad (14.17)$$

where m ($=b/a$) is the aspect ratio of the ellipse and $X=x/a$, $Y=y/a$. It is interesting to note that when the ϕ_i's are orthonormal with respect to the weight function st in Equation 14.15, Equation 14.13 reduces to

$$\sum_{j=1}^{N} (a_{ij} - \lambda^2 \delta_{ij})c_j = 0, \quad i = 1, 2, \dots, N \qquad (14.18)$$

where

$$\delta_{ij} = \begin{cases} 0, & \text{if } i \ne 0 \\ 1, & \text{if } i = j \end{cases} \qquad (14.19)$$

14.3   Generation of BCOPs

The present orthogonal polynomials are generated in the same way as given in Singh and Chakraverty (1991, 1992a, 1992b, 1999) by considering $st = (1+\delta_1 x + \delta_2 x^2)(1 + \psi_1 x + \psi_2 x^2)$ as the weight function. For this, suitable sets of linearly independent functions are chosen to start as

$$F_i(x,y) = f(x,y)f_i(x,y), \quad i = 1, 2, \dots, N \qquad (14.20)$$

where $f(x,y)$ satisfies the essential boundary conditions and $f_i(x,y)$ are linearly independent functions involving products of non-negative integral powers of x and y. The function $f(x,y)$ is defined by

$$f(x,y) = (1 - x^2 - y^2/m^2)^u \tag{14.21}$$

where u is 0, 1, or 2 according to the plate being subjected to free (F), simply supported (S), or clamped (C) edge condition. Then, the orthonormal set from $F_i(x,y)$ has been generated by the well-known Gram–Schmidt process as mentioned in the previous chapters (Singh and Chakraverty 1991, 1992a,b).

14.4 Some Numerical Results and Discussions

Although numerical results for various values of the parameters may be worked out, only few are reported here. Results have been computed for various orthotropic materials, the values of which are taken from Chakraverty and Petyt (1999). The parameters are considered to be governed by the Equations 14.4 through 14.8. Results are given when frequency parameter λ converges to at least four significant digits for all the boundary conditions, viz., C, S, and F: i.e., clamped, simply supported, and free, respectively. The order of approximation is taken as 15 to ensure convergence, and computations are all carried out in double precision arithmetic. Table 14.1 and Figure 14.1 show the convergence for $m = 0.5$. For the sake of comparison, results were also derived for homogeneous orthotropic and isotropic plates with variable as well as constant thickness. Table 14.2 shows the results for fundamental frequency parameters for orthotropic, clamped elliptic plates for various materials M1, M2, M3, and M4 the properties of which are

M1(Glass epoxy): $D_{x0}/H_0 = 3.75$, $D_{y0}/H_0 = 0.8$, $v_x = 0.26$
M2(Boron epoxy): $D_{x0}/H_0 = 13.34$, $D_{y0}/H_0 = 1.21$, $v_x = 0.23$
M3(Carbon epoxy): $D_{x0}/H_0 = 15.64$, $D_{y0}/H_0 = 0.91$, $v_x = 0.32$
M4(Kevlar): $D_{x0}/H_0 = 2.60$, $D_{y0}/H_0 = 2.60$, $v_x = 0.14$

The results are found to tally excellently with the recent results of Kim (2003), and in good agreement with Chakraverty and Petyt (1999), Sakata (1976), Rajappa (1963), and Mcnitt (1962). Table 14.3 shows the comparison of the first four frequencies with Chakraverty and Petyt (1997) for isotropic plates with variable thickness taking $\alpha_1 = \alpha_2 = 0.1$. A very close agreement is found in most of the cases.

TABLE 14.1

Convergence Table for $m = 0.5$, $D_{x0}/H_0 = 13.34$, $D_{y0}/H_0 = 1.21$, $\nu_X = 0.23$ (Boron–Epoxy), $\psi_1 = -0.2$, $\psi_2 = 0.6$, $\delta_1 = 0.6$, $\delta_2 = 0.2$, $\alpha_1 = 0.2$, $\alpha_2 = 0.2$, $\beta_1 = 0.6$, $\beta_2 = 0$, $\gamma_1 = 0.6$, and $\gamma_2 = 0.2$

B.C.	N	λ_1	λ_2	λ_3	λ_4	λ_5
C	5	44.8967	88.3861	90.8014	137.8200	170.0498
	10	44.8789	88.3219	90.6884	137.5861	152.6673
	12	44.8782	88.3203	90.6829	137.5766	152.4961
	13	44.8782	88.3203	90.6829	137.5764	152.4866
	14	44.8782	88.3203	90.6829	137.5763	152.4816
	15	44.8782	88.3203	90.6829	137.5763	152.4802
S	5	20.2775	54.3775	59.2479	99.3462	122.4417
	10	20.2412	54.0750	58.8974	98.34185	110.9925
	12	20.2401	54.0636	58.8919	98.3145	110.1100
	13	20.2401	54.0635	58.8916	98.3128	110.1091
	14	20.2401	54.0635	58.8916	98.3127	110.1058
	15	20.2401	54.0635	58.8916	98.3127	110.1029
F	5	12.3835	23.9290	40.3459	42.8981	57.6157
	10	12.3717	23.0814	35.4593	41.6875	53.8372
	12	12.37155	23.0300	35.4197	41.6091	53.7379
	13	12.37155	23.0300	35.3998	41.6077	53.7138
	14	12.37155	23.0298	35.3901	41.6075	53.7008
	15	12.37155	23.0298	35.3900	41.6075	53.7008

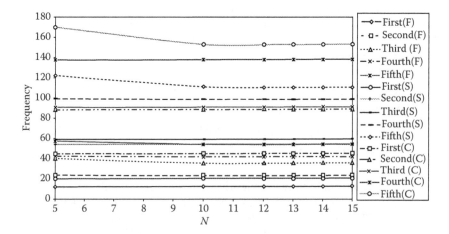

FIGURE 14.1
Convergence for $m = 0.5$ and the other parameters as per Table 14.1 (boron–epoxy).

TABLE 14.2

Comparison of Fundamental Frequency Parameters for Orthotropic Elliptic Plates

Material	m	Present	Kim (2003)	Chakraverty and Petyt (1999)	Sakata (1976)	Rajappa (1963)	McNitt (1962)
M1	0.2	136.2919	136.29	136.51	147.24	144.52	143.11
	0.5	27.3448	27.345	27.376	28.280	27.757	27.485
	0.8	16.23414	16.234	16.242	16.742	16.433	16.272
	1.0	14.2220	14.222	14.225	14.720	14.448	14.306
M2	0.2	168.5974	168.60	168.53	180.60	177.27	175.53
	0.5	37.1926	37.193	37.181	38.308	37.601	37.232
	0.8	25.9129	25.913	25.907	26.823	26.327	26.069
	1.0	24.0624	24.062	24.057	25.130	24.666	24.424
M3	0.2	148.555	148.56	148.67	158.10	155.18	153.66
	0.5	35.8641	35.864	35.876	36.953	36.270	35.915
	0.8	26.9579	26.958	26.959	28.016	27.498	27.229
	1.0	25.4433	25.443	25.442	26.735	26.241	25.984
M4	0.2	236.6095	236.60	236.43	261.08	256.26	253.75
	0.5	42.0735	42.074	42.043	44.080	43.266	42.842
	0.8	19.7401	19.740	19.726	20.351	19.975	19.779
	1.0	15.15248	15.152	15.142	15.596	15.308	15.158

Tables 14.4 through 14.7 and Figures 14.2 through 14.9 summarize results of the first five frequencies corresponding to the variation of the different parameters. Numerical results have been obtained for all the three boundary conditions: clamped (C), simply supported (S), and free (F).

Table 14.4 and Figures 14.2 through 14.5 show the first five frequency parameters for glass–epoxy material with $\alpha_1 = 0.5$, $\alpha_2 = -0.3$, $\beta_1 = -0.2$, $\beta_2 = 0.6$, $\gamma_1 = 0.3$, $\gamma_2 = -0.3$, and $m = 0.5$ for various values of density parameters δ_1, δ_2 and thickness parameters ψ_1, ψ_2. The above figures show the variations of the first five frequencies with δ_1, δ_2, ψ_1, and ψ_2, respectively. Trends of all the boundary conditions C, S, and F are shown in these figures. The following observations can be made:

Keeping other parameters constant,

1. If density parameter δ_2 is increased from -0.2 to 0.6, the frequency may be seen to decrease for various sets of parameters as mentioned above.

2. Similarly, it may be observed that taking $\psi_1 = 0.6$, $\psi_2 = 0.6$, the frequency decreases if density parameter δ_1 is increased from 0.2 to 0.6. The same trends as discussed above may be observed for all three boundary conditions.

3. As the thickness parameter ψ_1 is decreased from 0.6 to -0.2, frequency parameter too decreases. The exception to this trend may

TABLE 14.3

Comparison of Frequencies with Chakraverty and Petyt (1997) for Isotropic Plates with Variable Thickness ($\alpha_1 = 0.1$ and $\alpha_2 = 0.1$)

Edge Condition	m	x_1		x_2		x_3		x_4	
		Chakraverty and Petyt (1997)	Present	Chakraverty and Petyt (1997)	Present	Chakraverty and Petyt (1997)	Present	Chakraverty and Petyt (1997)	Present
Free	1.0	5.444	5.445	9.114	9.113	12.78	12.78	20.82	20.82
	0.8	6.195	6.195	6.767	6.767	12.40	12.40	15.23	15.23
	0.5	6.628	6.628	10.60	10.60	17.19	17.19	22.51	22.50
	0.2	6.718	6.718	17.62	17.614	25.70	25.70	35.61	33.655
Simply supported	1.0	5.035	5.035	14.18	14.18	14.26	14.26	26.27	26.28
	0.8	6.535	6.535	16.03	16.03	20.36	20.36	30.06	30.04
	0.5	13.49	13.497	24.27	24.27	39.56	39.499	46.67	46.67
	0.2	70.31	70.31	90.42	90.35	116.4	114.206	146.8	142.037
Clamped	1.0	10.60	10.60	21.68	21.68	22.15	22.156	35.84	35.84
	0.8	13.61	13.61	25.30	25.30	30.77	30.77	41.52	41.52
	0.5	27.78	27.78	40.56	40.56	57.85	57.82	70.42	70.428
	0.2	150.3	150.3	173.0	172.91	203.0	199.773	236.9	230.62

TABLE 14.4

Values of λ for $D_{x0}/H_0 = 3.75$, $D_{y0}/H_0 = 0.8$, $\nu_X = 0.26$ (Glass–Epoxy), $\alpha_1 = 0.5$, $\alpha_2 = -0.3$, $\beta_1 = -0.2$, $\beta_2 = 0.6$, $\gamma_1 = 0.3$, $\gamma_2 = -0.3$, $m = 0.5$

| $\delta_2 \rightarrow$ | | 0.6 | | | 0.2 | | | −0.2 | | |
(ψ_1, ψ_2)	δ_1	F	S	C	F	S	C	F	S	C
(0.6, 0.6)	0.6	11.5144	15.3277	33.3334	11.7869	15.6365	33.7275	13.1553	15.9629	34.1320
		11.8482	33.6793	58.8467	12.1613	35.0612	60.5557	13.3832	36.6171	62.4019
		30.3285	48.5410	72.2294	30.5901	49.1266	72.8622	31.3832	49.7236	73.5000
		31.3236	64.9629	95.0333	32.0006	67.7378	98.5315	34.9684	71.0286	102.4865
		33.4278	74.5107	103.5190	34.2047	76.7327	105.9248	37.6184	79.1263	108.4606
	0.2	11.7869	15.4860	33.5474	12.5723	15.8038	33.9474	13.6603	16.1399	34.3577
		12.1613	34.3496	59.7171	12.9197	35.8150	61.5073	13.8370	37.47175	63.4331
		30.5901	48.8585	72.5821	31.1021	49.4507	73.2181	31.6382	50.0541	73.8590
		32.0006	66.2267	96.7299	33.7807	69.2321	100.4504	36.06363	72.8286	104.6724
		34.2047	75.6502	104.7969	36.2317	77.9619	107.2739	38.8726	80.4542	109.9849
(0.6, 0.2)	0.6	11.4553	14.1986	30.9581	12.1126	14.4799	31.3414	12.8955	14.77695	31.7354
		11.7716	31.5628	86.9457	12.4179	32.8556	55.4407	13.3556	34.3111	57.1999
		28.0893	46.1520	97.2732	28.5695	46.7356	70.2514	29.0728	47.3307	70.9061
		29.4457	60.3926	69.6015	30.9863	62.9932	90.2249	32.9164	66.0727	93.9497
		31.7345	69.5384	53.8182	33.4871	71.6833	99.6534	35.6896	74.0009	102.1692
(0.6, −0.2)	0.6	11.1147	13.1909	28.6651	11.7268	13.4467	29.0391	12.4504	13.7165	29.4238
		12.0186	29.4439	48.6948	12.7240	30.6431	50.2232	13.6429	31.9924	51.8873
		26.0248	43.8472	66.9615	26.4857	44.4302	67.6358	26.9717	45.0250	68.3143
		27.8135	55.8309	78.6493	29.2736	58.2510	81.6907	31.0878	61.1117	85.1633
		29.9862	64.6030	90.8376	31.6183	66.6650	93.1928	33.6394	68.9006	95.6910

(continued)

TABLE 14.4 (continued)

Values of λ for $D_{x0}/H_0 = 3.75$, $D_{y0}/H_0 = 0.8$, $\nu_X = 0.26$ (Glass–Epoxy), $\alpha_1 = 0.5$, $\alpha_2 = -0.3$, $\beta_1 = -0.2$, $\beta_2 = 0.6$, $\gamma_1 = 0.3$, $\gamma_2 = -0.3$, $m = 0.5$

$\delta_2 \rightarrow$		0.6			0.2			-0.2	
(ψ_1, ψ_2)	δ_1	F	S	C	F	S	C	F	S	C
(0.2, 0.6)	0.6	11.1883	14.5106	31.1375	11.8964	14.8162	31.5361	12.7552	15.1397	31.9456
		11.2633	31.3910	53.7842	12.0257	32.7786	55.4924	13.0831	34.3453	57.3503
		28.6408	46.5828	69.8686	29.1414	47.1813	70.5301	29.6670	47.7909	71.1959
		29.0838	59.4088	72.2692	30.7532	62.2082	89.7360	32.9015	65.5743	93.7385
		30.9053	70.2326	86.2388	32.8022	72.4875	100.2072	35.2837	74.9269	102.8181
(−0.2, 0.6)	0.6	11.1328	14.5779	30.4626	11.8846	14.8901	30.8694	12.8085	15.2206	31.2874
		11.3086	31.1300	52.0610	12.1641	32.5880	53.8278	13.3963	34.2540	55.7584
		28.6549	46.5366	69.6949	29.1583	47.1344	70.3617	29.6892	47.7427	71.0320
		28.7093	57.6043	82.7044	30.5221	60.5984	86.3724	32.9080	64.2470	90.6134
		30.1911	70.3701	97.3500	32.2458	72.6970	99.8968	35.0430	75.2148	102.5891

TABLE 14.5

Values of λ for $D_{x0}/H_0 = 13.9$, $D_{y0}/H_0 = 0.79$, $\nu_X = 0.28$ (Graphite–Epoxy), $\psi_1 = 0.6$, $\psi_2 = 0.2$, $\delta_1 = 0.6$, $\delta_2 = -0.2$, $\beta_1 = -0.3$, $\beta_2 = 0.2$, $\gamma_1 = 0.6$, and $\gamma_2 = -0.3$

$\alpha_2 \rightarrow$		0.7			0.3			-0.4		
m	α_1	F	S	C	F	S	C	F	S	C
0.5	0.7	14.5123	18.4121	46.2358	14.4666	18.3251	44.8248	14.3653	18.1612	42.0905
		24.3458	50.5667	79.7889	24.1819	50.4999	78.9651	23.8551	50.3563	77.4634
		31.8369	59.3928	101.0664	31.7310	58.0448	97.6467	31.5416	55.4518	90.8490
		46.1019	94.8389	133.6405	45.9227	93.9421	133.0891	45.5870	92.2951	130.2142
		49.0588	99.5743	137.7193	48.7878	99.3645	135.1186	48.2210	98.8778	132.0123
	0.3	14.4505	18.2755	44.6625	14.3945	18.1819	43.1484	14.2647	18.0032	40.1538
		24.0882	50.4731	78.8028	23.9014	50.3944	77.9534	23.5185	50.2152	76.3980
		31.6710	57.4716	97.0878	31.5636	56.0084	93.3453	31.3687	53.1322	85.6794
		45.8140	93.4824	132.9471	45.6233	92.5463	131.7694	45.2559	90.8002	126.5156
		48.6710	99.2912	134.5230	48.3515	99.0239	132.3524	27.6648	98.3023	131.0619
	-0.7	14.2236	17.8864	40.1965	14.1120	17.7664	38.2608	13.7380	17.4852	33.7332
		23.2673	50.1583	76.2004	22.9716	49.9351	75.2680	22.1861	45.3727	65.8846
		31.2465	51.9062	85.3595	31.1269	50.0105	80.1447	30.8669	49.5060	73.5268
		45.0075	89.8074	125.6796	44.7530	88.6933	122.3650	44.0413	86.1474	110.8329
		47.3346	98.0124	130.8479	46.7483	96.8389	129.5661	44.5351	86.6678	115.6350
1.0	0.7	7.3963	13.0631	30.3830	7.3881	12.9292	29.1481	7.3693	12.6670	26.4812
		7.5853	21.4972	30.8958	7.5818	21.4046	29.92875	7.5743	21.2310	27.9790
		17.8903	34.3631	35.2463	17.8574	34.2970	34.2151	17.7856	34.1724	32.3431
		18.5187	51.8918	41.2408	18.5053	51.8330	40.4730	18.4783	49.3417	39.0279
		25.7125	53.5518	49.8189	25.48915	52.1268	49.1843	25.0639	51.7160	47.9735

(continued)

TABLE 14.5 (continued)

Values of λ for $D_{x0}/H_0 = 13.9$, $D_{y0}/H_0 = 0.79$, $\nu_X = 0.28$ (Graphite–Epoxy), $\psi_1 = 0.6$, $\psi_2 = 0.2$, $\delta_1 = 0.6$, $\delta_2 = -0.2$, $\beta_1 = -0.3$, $\beta_2 = 0.2$, $\gamma_1 = 0.6$, and $\gamma_2 = -0.3$

$\alpha_2 \rightarrow$		0.7			0.3			-0.4	
m / α_1	F	S	C	F	S	C	F	S	C
0.3	7.3852	12.8582	28.9780	7.3749	12.7092	27.5511	7.3497	12.4101	24.3722
	7.5802	21.3494	29.7412	7.5761	21.2501	28.6688	7.5664	21.0611	26.4486
	17.8446	34.2570	34.0840	17.8046	34.1864	33.0230	17.7130	34.0487	31.0494
	18.4980	51.5495	40.3599	18.4827	49.9845	39.5525	18.4507	46.8375	37.9964
	25.3605	51.8015	49.0915	25.1178	51.7359	48.4189	24.6441	59.1476	47.0926
-0.7	7.3417	12.2364	24.4784	7.3181	12.0175	22.1931	7.2121	11.4740	13.0793
	7.5623	20.9296	26.2769	7.5536	20.8028	24.7481	7.4755	20.4963	20.6901
	17.6789	33.9583	30.9682	17.5991	33.8609	29.7183	17.1527	33.5350	22.5913
	18.4317	45.6111	37.9382	18.4065	43.3770	36.9366	18.3279	36.7685	26.0579
	24.3310	57.8933	47.0578	23.9933	56.0469	46.1734	23.1991	51.9341	34.08045

TABLE 14.6

Values of λ for $D_{x0}/H_0 = 3.75$, $D_{y0}/H_0 = 0.8$, $\nu_X = 0.26$ (Glass–Epoxy), $\psi_1 = 0.6$, $\psi_2 = 0.2$, $\delta_1 = 0.6$, $\delta_2 = -0.2$, $\alpha_1 = 0.6$, $\alpha_2 = -0.2$, $\beta_1 = -0.4$, $\gamma_1 = -0.2$, and $\gamma_2 = 0.4$

m	β_2	B.C.	λ_1	λ_2	λ_3	λ_4	λ_5
0.5	-0.9	F	12.2493	13.4561	28.2858	30.9824	36.0650
		S	14.4990	34.4790	45.1133	66.8403	71.6075
		C	30.2631	56.4692	66.6046	94.0199	98.1954
	0.3	F	12.2678	13.4577	28.5736	31.0534	36.1208
		S	14.6780	34.578	46.6311	66.9669	73.1080
		C	31.3951	57.2317	69.7975	94.6653	100.8950
	0.9	F	12.2754	13.4580	28.7074	31.0834	36.1460
		S	14.7607	34.6260	47.3323	67.0198	73.7262
		C	31.9185	57.5961	71.2485	94.9589	102.1546
1.0	-0.9	F	6.2585	6.9569	14.0280	15.3078	15.9679
		S	7.3941	14.9459	25.7862	27.2673	37.8513
		C	18.5800	26.1417	37.2755	44.7857	51.9303
	0.3	F	6.3379	7.3291	14.0759	16.2746	17.0901
		S	7.4820	16.2725	27.3404	29.7066	38.6075
		C	18.9655	28.0342	42.1172	44.9657	56.7238
	0.9	F	6.3610	7.4644	14.0936	16.3840	17.6591
		S	7.5190	16.8064	27.3638	31.1536	38.9378
		C	19.1428	28.8250	43.9678	45.0518	57.2040

TABLE 14.7

Values of λ for $D_{x0}/H_0 = 13.34$, $D_{y0}/H_0 = 1.21$, $\nu_X = 0.23$ (Boron–Epoxy), $\psi_1 = 0.6$, $\psi_2 = 0.2$, $\delta_1 = -0.2$, $\delta_2 = 0.6$, $\alpha_1 = -0.2$, $\alpha_2 = 0.2$, $\beta_1 = -0.2$, $\beta_2 = 0.6$, $\gamma_2 = 0.5$, $m = 0.5$

γ_1	B.C.	λ_1	λ_2	λ_3	λ_4	λ_5
0.6	F	13.0147	22.4816	35.1973	42.4465	50.5541
	S	19.7098	51.5096	58.7874	95.9376	104.1882
	C	43.3165	85.2458	88.7808	134.3557	145.9498
0.4	F	12.7003	22.4801	35.0383	41.8330	50.4818
	S	19.6902	51.4809	58.6583	95.6622	104.1151
	C	43.2126	85.0651	88.6227	134.0053	145.7313
0.2	F	12.3651	22.4785	34.8695	41.1925	50.4080
	S	19.6702	51.4517	58.5284	95.3841	104.0411
	C	43.1082	84.8836	88.4640	133.6526	145.5114
-0.2	F	11.6199	22.4751	34.4981	39.8201	50.2558
	S	19.6291	51.3919	58.2659	94.8189	103.8901
	C	42.8980	84.5180	88.1445	132.9401	145.0678
-0.4	F	11.2019	22.4733	34.2929	39.0824	50.1772
	S	19.6079	51.3611	58.1332	94.5318	103.8130
	C	42.7922	84.3339	87.9838	132.5803	144.8436
-0.8	F	10.2469	22.4694	33.8351	37.4856	50.0143
	S	19.5642	51.2980	57.8650	93.9478	103.6554
	C	42.5790	83.9632	87.6602	131.8533	144.3916

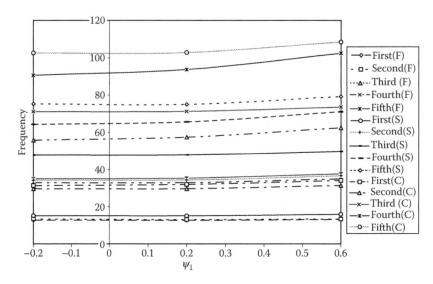

FIGURE 14.2
Variation of first five frequencies with linear thickness variation parameter ($m = 0.5$, glass–epoxy).

be observed in the second frequency parameter of free boundary condition as ψ_2 decreases from 0.2 to -0.2.

4. However, no trend in frequency parameters is observed from decrease in thickness parameter ψ_2 from 0.6 to -0.2.

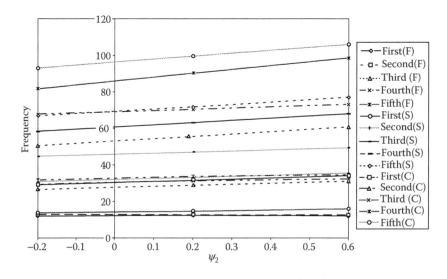

FIGURE 14.3
Variation of first five frequencies with quadratic thickness variation parameter ($m = 0.5$, glass–epoxy).

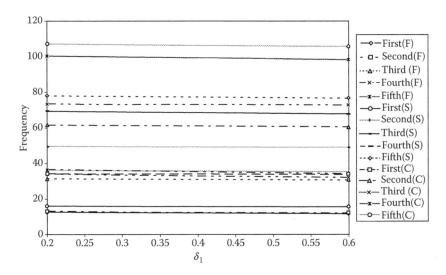

FIGURE 14.4
Variation of first five frequencies with linear density variation parameter ($m = 0.5$, glass–epoxy).

Next, Table 14.5 investigates the effect of variation of E_x parameters α_1 and α_2 on frequency parameter λ for graphite–epoxy material (Chakraverty and Petyt (1999)) with $\psi_1 = 0.6$, $\psi_2 = 0.2$, $\delta_1 = 0.6$, $\delta_2 = -0.2$, $\beta_1 = -0.3$, $\beta_2 = 0.2$, $\gamma_1 = 0.6$, $\gamma_2 = -0.3$, and $m = 0.5$ and 1.0. Consequently, Figures 14.6 and 14.7

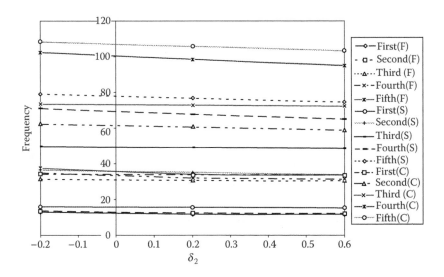

FIGURE 14.5
Variation of first five frequencies with quadratic density variation parameter ($m = 0.5$, glass–epoxy).

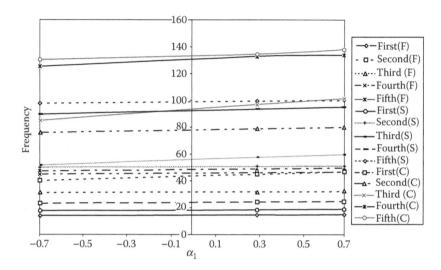

FIGURE 14.6
Variation of first five frequencies with linear variation of Young's modulus along *x*-direction ($m = 0.5$, graphite–epoxy).

depict the variation of the first five frequency parameters with α_1 and α_2, respectively. Here, the graphs in each case of the boundary conditions, viz., F, S, and C are again given.

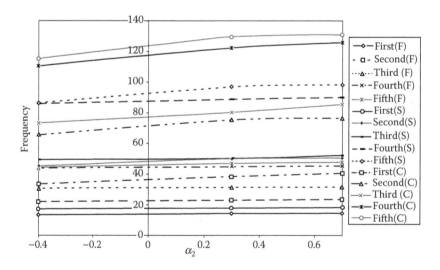

FIGURE 14.7
Variation of first five frequencies with quadratic variation of Young's modulus along *x*-direction ($m = 0.5$, graphite–epoxy).

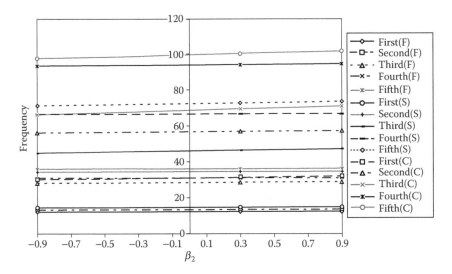

FIGURE 14.8
Variation of first five frequencies with quadratic variation of Young's modulus along y-direction ($m = 0.5$, glass–epoxy).

The following inferences may be drawn from Table 14.5:

1. As α_1 decreases from 0.7 to -0.7, frequency parameter λ decreases.
2. Frequencies also decrease for decrease in value of α_2 from 0.7 to -0.4.

Table 14.6 includes the variation of λ with respect to E_y parameter β_2 only for glass–epoxy material. The other parameters are taken as $\psi_1 = 0.6$, $\psi_2 = 0.2$, $\delta_1 = 0.6$, $\delta_2 = -0.2$, $\alpha_1 = 0.6$, $\alpha_2 = -0.2$, $\beta_1 = -0.4$, $\gamma_1 = -0.2$, $\gamma_2 = 0.4$, and $m = 0.5$ and 1.0. Here, the frequencies may be seen to decrease with decrease in value of β_2 from 0.9 to -0.9, keeping all other parameters fixed. Figure 14.8 depicts the variation of first five frequencies with β_2 for $m = 0.5$ for all the boundary conditions.

In Tables 14.5 and 14.6, the same trends as discussed above may be observed for all three boundary conditions and for both elliptic ($m = 0.5$) as well as circular ($m = 1.0$) plate. It is quite evident from the above tables that the values of λ are considerably lower for a circular plate when compared with an elliptic plate.

Finally, Table 14.7 shows frequency parameters for various values of shear modulus parameter, γ_1. The computations have been undertaken for boron–epoxy material, where $\psi_1 = 0.6$, $\psi_2 = 0.2$, $\delta_1 = -0.2$, $\delta_2 = 0.6$, $\alpha_1 = -0.2$, $\alpha_2 = 0.2$, $\beta_1 = -0.2$, $\beta_2 = 0.6$, $\gamma_2 = 0.5$, and $m = 0.5$. It may be seen that as γ_1 decreases, the frequency parameter too decreases for free, simply supported, and clamped boundary conditions. Again Figure 14.9 shows the variation of the first five frequency parameters with γ_1 for all the three boundary conditions.

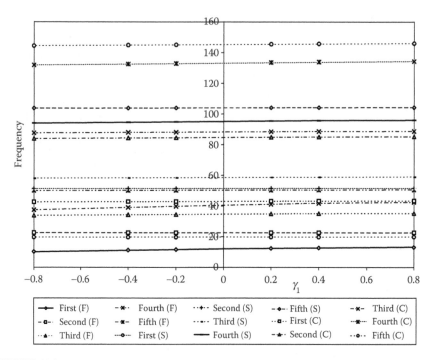

FIGURE 14.9
Variation of first five frequencies with linear variation of shear modulus ($m = 0.5$, boron-epoxy).

14.5 Conclusion

The effect of nonhomogeneity and variable thickness on the natural frequencies of circular and elliptic plates with rectangular orthotropic material has been studied by using BCOPs in the Rayleigh–Ritz method. A desired frequency can be obtained by a proper choice of the various plate parameters considered here, which would be beneficial for the design engineers.

Bibliography

Chakraverty, S. and Petyt, M. 1997. Natural frequencies for free vibration of nonhomogeneous elliptic and circular plates using two-dimensional orthogonal polynomials. *Applied Mathematical Modelling*, 21: 399–417.

Chakraverty, S. and Petyt, M. 1999. Free vibration analysis of elliptic and circular plates having rectangular orthotropy. *Structural Engineering and Mechanics*, 7(1): 53–67.

Chakraverty, S., Jindal Ragini, and Agarwal, V.K. 2007. Vibration of non-homogenous orthotropic elliptic and circular plates with variable thickness, *Transaction of the American Society of Mechanical Engineers*, 129, 256–259.

Chakraverty, S., Jindal Ragini, and Agarwal, V.K. 2007. Effect of non-homogeneity of natural frequencies of vibration of elliptic plates. *Mechanics*, 42, 585–599.

Gupta, A.P. and Bhardwaj, N. 2004. Vibration of rectangular orthotropic elliptic plates of quadratically varying thickness resting on elastic foundation. *Journal of Vibration and Acoustics*, 126(1): 132–140.

Hassan, S. and Makray, M. 2003. Transverse vibrations of elliptical plate of linearly varying thickness with half of the boundary clamped and the rest simply supported. *International Journal of Mechanical Science*, 45: 873–890.

Kim, C.S. 2003. Natural frequencies of orthotropic, elliptical and circular plates. *Journal of Sound and Vibration*, 259(3): 733–745.

Kim, C.S. and Dickinson, S.M. 1989. On the lateral vibration of thin annular and circular composite plates subject to certain complicating effects. *Journal of Sound and Vibration*, 130(3): 363–377.

Lal, R. and Sharma, S. 2004. Axisymmetric vibrations of non-homogeneous polar orthotropic annular plates of variable thickness. *Journal of Sound and Vibration*, 272: 245–265.

Laura, P.A.A. and Gutierrez, R.H. 2002. Transverse vibrations of rectangular plates of generalized anisotropy and discontinuously varying thickness. *Journal of Sound and Vibration*, 250(3): 569–574.

Leissa, A.W. 1969. *Vibration of Plates*. NASA SP-160, Washington, DC.

Leissa, A.W. 1977. Recent research in plate vibrations, 1973–1976: Classical theory. *Shock and Vibration Digest*, 9(10): 3–24.

Leissa, A.W. 1978. Recent research in plate vibrations: 1973–1976: Complicating effects. *Shock and Vibration Digest*, 10(12): 21–35.

Leissa, A.W. 1981a. Plate vibration research, 1976–1980: Classical theory. *Shock and Vibration Digest*, 13(9): 11–22.

Leissa, A.W. 1981b. Plate vibration research, 1976–1980: Complicating effects. *Shock and Vibration Digest*, 13(10): 19–36.

McNitt, R.P. 1962. Free vibration of a clamped elliptical plate. *Journal of Aeronautical Science*, 29: 1124–1125.

Narita, Y. 1986. Free vibration analysis of orthotropic elliptical plates resting on arbitrary distributed point support. *Journal of Sound and Vibration*, 108: 1–10.

Rajappa, N.R. 1963. Free vibration of rectangular and circular orthotropic plates. *AIAA J.1*, 1194–1195.

Sakata, T. 1976. A reduction method for problems of vibration of orthotropic plates. *Journal of Sound and Vibration*, 48: 405–412.

Singh, B. and Chakraverty, S. 1991. Transverse vibrations of completely free elliptic and circular plates using orthogonal polynomials in Rayleigh–Ritz method. *International Journal of Mechanical Science*, 33(9): 741–751.

Singh, B. and Chakraverty, S. 1992a. Transverse vibration of simply-supported elliptic and circular plates using orthogonal polynomials in two variables. *Journal of Sound and Vibration*, 152(1): 149–155.

Singh, B. and Chakraverty, S. 1992b. On the use of orthogonal polynomials in Rayleigh–Ritz method for the study of transverse vibration of elliptic plates. *Computers and Structures*, 43: 439–443.

Singh, B. and Chakraverty, S. 1994. Use of characteristic orthogonal polynomials in two dimensions for transverse vibration of elliptic and circular plates with variable thickness. *Journal of Sound and Vibration*, 173(3): 289–299.

Singh, B. and Tyagi, D.K. 1985. Transverse vibration of elliptic plates with variable thickness. *Journal of Sound and Vibration*, 99(3): 379–391.

Tomar, J.S., Gupta, D.C, and Jain, N.C. 1982. Axisymmetric vibrations of an isotropic elastic nonhomogeneous circular plates of linearly varying thickness. *Journal of Sound and Vibration*, 85(3): 365–370.

Index

Milton Keynes UK
Ingram Content Group UK Ltd.
UKHW021833071024
449327UK00021B/1491